AF692744

Recent Process Design and Development Strategies for Dental Materials

Recent Process Design and Development Strategies for Dental Materials

Editor

Mary Anne Melo

MDPI

Basel • Beijing • Wuhan • Barcelona • Belgrade • Novi Sad • Cluj • Manchester

Editor
Mary Anne Melo
Comprehensive Dentistry
University of Maryland
Baltimore
United States

Editorial Office
MDPI
St. Alban-Anlage 66
4052 Basel, Switzerland

This is a reprint of articles from the Special Issue published online in the open access journal *International Journal of Molecular Sciences* (ISSN 1422-0067) (available at: www.mdpi.com/journal/ijms/special_issues/1L43R8082K).

For citation purposes, cite each article independently as indicated on the article page online and as indicated below:

Lastname, A.A.; Lastname, B.B. Article Title. *Journal Name* **Year**, *Volume Number*, Page Range.

ISBN 978-3-7258-1076-5 (Hbk)
ISBN 978-3-7258-1075-8 (PDF)
doi.org/10.3390/books978-3-7258-1075-8

Cover image courtesy of Mary Anne Melo

© 2024 by the authors. Articles in this book are Open Access and distributed under the Creative Commons Attribution (CC BY) license. The book as a whole is distributed by MDPI under the terms and conditions of the Creative Commons Attribution-NonCommercial-NoDerivs (CC BY-NC-ND) license.

Contents

About the Editor

Mary Anne Melo

Mary Anne Melo (DDS, MSc, Ph.D., Fellow ADM) is a clinical professor at the University of Maryland School of Dentistry and currently serves as chair of the Department of General Dentistry. Dr. Melo applies her experience as a dentist and dental materials researcher to advance the development of smart and bioactive restorative materials. Her clinical areas of interest focus on minimally invasive dentistry, management of high-caries-risk patients and esthetic dentistry. Her research has focused chiefly on anticaries approaches for caries-inhibiting, antibacterial or remineralization functionalities. Her research group has pioneered the investigations of antibacterial and remineralizing dental adhesives and resin composites. Dr. Melo is a current member of the Academy of Operative Dentistry, the International Association for Dental Research, The American Academy of Cariology and the American Academy of Cosmetic Dentistry. Dr. Melo is a co-inventor of two patents, has edited three books and has published more than 200 papers in the area of dental materials. She lectures nationally and internationally on diverse topics of restorative dentistry.

Preface

Through the exploration of the rapidly evolving field of dental materials, this reprint aims to provide a comprehensive examination of the latest advancements and innovations in restorative dentistry. Inspired by the current Special Issue of the *International Journal of Molecular Sciences*, we delve into the processes pushing the boundaries of dental materials and techniques. Our goal with this reprint is to build on the core research featured in the referenced editorial and pave the way for new advances in dental materials. It underscores the vital importance of collaborative efforts that merge materials science, engineering and clinical aspects to tackle ongoing challenges and explore exciting new opportunities in restorative dentistry.

Mary Anne Melo
Editor

International Journal of
Molecular Sciences

Article

Effects of TiO_2 Nanotubes and Reduced Graphene Oxide on *Streptococcus mutans* and Preosteoblastic Cells at an Early Stage

Min-Kyung Ji [1,†], Hyeonji Kim [2,†], Geonwoo Jeong [3], Won-Jae Kim [4], Je-Hwang Ryu [5], Hoonsung Cho [3,*] and Hyun-Pil Lim [1,2,*]

1 Dental 4D Research Center, Chonnam National University, 33 Yongbong-ro, Buk-gu, Gwangju 61186, Republic of Korea; perizimin@gmail.com
2 Department of Prosthodontics, School of Dentistry, Chonnam National University, 33 Yongbong-ro, Buk-gu, Gwangju 61186, Republic of Korea; otwohh@naver.com
3 Department of Materials Science & Engineering, Chonnam National University, 77 Yongbong-ro, Buk-gu, Gwangju 61186, Republic of Korea; rjsdn7927@daum.net
4 Department of Oral Physiology, School of Dentistry, Chonnam National University, 33 Yongbong-ro, Buk-gu, Gwangju 61186, Republic of Korea; wjkim@jnu.ac.kr
5 Department of Pharmacology and Dental Therapeutics, School of Dentistry, Chonnam National University, 77 Yongbong-ro, Buk-gu, Gwangju 61186, Republic of Korea; jesryu@jnu.ac.kr
* Correspondence: cho.hoonsung@jnu.ac.kr (H.C.); mcnihil@jnu.ac.kr (H.-P.L.); Tel.: +82-62-530-1717 (H.C.); +82-62-530-5577 (H.-P.L.)
† These authors contributed equally to this work.

Abstract: The effects of TiO_2 nanotube (TNT) and reduced graphene oxide (rGO) deposition onto titanium, which is widely used in dental implants, on *Streptococcus mutans* (*S. mutans*) and preosteoblastic cells were evaluated. TNTs were formed through anodic oxidation on pure titanium, and rGO was deposited using an atmospheric plasma generator. The specimens used were divided into a control group of titanium specimens and three experimental groups: Group N (specimens with TNT formation), Group G (rGO-deposited specimens), and Group NG (specimens under rGO deposition after TNT formation). Adhesion of *S. mutans* to the surface was assessed after 24 h of culture using a crystal violet assay, while adhesion and proliferation of MC3T3-E1 cells, a mouse preosteoblastic cell line, were evaluated after 24 and 72 h through a water-soluble tetrazolium salt assay. TNT formation and rGO deposition on titanium decreased *S. mutans* adhesion ($p < 0.05$) and increased MC3T3-E1 cell adhesion and proliferation ($p < 0.0083$). In Group NG, *S. mutans* adhesion was the lowest ($p < 0.05$), while MC3T3-E1 cell proliferation was the highest ($p < 0.0083$). In this study, TNT formation and rGO deposition on a pure titanium surface inhibited the adhesion of *S. mutans* at an early stage and increased the initial adhesion and proliferation of preosteoblastic cells.

Keywords: TiO_2 nanotubes; reduced graphene oxide; nonthermal atmospheric plasma; *Streptococcus mutans*; preosteoblastic cells

check for updates

Citation: Ji, M.-K.; Kim, H.; Jeong, G.; Kim, W.-J.; Ryu, J.-H.; Cho, H.; Lim, H.-P. Effects of TiO_2 Nanotubes and Reduced Graphene Oxide on *Streptococcus mutans* and Preosteoblastic Cells at an Early Stage. *Int. J. Mol. Sci.* **2024**, *25*, 1351. https://doi.org/10.3390/ijms25021351

Academic Editor: Luigi Canullo

Received: 23 November 2023
Revised: 12 January 2024
Accepted: 17 January 2024
Published: 22 January 2024

Copyright: © 2024 by the authors. Licensee MDPI, Basel, Switzerland. This article is an open access article distributed under the terms and conditions of the Creative Commons Attribution (CC BY) license (https://creativecommons.org/licenses/by/4.0/).

1. Introduction

The loss of supporting tissues around dental implants caused by bacterial infection is one of the many causes of implant failure [1,2]. Simonis et al. [3] reported in a follow-up study over 10 to 16 years that, despite the implant success rate of 82.94%, the biological failure rate of the implants was 16.94% and the technical failure rate was 31.09. Peri-implantitis is a term used to describe cases in which inflammation generated in the tissues surrounding osseointegrated implants has destructively progressed to the extent that it affects implant stability [4]. The bacterial colonization pattern in the biofilm of natural teeth is similar to that of dental implants with peri-implantitis [5].

Streptococcus mutans (*S. mutans*) is a bacterium that easily attaches to the tooth surface and tissue, making large contributions to initial biofilm formation [6]. *S. mutans* induces the adhesion and proliferation of bacteria that otherwise lack adhesion mechanisms, causing periodontal disease and increasing the likelihood of recurrence after treatment [7]. The initial biofilm formed by *S. mutans* facilitates late bacterial colonization through interactions with other bacteria, including *Aggregatibacter actinomycetemcomitans* (*Aa*), *Fusobacterium nucleatum* (*Fn*), and *Porphyromonas gingivalis* (*P. gingivalis*), and this biofilm accumulates down to the subgingival area [8]. Therefore, to ensure successful implantation, it is important to promote the osseointegration of implant surfaces by suppressing the adhesion of bacteria, such as *S. mutans*, onto the implant surfaces after implantation. It has been reported that the topography of implant surfaces affects not only the adhesion of cells but also that of bacteria [9,10]. Various attempts at surface modification to increase the degree of bonding between implant surfaces and bone tissues have been made [11]. On titanium surfaces exposed to air, oxide films such as TiO, amorphous TiO_2, TiO_2 (anatase), TiO_2 (brookite), TiO_2 (rutile), Ti_2O_3, and Ti_xO_y are formed [12]. The titanium oxide film promotes osseointegration formation and corrosion resistance, but the naturally occurring oxide film is not dense, and the film layer is thin. Anodic oxidation, one of the methods for modifying the titanium surface, is a method of electrochemically forming a thin, rough, porous amorphous TiO_2 oxide film on the titanium surface [13]. The formation of amorphous TiO_2 nanotubes (TNTs) on titanium surfaces by anodic oxidation increased the surface area of titanium, leading to enhanced cell adhesion, proliferation, and differentiation. This result has been reported to improve the bonding strength between the dental implant and the surrounding bone and promote osteogenesis [14–16].

Graphene oxide (GO) nanomaterials have been increasingly studied for biomedical applications including drug delivery carriers, imaging agents, biosensors, and tissue engineering scaffolds, due to their outstanding physicochemical, optical, electrical, and mechanical properties [17,18]. In particular, the potential of graphene and its derivatives as 2D culture platforms for the differentiation of various types of stem cells towards osteogenesis has attracted considerable attention [19]. GO is a material with a monolayer structure in which carbon atoms are bonded covalently in a hexagonal honeycomb shape. Since GO has various functional groups on its surface, including epoxy, hydroxyl, carbonyl, and carboxylic acid, it was reported that it could serve as a GO-based composite [20–22]. GO is typically synthesized by mechanically or chemically exfoliating bulk graphite and then depositing it on a metal catalyst using chemical vapor deposition (CVD) and Hummer's technique [23–25]. GO is hydrophilic due to functional groups such as carboxyl, hydroxyl, and epoxy and has a high mechanical flexibility. However, these GO deposition methods have disadvantages that include the potential for contamination by residue from the solution used during manufacturing and the generation of toxic gas [26–28].

Plasma is a partially ionized gas containing highly reactive particles such as electrically excited atoms, molecules, and free radical species [29]. Various gases, through atmospheric plasma, break down carbon molecules into carbon atoms and create GO [30,31]. GO deposition using atmospheric plasma is simple and cost-effective and has the advantage that no other additives or by-products are produced during the production of GO [32]. In this study, reduced GO (rGO) was synthesized using the atmospheric plasma method, which allows the synthesis of rGO without using any catalyst for effective deposition on various substrates. GO becomes rGO through a reduction process, which is hydrophobic. Hydrophobic rGO was synthesized by direct deposition using a mixture of methane and argon gas while super hydrophobic [33].

Most studies on graphene have focused on whether it is toxic in vitro and in vivo [34,35]. On the other hand, the in vivo bioactive potential of graphene and related materials remains to be studied. Although graphene-based materials have shown appropriate biocompatibility when used in orthopedic implants, little research has been performed to specifically test the biocompatibility of graphene for dental applications. Lee et al. [36] reported that rGO-based composite materials fabricated to accelerate bone regeneration have the potential to stimulate osteogenesis. Williams et al. [37] reported that particle size, shape, and concentration of graphene-based materials for applications are major factors affecting cytotoxicity, antibacterial properties, and cell differentiation ability.

This study aimed to determine the biofilm formation of *S. mutans* and the activity of MC3T3-E1 cells, a mouse preosteoblastic cell line, on titanium surfaces fabricated with TNTs and rGO.

2. Results

2.1. Surface Characteristics

In the scanning electron microscope (SEM) observations, TNTs were formed in Group N. It was also observed that, in Group G, rGO was deposited in the form of a cloud, and that, in Group NG, TNTs that formed on the surface were covered by the deposited rGO (Figure 1).

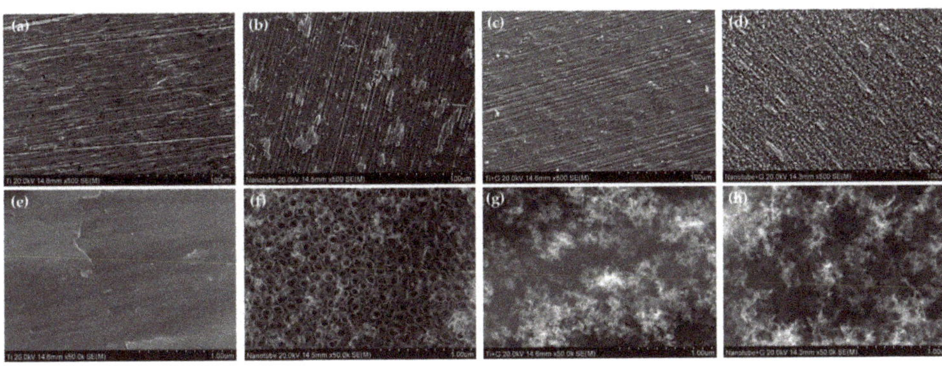

Figure 1. Surface morphology of (**a**,**e**) polished titanium (control group), (**b**,**f**) TiO_2 nanotube titanium (Group N), (**c**,**g**) reduced graphene oxide-deposited titanium (Group G), and (**d**,**h**) reduced graphene oxide-deposited TiO_2 nanotube titanium (Group NG) ((**a**–**d**): magnification = 500, (**e**–**h**): magnification = 50,000).

Surface roughness was measured to be 0.145 ± 0.036 μm for the control group, 0.204 ± 0.030 μm for Group N, 0.245 ± 0.032 μm for Group G, and 0.289 ± 0.021 μm for Group NG. The control group showed the lowest roughness, followed by Group N, Group G, and finally, Group NG, which had the highest roughness. The change in contact angle according to TNT formation and rGO deposition is shown in Figure 2. The contact angle was measured to be 70.26 ± 0.11° for the control group, 58.21 ± 4.32° for Group N, 97.10 ± 3.90° for Group G, and 99.92 ± 1.70° for Group NG. The control group showed the lowest contact angle, followed by Groups N, G, and NG.

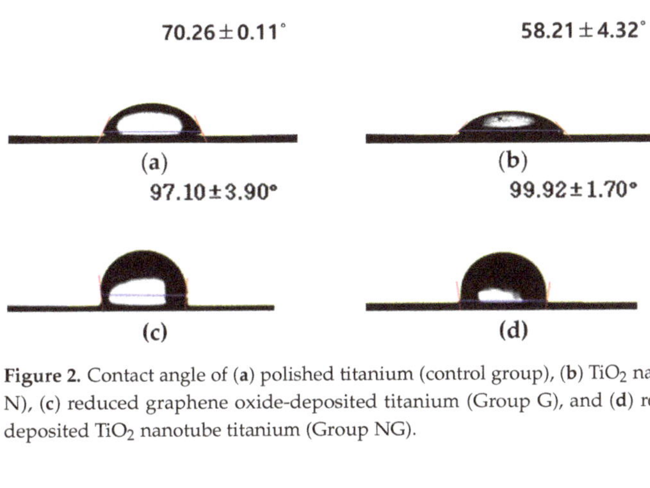

Figure 2. Contact angle of (**a**) polished titanium (control group), (**b**) TiO_2 nanotube titanium (Group N), (**c**) reduced graphene oxide-deposited titanium (Group G), and (**d**) reduced graphene oxide-deposited TiO_2 nanotube titanium (Group NG).

The results of a chemical composition analysis of the rGO deposits are shown in Figure 3. Of the three C1 peaks (284.6, 285.7, and 287.8 eV) detected in the titanium samples, the 284.6 eV peak reflects C-C bonded carbon, while the 285.7 eV peak reflects C-O bonded carbon, and the 287.8 eV peak reflects C=O bonded carbon. In the case of Ti2p, two distinct peaks were observed at 463.8 and 458.1 eV, indicating TiO_2.

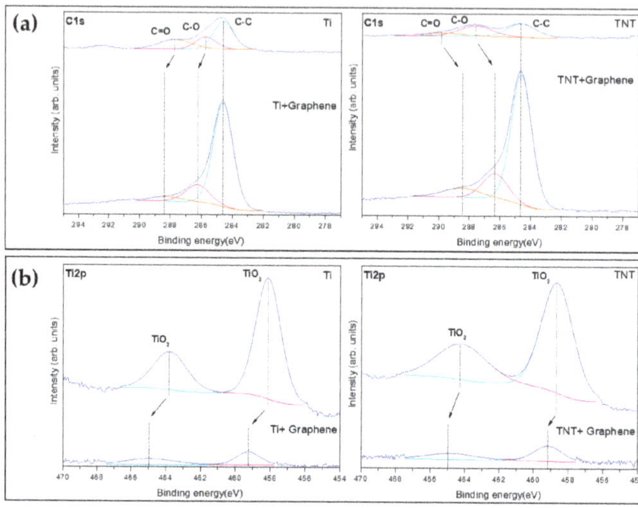

Figure 3. Chemical composition analysis of the reduced graphene oxide deposits. XPS high-resolution spectra at (**a**) C1s and (**b**) Ti2p. The Cls peak has decomposed into ClsA (cyan line), C1sB (magenta line), C1sC (orange line) and the Ti2p peak has decomposed into Ti2p1 (magenta line), Ti2p3 (cyan line). Dashed lines indicate that the binding energy of each peak has shifted.

After rGO deposition, the C1 peaks of the Ti samples increased, and the low peaks between 288.3 and 286.2 eV were slightly shifted to a higher bonding energy. The C1 peaks of TNT also increased, while the low peaks between 289.7 and 287.6 eV were slightly shifted to a lower bonding energy. The Ti2p peaks were slightly shifted to a higher bonding energy.

Based on the peaks shown in XPS survey profiles, the atomic percentage of each control and experimental group was graphed (Figure 4). For the control group, the carbon content of the surface was 32.99 at%, the oxygen content was 56.04 at%, and the titanium content was 10.96 at%. For Group N, the carbon content of the surface was 38.77 at%, the oxygen content was 48.77 at%, and the titanium content was 12.46 at%. For Group G, it was observed that the carbon content of the surface was significantly increased to 94.69 at%, the oxygen content was 5.00 at%, and the titanium content was significantly decreased to 0.32 at%. For Group NG, the carbon content of the surface was 94.79 at%, the oxygen content was 5.02 at%, and the titanium content was 0.19 at%.

The samples were analyzed via Raman spectroscopy to validate the presence of rGO. The D, G, and 2D bands in Group G occurred at 1350, 1593, and 2683 cm^{-1}, respectively, indicating that rGO was successfully synthesized (Figure 5a). Similarly, a major band position at 1348, 1583, and 2683 cm^{-1}, corresponding to rGO, was detected in Group NG (Figure 5b). The relative intensity ratio of I_{2D}/I_G can be used to distinguish the number of layers of graphene. The intensities of Group G, as shown in Figure 5a, are 550, 496, and 448 corresponding to the D, G, and 2D bands, respectively, and the intensity ratio for I_{2D}/I_G was 0.9, showing a multi-layered rGO. The intensities of Group NG, shown in Figure 5b, are 810, 606, and 453 for the D, G, and 2D peaks, respectively, and its ratio for I_{2D}/I_G is 0.74, which corresponds to a multi-layered rGO. The I_D/I_G ratio of Group G and NG is 1.1 and 1.3, respectively.

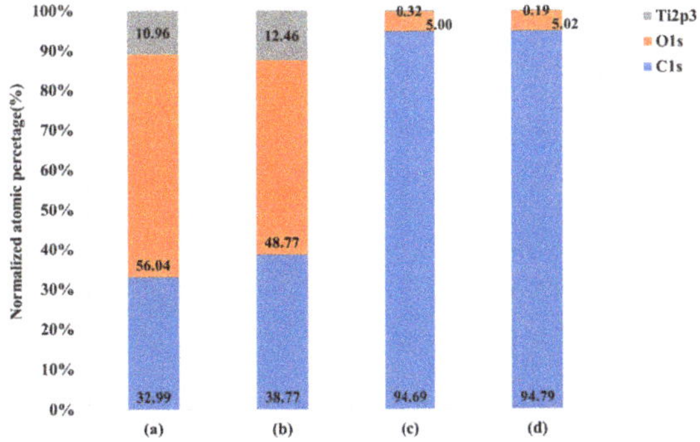

Figure 4. Normalized atomic percentage of each element on (**a**) polished titanium (control group), (**b**) TiO_2 nanotube titanium (Group N), (**c**) reduced graphene oxide-deposited titanium (Group G), and (**d**) reduced graphene oxide-deposited TiO_2 nanotube titanium (Group NG).

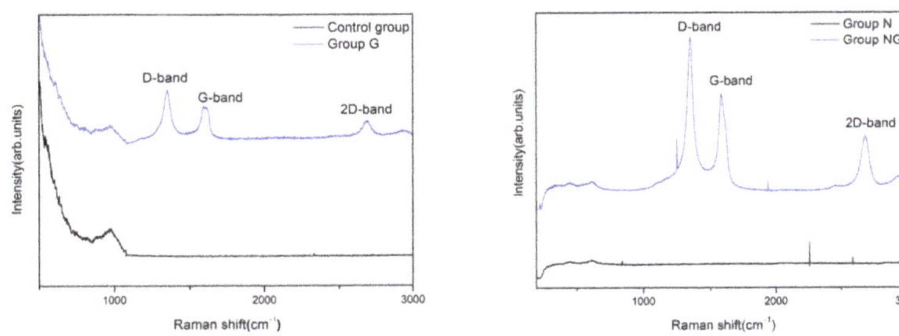

(**a**) (**b**)

Figure 5. Characterization of reduced graphene oxide-deposited surfaces using Raman spectroscopy on (**a**) polished titanium (control group) and reduced graphene oxide-deposited titanium (Group G) and (**b**) TiO_2 nanotube titanium (Group N) and reduced graphene oxide-deposited, TiO_2 nanotube titanium (Group NG).

2.2. *Assessment of the Ability to Inhibit Biofilm Formation*

By assessing the adhesion of *S. mutans* to specimens, it was found that adhesion was significantly reduced in all the experimental groups (Groups N, G, and NG) compared with the control group; further, the adhesion of *S. mutans* in Group NG was significantly lower than in Group N ($p < 0.05$; Figure 6).

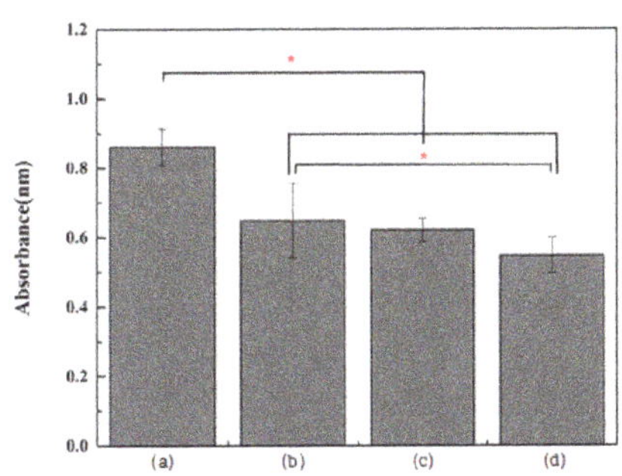

Figure 6. Absorbance level of *S.mutans* after incubation for 24 h on (**a**) polished titanium (control group), (**b**) TiO_2 nanotube titanium (Group N), (**c**) reduced graphene oxide-deposited titanium (Group G), and (**d**) reduced graphene oxide-deposited TiO_2 nanotube titanium (Group NG) (the result of one-way ANOVA test, *: significant at $p < 0.05$).

2.3. Assessment of Osteoblastic Activity

In total, 4×10^4 cells/mL of MC3T3-E1 cells were dispensed onto the specimen, and after culturing for 4 h, the adhesion of the cells was observed using an SEM. In all specimens, MC3T3-E1 cells were observed to be well attached. Compared to the control group, the number of cells attached to the surface was higher in Group N, G, and NG. No obvious differences were observed between the TNT-formed surface and the rGO-deposited surface (Figure 7).

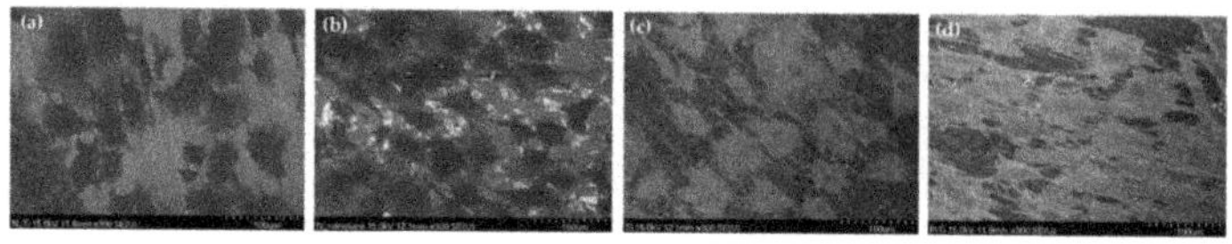

Figure 7. SEM images of MC3T3-E1 cells after incubation for 4 h on (**a**) polished titanium (control group), (**b**) TiO_2 nanotube titanium (Group N), (**c**) reduced graphene oxide-deposited titanium (Group G), and (**d**) reduced graphene oxide-deposited TiO_2 nanotube titanium (Group NG) ((**a**–**d**): magnification = 300).

The viability of MC3T3-E1 cells was significantly increased in all three experimental groups compared with the control group ($p < 0.0083$). The surfaces of the two rGO-coated groups (Group G and NG) were associated with significantly higher MC3T3-E1 cell viability compared with those on which TNTs were formed (Group N) ($p < 0.0083$; Figure 8). The proliferation of MC3T3-E1 cells was significantly increased in all three experimental groups compared with the control group ($p < 0.0083$). In Group NG, cell proliferation was significantly higher than that of surfaces on which TNTs were formed (Group N) and those on which rGO was coated (Group G) ($p < 0.0083$; Figure 9).

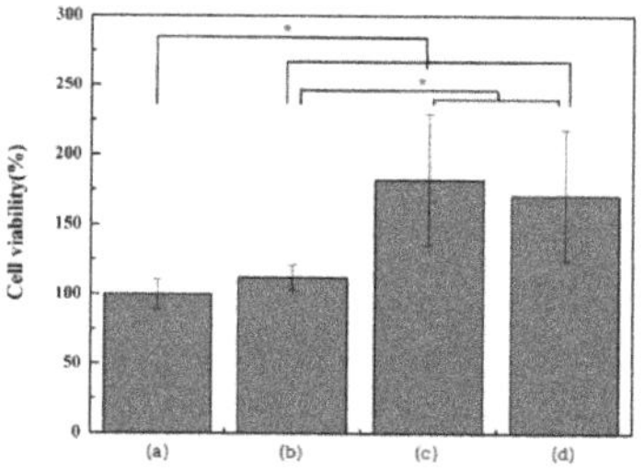

Figure 8. Evaluation of MC3T3-E1 cell viability after incubation for 24 h on (**a**) polished titanium (control group), (**b**) TiO_2 nanotube titanium (Group N), (**c**) reduced graphene oxide-deposited titanium (Group G), and (**d**) reduced graphene oxide-deposited TiO_2 nanotube titanium (Group NG) (the result of Kruskal–Wallis test, *: significant at $p < 0.0083$).

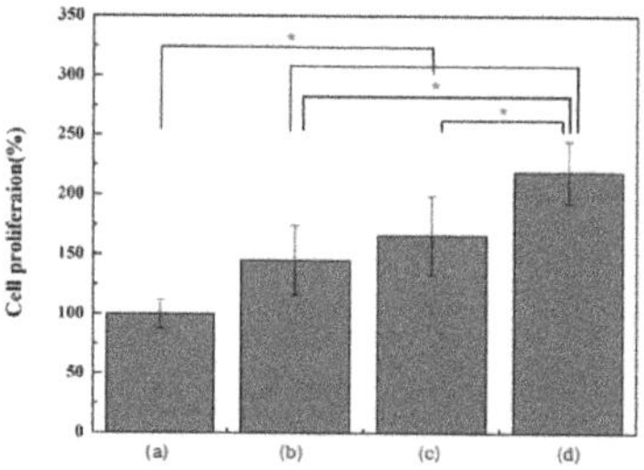

Figure 9. Evaluation of MC3T3-E1 cell proliferation after incubation for 72 h on (**a**) polished titanium (control group), (**b**) TiO_2 nanotube titanium (Group N), (**c**) reduced graphene oxide-deposited titanium (Group G), and (**d**) reduced graphene oxide-deposited TiO_2 nanotube titanium (Group NG) (the result of Kruskal–Wallis test, *: significant at $p < 0.0083$).

3. Discussion

Bacterial adhesion and osteoblastic activity are significantly affected by the surface morphology and roughness of the implants. This study sought to examine the extent to which TNT formation and rGO deposition onto titanium surfaces inhibit the adhesion of *S. mutans* and activate osteoblasts. High surface roughness increases the accumulation of biofilm [7], as well as the initial adhesion of osteoblasts [38,39]. In the crystal violet assay conducted in this study, it was observed that as surface roughness increased—in the order of the control group (0.145 $\pm$ 0.036 μm), Group N (0.204 $\pm$ 0.030 μm), Group G (0.245 $\pm$ 0.032 μm), and Group NG (0.289 $\pm$ 0.021 μm)—the adhesion of *S. mutans* decreased. This result contradicts the findings of other studies, which have reported that increases in surface roughness facilitated bacterial colonization, resulting in increased bacterial adhesion [40,41]. Bacterial behavior varies depending on the size and topographic properties of TNTs. Shi et al. [42] reported that the number of bacteria cultured on the relatively rough surface of TNTs was significantly lower than the number of bacteria cultured on smooth Ti surfaces. The roughened nanointerface properties of TiO_2 cause a stress response in some bacteria, leading to the rupture of the bacterial cell membrane and apoptosis. In addition, the adhesion of *S. mutans* onto rGO-deposited surfaces was further reduced compared with adhesion on the control group and TNT surfaces. The known antibacterial mechanisms of GO are as follows: (1) physical direct interaction of extremely sharp edges of nanomaterials with cell wall membrane [43], (2) ROS generation [44], (3) trapping the bacteria within the aggregated nanomaterials [45], (4) oxidative stress [46], (5) interruption in the glycolysis process of the cells, (6) DNA damaging [47], (7) metal ion release [48], and (8) contribution in generation/explosion of nanobubbles [49]. Bacterial adhesion was further reduced on the rGO-deposited surfaces after TNT formation, even more so than on the rGO-deposited surfaces alone. After the surface roughness increased with TNT formation, the bacterial cell wall membrane was directly damaged by the sharp edges of the deposited rGO, which appears to have further enhanced the antibacterial effect. In this study, it was found that rGO deposition decreased hydrophilicity and increased hydrophobicity. It has been reported that bacteria adhere better to hydrophobic surfaces than to hydrophilic ones [50]. The bacterium *S. mutans* used in this study is reportedly hydrophobic [51]. It can be expected that the surface rGO coating reduces hydrophilicity, thereby increasing the adhesion of the hydrophobic *S. mutans*. However, in this study, when comparing the control group and the rGO-deposited groups (Groups G and NG), it was observed that bacterial adhesion was significantly reduced in the rGO-deposited groups. Oxidative stress caused by rGO functional groups may damage biofilms on the specimens. Although the rGO coating reduced surface hydrophilicity and increased surface hydrophobicity, bacterial adhesion was reduced, as the increased hydrophobicity was offset by the physical properties of rGO.

Raman spectroscopy is widely used to characterize crystal structure, disorder, and defects in graphene-based materials [52]. In this study, we analyzed the samples using Raman spectroscopy to confirm the presence of rGO. Figure 5 shows the Raman spectra of rGO. The peak at 1350 and 1348 cm^{-1} in Group G and Group NG, respectively, corresponds to the D band, caused by first-order Raman scattering due to the E_{2g} mode following the Raman selection rule among Γ modes. This band is present in all sp^2 carbon systems forming a C-C bond. The peak at 1593 and 1583 cm^{-1} in Group G and Group NG, respectively, corresponds to the G band, caused by the A_{1g} vibration mode, and its intensity can increase due to disorder structure of the plane. The peak at 2683 cm^{-1} in both Group G and Group NG represents the 2D band, indicating secondary scattering in which two phonons of D mode are emitted [53]. Figure 5 shows that the I_D/I_G ratio of rGO shows an increasing trend in Group G (1.1) compared with in Group NG (1.3). This suggest that more graphitic domains are formed and the sp^2 cluster number is increased [54]. Additionally, the number of graphene layers can be estimated through the I2D/G ratio. A 2D/G ratio of >2, 1~2, and <1 corresponds to single-layered, double-layered, and multi-layered graphene, respectively [53]. Therefore, it is evident that both Group G and Group NG are deposited with multi-layers of rGO, attributed to the synthesis method using nonthermal atmospheric plasma.

As for the changes in osteoblast activity, the adhesion and proliferation of MC3T3-E1 cells increased significantly in all three experimental groups compared with the control group. TNT formation on titanium surfaces also exhibits a higher proliferation rate than the control group. Oh et al. [55] reported that TNT arrays on titanium induced osteoblast proliferation by 300–400% compared to unmodified titanium surfaces. The reason for this is that the lateral spacing of nanoscale features can influence and change cell behavior [56]. The improved surface roughness of current bioactive implants is one of the important factors that provide appropriate clues to a good cellular response to the implanted material. Many studies on the effects of macro- and micro-roughness on cellular response and tissue formation have been inconclusive. However, it is reported that the increased roughness and surface area on the nanotube surface compared to unmodified titanium surfaces can influence and change certain cell behavior [14,57]. TNTs are known to have a greater surface energy than the untreated Ti [58]. It is reported that the contact angle, which means the wettability of the surface, is improved to be more hydrophilic on the nanotube surface, which is advantageous for improving protein adsorption and cell adhesion [59]. However, it has been reported that in the case of bacteria, TNTs can significantly reduce the number of cells attached to the surface [60]. Currently, opinions are inconsistent on how the hydrophilic/free properties of materials regulate bacterial cell behavior. Previous studies have reported that improving the hydrophilicity of the surface can promote bacterial attachment and proliferation, and that bacteria grow as the diameter of TNT increases [61,62]. However, Xiaoguo et al. [42] reported that as the diameter of TNTs increased, *P.gingivalis* tended to decrease and then increase again, indicating that there are complex factors controlling bacterial behavior at the biocompatible interface.

Item et al. [63] formed TNT on the surface of Ti6Al4V-ELI, electrophoretically deposited rGO, and reported the investigation of the antibacterial activity of *Staphylococcus aureus* and biocompatibility of L-929 fibroblast cells. TNT-rGO enhanced antibacterial activity without causing any morphological damage to bacteria, and for L-929 fibroblasts, rGO had a positive effect on cell adhesion and proliferation. In particular, the increase within the rGO-deposited groups was higher and statistically significant. The rGO formed in this study was deposited in the form of a cloud. Although it has a different shape from the surface utilized by Item et al. [63], the reason for the same results is thought to be because rGO has properties that promote the adhesion, proliferation, and differentiation of osteoblasts [58–60].

However, this study has some limitations in that all experiments were conducted in vitro. Although the roughness and surface area of TNTs can influence and change the behavior of bacteria and cells, this study tested only one size of TNT and rGO surfaces deposited under one condition. It is necessary to investigate TNT and rGO surfaces formed under various conditions and to evaluate changes in the properties of these surfaces and their activity against bacteria and cells. The results could differ if it was conducted in an oral environment with saliva. Also, only a single strain of *S. mutans* was used in this study, but the properties of bacteria itself can also influence the result [64]. The cell wall composition of Gram-negative bacteria is different from that of *S. mutans*, which indicates the different surface properties of bacteria. Hence, further studies are needed considering these limitations.

4. Materials and Methods

4.1. Samples

Pure titanium (ASTM Grade IV, Kobe Steel, Kobe, Japan) was prepared as disks with a thickness of 3 mm and a diameter of 15 mm. With a polishing machine (Labopol-5, Struers, Copenhagen, Denmark), the surface of each specimen was polished with 600-grit SiC abrasive paper under running water and sequentially up to 2000-grit SiC abrasive paper to finish. After the specimens were polished, they were ultrasonically cleaned using acetone, ethanol, and distilled water for 15 min each and dried. All specimens were sterilized with ethylene oxide (EO) gas.

4.2. Surface Treatment

4.2.1. Anodic Oxidation

Anodic oxidation was performed using a DC power supply (Fine Power F-3005, SG EMD, Anyang, Republic of Korea). The electrolyte solution was prepared by adding 1 M phosphoric acid and 1.5 wt% hydrofluoric acid to distilled water. A platinum plate was connected to the cathode, and a titanium specimen was connected to the anode. The specimen and the platinum plate were dipped in the electrolyte solution, after which a voltage of 20 V was applied for ten minutes.

4.2.2. Reduced Graphene Oxide Deposition

Deposition of rGO onto the specimens was performed using a nonthermal atmospheric plasma generator (PGS-300, Expantech Co., Suwon, Republic of Korea). Argon gas (4 L/min) and methane gas (3.5 L/min) were mixed in a quartz tube to generate plasma, and this plasma was then applied to the specimens at a rate of 10 L/min with a power of 300 W using a plasma generator with a high-frequency (900 MHz) resonator (Table 1). The distance between the plasma flame and the specimens was maintained at 15 mm, and the plasma flame was moved from left to right while the specimens were rotated at 180 rpm so that the rGO could be evenly deposited. Plasma was applied for a total of six minutes per specimen by setting the plasma to reciprocate 24 times at 15 s per reciprocation. Figure 10 shows a schematic diagram of the atmospheric plasma-based rGO deposition.

Table 1. Parameters of the atmospheric plasma generator.

Parameter	Value
Average working power (W)	300
Voltage (V)	27
Frequency (MHz)	900
Atmospheric pressure (Torr)	760
Plasma density (cm^3)	10^{15}

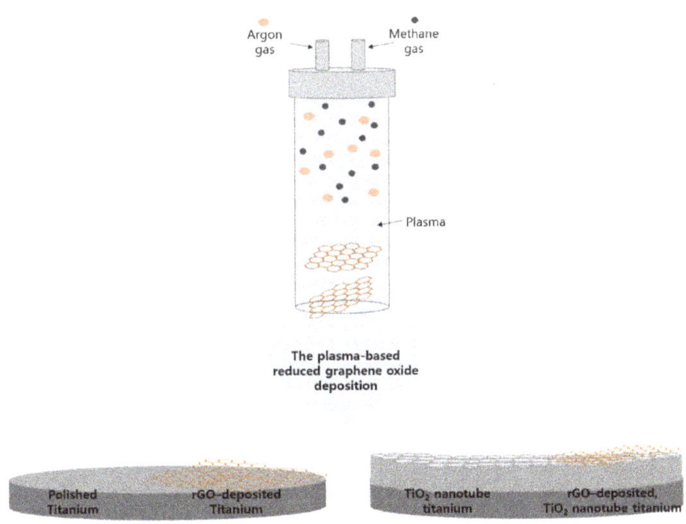

Figure 10. Schematic diagram of the atmospheric plasma-based reduced graphene oxide deposition.

4.3. Classification of Experimental Groups

The specimens used in this study were divided into a control group (polished titanium) and three experimental groups: Group N (TiO_2 nanotube titanium), Group G (rGO-deposited titanium), and Group NG (rGO-deposited TiO_2 nanotube titanium) (Table 2).

Table 2. Experimental groups in this study.

Group	Coating Condition
Control	Polished titanium
N	TiO_2 nanotube titanium
G	Reduced graphene oxide-deposited titanium
NG	Reduced graphene oxide-deposited TiO_2 nanotube titanium

4.4. Assessment of Surface Characteristics

The surface structures of TNTs and rGO formed on the titanium specimens were observed using a field emission scanning electron microscope (S-4700, Hitachi, Horiba, Osaka, Japan). The surface roughness of the control and three experimental groups (N, G, and NG) was measured using a 2D contact stylus profilometer (DIAVITE DH-7, Asmeto AG, Schwyz, Switzerland). Surface roughness was measured at three points on each specimen, and the average of these three values was used as the specimen's average roughness (Ra). To compare changes in the surface hydrophilicity of the specimens, 4 μL of distilled water was dropped on each specimen's surface. The angle between the surface and the solution was measured after ten seconds using a video contact angle measuring device (Phoenix 300, SEO Co., Suwon, Republic of Korea). For each group, the contact angle of the specimens was measured and averaged. X-ray photoemission spectroscopy (XPS; MultiLab 2000, Thermo Electron Corporation, Warwickshire, UK) was performed to assess the elemental changes of the surfaces after rGO deposition. The area values of each peak for the detected elements were normalized and expressed as a quantitative ratio. Laser Raman spectroscopy (NRS-5100, JASCO, Tokyo, Japan) was performed to determine the rGO on the surface at 532.13 nm.

4.5. Assessment of the Ability to Inhibit Biofilm Formation

The ability of the surfaces to inhibit biofilm formation was assessed using a Gram-positive, facultative anaerobic bacterium, *S. mutans* (KCOM 1504), which causes early biofilm formation. *S. mutans* strains were purchased from the Korean Collection for Oral Microbiology (KCOM, Gwangju, Republic of Korea) and cultured in brain heart infusion (BHI; Becton, Dickinson and Company, Sparks, MD, USA) medium. A single colony of *S. mutans* formed on a solid medium was transferred into a liquid medium and cultured in an incubator (LIB-150M, DAIHAN Labtech Co., Namyangju, Republic of Korea) at 37 °C. Biofilm formation was assessed via a crystal violet staining assay. Bacteria at a concentration of 1.5×10^7 CFU/mL were inoculated on the specimens, which were then cultured. After *S. mutans* was cultured for 24 h, the specimens were carefully washed twice with phosphate-buffered saline (PBS) to remove bacteria that had failed to attach to the specimens. A 0.3% crystal violet solution was dispensed onto the specimens, staining them for ten minutes. After removing the crystal violet solution using suction, the specimens were washed three times with PBS and dried for 15 min. A destaining solution (80% ethyl alcohol and 20% acetone) was dispensed onto the dried specimens and stirred for one hour. Then, 200 μL of the destaining solution was placed in a 96-well plate, and the absorbance was measured at 595 nm using a VersaMax ELISA microplate reader (Molecular Devices, San Jose, CA, USA).

4.6. Osteoblastic Activity

MC3T3-E1 mouse (*Mus musculus*) preosteoblastic cells were purchased from the American Type Culture Collection (ATCC; Manassas, VA, USA). The cells were cultured in an α-Minimum Essential Medium (α-MEM; Gibco-BRL, Grand Island, NY, USA) containing 10% fetal bovine serum (FBS) and 100 U/mL penicillin at 37 °C in a 5% CO_2 culture incubator (FormaSeries II 3111 Water Jacketed CO_2 Incubator, Thermo Scientific, Waltham, MA, USA). Culture media were replaced every three days, and the cells were subcultured until the number of cells was sufficient for the necessary tests. Four to seven generations of cells were used in this study. Ten specimens per group were placed in a 24-well plate, and 4×10^4 cells/mL of MC3T3-E1 cells were dispensed onto the specimens and cultured in an incubator set at 5% CO_2 and 37 °C for 24 and 72 h. After both incubation periods, to assess cell viability and proliferation, the WST-8 reagent (EZ-Cytox, Itsbio, Inc., Seoul, Republic of Korea) was dispensed into each well, and the plate was put into an incubator set at 37 °C and 5% CO_2 for the reaction. When the orange color was developed by the WST-8 reagent, 100 μL of the culture medium containing the reagent was transferred from each well into a 96-well plate, and the absorbance was measured at 450 nm using a VersaMax ELISA microplate reader (Molecular Devices, San Jose, CA, USA).

The adhesion of MC3T3-E1 cells was observed using a scanning electron microscope (SEM). Two specimens per group were placed in a 24-well plate, and 4×10^4 cells/mL of MC3T3-E1 cells were dispensed onto the specimens and cultured in an incubator set at 5% CO_2 and 37 °C for 4 h. After 4 h of cell culture, the cells on the specimens were fixed in 2.5% glutaraldehyde for 2 h. After carefully washing with PBS solution twice, the cells were dehydrated in an ethanol gradient in the order of 40%, 50%, 60%, 70%, 80%, 90%, and 100% for 10 min at each concentration. The sample was dried on a clean bench for 2 h after the dehydration process. The cell morphology was observed using an SEM.

4.7. Statistical Analysis

For the statistical analyses conducted in this study, SPSS Statistics V21.0 (SPSS Inc., Chicago, IL, USA) was used. Data collected to evaluate biofilm-inhibition ability met the assumption of normality according to a Shapiro–Wilk test. Since the assumption of homogeneity of variance was also not violated, statistical analyses were performed using a parametric ANOVA and a post hoc Tukey test. The significance of all the results was tested at the level of a p-value less than 0.05. Since osteoblast adhesion and proliferation data did not satisfy the assumption of normality in the Shapiro–Wilk test, these data were statistically analyzed using a Kruskal–Wallis test, a nonparametric ANOVA method. After a Mann–Whitney U test was performed, the type I error was corrected using Bonferroni's method to find groups of different sizes so that the significance was tested at a p-value less than 0.0083.

5. Conclusions

TNT formation and rGO deposition on a pure titanium surface decreased the adhesion of *S. mutans* at an early stage of 24 h ($p < 0.05$) and increased the adhesion and proliferation of MC3T3-E1 cells ($p < 0.0083$). The rGO-deposited surface with TNTs showed the lowest adhesion rate of *S. mutans* ($p < 0.05$) and the best proliferation of MC3T3-E1 cells ($p < 0.0083$).

Within the limits of this study, we suggest that TNT formation and rGO deposition on pure titanium surfaces can reduce the adhesion of *S. mutans* at an early stage. This surface modification is expected to improve the proliferation of preosteoblastic cells.

Author Contributions: Conceptualization, M.-K.J., H.K. and H.-P.L.; methodology, H.C. and H.-P.L.; software and validation and formal analysis, G.J.; investigation and data curation, M.-K.J. and H.K.; writing—original draft preparation, M.-K.J. and H.K.; writing—review and editing, M.-K.J. and G.J.; visualization and supervision, H.C. and H.-P.L.; project administration, J.-H.R. and W.-J.K.; funding acquisition, J.-H.R. and W.-J.K.; All authors have read and agreed to the published version of the manuscript.

Funding: This work was supported by the National Research Foundation of Korea (NRF), a grant funded by the Korean government (MSIT) (No. 2020R1C1C1005683, No. 2022R1A4A1029312), and by the Korean Fund for Regenerative Medicine (KFRM) grant (Ministry of Science and ICT, Ministry of Health & Welfare, 22A0104L1).

Institutional Review Board Statement: Not applicable.

Informed Consent Statement: Not applicable.

Data Availability Statement: Data are contained within the article.

Conflicts of Interest: The authors declare no conflicts of interest.

References

1. Quirynen, M.; De Soete, M.; Van Steenberghe, D. Infectious risks for oral implants a review of the literature. *Clin. Oral Implant. Res. Rev. Artic.* **2002**, *13*, 1–19. [CrossRef]
2. Mombelli, A.; Müller, N.; Cionca, N. The epidemiology of peri-implantitis. *Clin. Oral Implant. Res.* **2012**, *23*, 67–76. [CrossRef]
3. Simonis, P.; Dufour, T.; Tenenbaum, H. Long-term implant survival and success: A 10–16-year follow-up of non-submerged dental implants. *Clin. Oral Implant. Res.* **2010**, *21*, 772–777. [CrossRef] [PubMed]
4. Sánchez Garcés, M.; Gay Escoda, C. Periimplantitis. *Med. Oral Patol. Oral Cir. Bucal* **2004**, *9*, 63–74.
5. Fürst, M.M.; Salvi, G.E.; Lang, N.P.; Persson, G.R. Bacterial colonization immediately after installation on oral titanium implants. *Clin. Oral Implant. Res.* **2007**, *18*, 501–508. [CrossRef] [PubMed]
6. Krzyściak, W.; Jurczak, A.; Kościelniak, D.; Bystrowska, B.; Skalniak, A. The virulence of *Streptococcus mutans* and the ability to form biofilms. *Eur. J. Clin. Microbiol. Infect. Dis.* **2014**, *33*, 499–515. [CrossRef]
7. Inoue, M.; Koga, T.; Sato, S.; Hamada, S. Synthesis of adherent insoluble glucan by the concerted action of the two glucosyltransferase components of *Streptococcus mutans*. *FEBS Lett.* **1982**, *143*, 101–104. [CrossRef]
8. Badihi Hauslich, L.; Sela, M.N.; Steinberg, D.; Rosen, G.; Kohavi, D. The adhesion of oral bacteria to modified titanium surfaces: Role of plasma proteins and electrostatic forces. *Clin. Oral Implant. Res.* **2013**, *24*, 49–56. [CrossRef] [PubMed]
9. Anselme, K.; Bigerelle, M. Effect of a gold–palladium coating on the long-term adhesion of human osteoblasts on biocompatible metallic materials. *Surf. Coat. Technol.* **2006**, *200*, 6325–6330. [CrossRef]
10. Le Guéhennec, L.; Soueidan, A.; Layrolle, P.; Amouriq, Y. Surface treatments of titanium dental implants for rapid osseointegration. *Dent. Mater.* **2007**, *23*, 844–854. [CrossRef]
11. Wennerberg, A.; Albrektsson, T. On implant surfaces: A review of current knowledge and opinions. *Int. J. Oral Maxillofac. Implant.* **2010**, *25*, 63–74.
12. Park, S.; Min, D.; Lim, H.; Yoon, D.; Lee, K. Effect of heat treatment on phase transition of nanotubular titanium oxide arrays. *J. Nanosci. Nanotechnol.* **2011**, *11*, 1476–1479. [CrossRef] [PubMed]
13. Zhao, J.; Wang, X.; Chen, R.; Li, L. Fabrication of titanium oxide nanotube arrays by anodic oxidation. *Solid State Commun.* **2005**, *134*, 705–710. [CrossRef]
14. Brammer, K.S.; Oh, S.; Frandsen, C.J.; Jin, S. Biomaterials and biotechnology schemes utilizing TiO_2 nanotube arrays. *Biomater. Sci. Eng.* **2011**, *9*, 193–210.
15. Lavenus, S.; Louarn, G.; Layrolle, P. Nanotechnology and dental implants. *Int. J. Biomater.* **2010**, *2010*, 915327. [CrossRef]
16. Ellingsen, J.E.; Johansson, C.B.; Wennerberg, A.; Holmén, A. Improved retention and bone-to-implant contact with fluoride-modified titanium implants. *Int. J. Oral Maxillofac. Implant.* **2004**, *19*, 659–666.
17. Zhang, Y.; Nayak, T.R.; Hong, H.; Cai, W. Graphene: A versatile nanoplatform for biomedical applications. *Nanoscale* **2012**, *4*, 3833–3842. [CrossRef]
18. Feng, L.; Wu, L.; Qu, X. New horizons for diagnostics and therapeutic applications of graphene and graphene oxide. *Adv. Mater.* **2013**, *25*, 168–186. [CrossRef]
19. Nayak, T.R.; Andersen, H.; Makam, V.S.; Khaw, C.; Bae, S.; Xu, X.; Ee, P.-L.R.; Ahn, J.-H.; Hong, B.H.; Pastorin, G. Graphene for controlled and accelerated osteogenic differentiation of human mesenchymal stem cells. *ACS Nano* **2011**, *5*, 4670–4678. [CrossRef]
20. Dreyer, D.R.; Park, S.; Bielawski, C.W.; Ruoff, R.S. The chemistry of graphene oxide. *Chem. Soc. Rev.* **2010**, *39*, 228–240. [CrossRef]
21. Loh, K.P.; Bao, Q.; Ang, P.K.; Yang, J. The chemistry of graphene. *J. Mater. Chem.* **2010**, *20*, 2277–2289. [CrossRef]
22. Zhou, D.; Cheng, Q.-Y.; Han, B.-H. Solvothermal synthesis of homogeneous graphene dispersion with high concentration. *Carbon* **2011**, *49*, 3920–3927. [CrossRef]
23. Yi, M.; Shen, Z. A review on mechanical exfoliation for the scalable production of graphene. *J. Mater. Chem. A* **2015**, *3*, 11700–11715. [CrossRef]
24. Brownson, D.A.; Banks, C.E. The electrochemistry of CVD graphene: Progress and prospects. *Phys. Chem. Chem. Phys.* **2012**, *14*, 8264–8281. [CrossRef]
25. Sali, S.; Mackey, H.R.; Abdala, A.A. Effect of graphene oxide synthesis method on properties and performance of polysulfone-graphene oxide mixed matrix membranes. *Nanomaterials* **2019**, *9*, 769. [CrossRef] [PubMed]
26. Brownson, D.A.; Metters, J.P.; Kampouris, D.K.; Banks, C.E. Graphene electrochemistry: Surfactants inherent to graphene can dramatically effect electrochemical processes. *Electroanalysis* **2011**, *23*, 894–899. [CrossRef]
27. Brownson, D.A.; Banks, C.E. Graphene electrochemistry: Surfactants inherent to graphene inhibit metal analysis. *Electrochem. Commun.* **2011**, *13*, 111–113. [CrossRef]
28. Brownson, D.A.; Banks, C.E. Fabricating graphene supercapacitors: Highlighting the impact of surfactants and moieties. *Chem. Commun.* **2012**, *48*, 1425–1427. [CrossRef]

29. Arora, V.; Nikhil, V.; Suri, N.; Arora, P. Cold atmospheric plasma (CAP) in dentistry. *Dentistry* **2014**, *4*, 1. [CrossRef]
30. Lambert, T.N.; Luhrs, C.C.; Chavez, C.A.; Wakeland, S.; Brumbach, M.T.; Alam, T.M. Graphite oxide as a precursor for the synthesis of disordered graphenes using the aerosol-through-plasma method. *Carbon* **2010**, *48*, 4081–4089. [CrossRef]
31. Phillips, J.; Luhrs, C.C.; Richard, M. Engineering particles using the aerosol-through-plasma method. *IEEE Trans. Plasma Sci.* **2009**, *37*, 726–739. [CrossRef]
32. Rho, K.; Park, C.; Alam, K.; Kim, D.; Ji, M.-K.; Lim, H.-P.; Cho, H. Biological effects of plasma-based graphene oxide deposition on titanium. *J. Nanomater.* **2019**, *2019*, 9124989. [CrossRef]
33. Alam, K.; Jo, Y.Y.; Park, C.-K.; Cho, H. Synthesis of graphene oxide using atmospheric plasma for prospective biological applications. *Int. J. Nanomed.* **2020**, *15*, 5813–5824. [CrossRef] [PubMed]
34. Sanchez, V.C.; Jachak, A.; Hurt, R.H.; Kane, A.B. Biological interactions of graphene-family nanomaterials: An interdisciplinary review. *Chem. Res. Toxicol.* **2012**, *25*, 15–34. [CrossRef] [PubMed]
35. Bianco, A. Graphene: Safe or toxic? The two faces of the medal. *Angew. Chem. Int. Ed.* **2013**, *52*, 4986–4997. [CrossRef] [PubMed]
36. Lee, J.H.; Shin, Y.C.; Lee, S.-M.; Jin, O.S.; Kang, S.H.; Hong, S.W.; Jeong, C.-M.; Huh, J.B.; Han, D.-W. Enhanced osteogenesis by reduced graphene oxide/hydroxyapatite nanocomposites. *Sci. Rep.* **2015**, *5*, 18833. [CrossRef] [PubMed]
37. Williams, A.; Moore, E.; Thomas, A.; Johnson, J. Graphene-Based Materials in Dental Applications: Antibacterial, Biocompatible, and Bone Regenerative Properties. *Int. J. Biomater.* **2023**, *2023*, 8803283. [CrossRef]
38. Chehroudi, B.; Gould, T.; Brunette, D. Effects of a grooved titanium-coated implant surface on epithelial cell behavior in vitro and in vivo. *J. Biomed. Mater. Res.* **1989**, *23*, 1067–1085. [CrossRef]
39. Ricci, J.; Spivak, J.; Blumenthal, N.; Alexander, H. Modulation of bone ingrowth by surface chemistry and roughness. In *The Bone-Biomaterial Interface*; University of Toronto Press: Toronto, ON, Canada, 1991; pp. 334–349.
40. Quirynen, M.; Bollen, C. The influence of surface roughness and surface-free energy on supra-and subgingival plaque formation in man: A review of the literature. *J. Clin. Periodontol.* **1995**, *22*, 1–14. [CrossRef]
41. Rimondini, L.; Farè, S.; Brambilla, E.; Felloni, A.; Consonni, C.; Brossa, F.; Carrassi, A. The effect of surface roughness on early in vivo plaque colonization on titanium. *J. Periodontol.* **1997**, *68*, 556–562. [CrossRef]
42. Shi, X.; Xu, Q.; Tian, A.; Tian, Y.; Xue, X.; Sun, H.; Yang, H.; Dong, C. Antibacterial activities of TiO_2 nanotubes on *Porphyromonas gingivalis*. *RSC Adv.* **2015**, *5*, 34237–34242. [CrossRef]
43. Akhavan, O.; Ghaderi, E. Toxicity of graphene and graphene oxide nanowalls against bacteria. *ACS Nano* **2010**, *4*, 5731–5736. [CrossRef] [PubMed]
44. Dutta, T.; Sarkar, R.; Pakhira, B.; Ghosh, S.; Sarkar, R.; Barui, A.; Sarkar, S. ROS generation by reduced graphene oxide (rGO) induced by visible light showing antibacterial activity: Comparison with graphene oxide (GO). *RSC Adv.* **2015**, *5*, 80192–80195. [CrossRef]
45. Akhavan, O.; Ghaderi, E.; Esfandiar, A. Wrapping bacteria by graphene nanosheets for isolation from environment, reactivation by sonication, and inactivation by near-infrared irradiation. *J. Phys. Chem. B* **2011**, *115*, 6279–6288. [CrossRef] [PubMed]
46. Liu, S.; Zeng, T.H.; Hofmann, M.; Burcombe, E.; Wei, J.; Jiang, R.; Kong, J.; Chen, Y. Antibacterial activity of graphite, graphite oxide, graphene oxide, and reduced graphene oxide: Membrane and oxidative stress. *ACS Nano* **2011**, *5*, 6971–6980. [CrossRef] [PubMed]
47. Pulingam, T.; Thong, K.L.; Appaturi, J.N.; Lai, C.W.; Leo, B.F. Mechanistic actions and contributing factors affecting the antibacterial property and cytotoxicity of graphene oxide. *Chemosphere* **2021**, *281*, 130739. [CrossRef] [PubMed]
48. Wang, Y.-W.; Cao, A.; Jiang, Y.; Zhang, X.; Liu, J.-H.; Liu, Y.; Wang, H. Superior antibacterial activity of zinc oxide/graphene oxide composites originating from high zinc concentration localized around bacteria. *ACS Appl. Mater. Interfaces* **2014**, *6*, 2791–2798. [CrossRef]
49. Jannesari, M.; Akhavan, O.; Hosseini, H.R.M.; Bakhshi, B. Oxygen-rich graphene/ZnO_2-Ag nanoframeworks with pH-switchable catalase/peroxidase activity as O_2 nanobubble-self generator for bacterial inactivation. *J. Colloid Interface Sci.* **2023**, *637*, 237–250. [CrossRef]
50. Liu, Z.; Robinson, J.T.; Sun, X.; Dai, H. PEGylated nanographene oxide for delivery of water-insoluble cancer drugs. *J. Am. Chem. Soc.* **2008**, *130*, 10876–10877. [CrossRef]
51. Nostro, A.; Cannatelli, M.; Crisafi, G.; Musolino, A.; Procopio, F.; Alonzo, V. Modifications of hydrophobicity, in vitro adherence and cellular aggregation of *Streptococcus mutans* by *Helichrysum italicum* extract. *Lett. Appl. Microbiol.* **2004**, *38*, 423–427. [CrossRef]
52. Sobon, G.; Sotor, J.; Jagiello, J.; Kozinski, R.; Zdrojek, M.; Holdynski, M.; Paletko, P.; Boguslawski, J.; Lipinska, L.; Abramski, K.M. Graphene oxide vs. reduced graphene oxide as saturable absorbers for Er-doped passively mode-locked fiber laser. *Opt. Express* **2012**, *20*, 19463–19473. [CrossRef] [PubMed]
53. Shen, Y.; Lua, A.C. A facile method for the large-scale continuous synthesis of graphene sheets using a novel catalyst. *Sci. Rep.* **2013**, *3*, 3037. [CrossRef] [PubMed]
54. Silva Filho, J.; Venancio, E.; Silva, S.; Takiishi, H.; Martinez, L.; Antunes, R. A thermal method for obtention of 2 to 3 reduced graphene oxide layers from graphene oxide. *SN Appl. Sci.* **2020**, *2*, 1450. [CrossRef]
55. Oh, S.; Daraio, C.; Chen, L.H.; Pisanic, T.R.; Fiñones, R.R.; Jin, S. Significantly accelerated osteoblast cell growth on aligned TiO_2 nanotubes. *J. Biomed. Mater. Res. Part A* **2006**, *78*, 97–103. [CrossRef] [PubMed]
56. Popat, K.C.; Daniels, R.H.; Dubrow, R.S.; Hardev, V.; Desai, T.A. Nanostructured surfaces for bone biotemplating applications. *J. Orthop. Res.* **2006**, *24*, 619–627. [CrossRef] [PubMed]

57. Zhou, H.; Gan, X.; Wang, J.; Zhu, X.; Li, G. Hemoglobin-based hydrogen peroxide biosensor tuned by the photovoltaic effect of nano titanium dioxide. *Anal. Chem.* **2005**, *77*, 6102–6104. [CrossRef] [PubMed]
58. Puckett, S.D.; Taylor, E.; Raimondo, T.; Webster, T.J. The relationship between the nanostructure of titanium surfaces and bacterial attachment. *Biomaterials* **2010**, *31*, 706–713. [CrossRef]
59. MacDonald, D.; Deo, N.; Markovic, B.; Stranick, M.; Somasundaran, P. Adsorption and dissolution behavior of human plasma fibronectin on thermally and chemically modified titanium dioxide particles. *Biomaterials* **2002**, *23*, 1269–1279. [CrossRef]
60. Colon, G.; Ward, B.C.; Webster, T.J. Increased osteoblast and decreased *Staphylococcus epidermidis* functions on nanophase ZnO and TiO_2. *J. Biomed. Mater. Res. Part A* **2006**, *78*, 595–604. [CrossRef]
61. Ji, J.; Zhang, W. Bacterial behaviors on polymer surfaces with organic and inorganic antimicrobial compounds. *J. Biomed. Mater. Res. Part A* **2009**, *88*, 448–453. [CrossRef]
62. Lee, J.H.; Lee, S.J.; Khang, G.; Lee, H.B. Interaction of fibroblasts on polycarbonate membrane surfaces with different micropore sizes and hydrophilicity. *J. Biomater. Sci. Polym. Ed.* **1999**, *10*, 283–294. [CrossRef] [PubMed]
63. Turu, I.C.; Bayraktar, S.; Akgul, B.; Ilhan-Sungur, E.; Abamor, E.S.; Cansever, N. Formation of TiO_2 nanotubes and deposition of silver nanoparticle and reduced graphene oxide: Antibacterial and biocompatibility behavior. *Surf. Coat. Technol.* **2023**, *470*, 129866. [CrossRef]
64. Murakami, A.; Arimoto, T.; Suzuki, D.; Iwai-Yoshida, M.; Otsuka, F.; Shibata, Y.; Igarashi, T.; Kamijo, R.; Miyazaki, T. Antimicrobial and osteogenic properties of a hydrophilic-modified nanoscale hydroxyapatite coating on titanium. *Nanomed. Nanotechnol. Biol. Med.* **2012**, *8*, 374–382. [CrossRef] [PubMed]

Disclaimer/Publisher's Note: The statements, opinions and data contained in all publications are solely those of the individual author(s) and contributor(s) and not of MDPI and/or the editor(s). MDPI and/or the editor(s) disclaim responsibility for any injury to people or property resulting from any ideas, methods, instructions or products referred to in the content.

International Journal of *Molecular Sciences*

Article

Disparity in the Influence of Implant Provisional Materials on Human Gingival Fibroblasts with Different Phases of Cell Settlement: An In Vitro Study

Takanori Matsuura [1,*,†], Stella Stavrou [1,†], Keiji Komatsu [1], James Cheng [1], Alisa Pham [1], Stephany Ferreira [2], Tomomi Baba [1], Ting-Ling Chang [1], Denny Chao [1] and Takahiro Ogawa [1]

1 Weintraub Center for Reconstructive Biotechnology, Division of Regenerative and Reconstructive Sciences, UCLA School of Dentistry, Los Angeles, CA 90095, USA; stella.c.stavrou@gmail.com (S.S.); jamescheng@dentistry.ucla.edu (J.C.); alisapham@dentistry.ucla.edu (A.P.); uclaosugi@gmail.com (T.B.); tlchang@dentistry.ucla.edu (T.-L.C.); denchao@ucla.edu (D.C.); togawa@dentistry.ucla.edu (T.O.)

2 West Los Angeles VA Medical Center, Los Angeles, CA 90073, USA; stephanyferreira3@gmail.com

* Correspondence: tmatsuura@g.ucla.edu

† These authors contributed equally to this work.

Abstract: The development of healthy peri-implant soft tissues is critical to achieving the esthetic and biological success of implant restorations throughout all stages of healing and tissue maturation, starting with provisionalization. The purpose of this study was to investigate the effects of eight different implant provisional materials on human gingival fibroblasts at various stages of cell settlement by examining initial cell attachment, growth, and function. Eight different specimens—bis-acrylic 1 and 2, flowable and bulk–fill composites, self-curing acrylic 1 and 2, milled acrylic, and titanium (Ti) alloy as a control—were fabricated in rectangular plates ($n = 3$). The condition of human gingival fibroblasts was divided into two groups: those in direct contact with test materials (contact experiment) and those in close proximity to test materials (proximity experiment). The proximity experiment was further divided into three phases: pre-settlement, early settlement, and late settlement. A cell culture insert containing each test plate was placed into a well where the cells were pre-cultured. The number of attached cells, cell proliferation, resistance to detachment, and collagen production were evaluated. In the contact experiment, bis-acrylics and composites showed detrimental effects on cells. The number of cells attached to milled acrylic and self-curing acrylic was relatively high, being approximately 70% and 20–30%, respectively, of that on Ti alloy. There was a significant difference between self-curing acrylic 1 and 2, even with the same curing modality. The cell retention ability also varied considerably among the materials. Although the detrimental effects were mitigated in the proximity experiment compared to the contact experiment, adverse effects on cell growth and collagen production remained significant during all phases of cell settlement for bis-acrylics and flowable composite. Specifically, the early settlement phase was not sufficient to significantly mitigate the material cytotoxicity. The flowable composite was consistently more cytotoxic than the bulk–fill composite. The harmful effects of the provisional materials on gingival fibroblasts vary considerably depending on the curing modality and compositions. Pre-settlement of cells mitigated the harmful effects, implying the susceptibility to material toxicity varies depending on the progress of wound healing and tissue condition. However, cell pre-settlement was not sufficient to fully restore the fibroblastic function to the normal level. Particularly, the adverse effects of bis-acrylics and flowable composite remained significant. Milled and self-curing acrylic exhibited excellent and acceptable biocompatibility, respectively, compared to other materials.

Keywords: implant provisional materials; peri-implant soft tissue; cytotoxicity; cytocompatibility; gingival fibroblasts

check for updates

Citation: Matsuura, T.; Stavrou, S.; Komatsu, K.; Cheng, J.; Pham, A.; Ferreira, S.; Baba, T.; Chang, T.-L.; Chao, D.; Ogawa, T. Disparity in the Influence of Implant Provisional Materials on Human Gingival Fibroblasts with Different Phases of Cell Settlement: An In Vitro Study. *Int. J. Mol. Sci.* **2024**, *25*, 123. https://doi.org/10.3390/ijms25010123

Academic Editor: Luigi Canullo

Received: 17 November 2023
Revised: 13 December 2023
Accepted: 19 December 2023
Published: 21 December 2023

Copyright: © 2023 by the authors. Licensee MDPI, Basel, Switzerland. This article is an open access article distributed under the terms and conditions of the Creative Commons Attribution (CC BY) license (https://creativecommons.org/licenses/by/4.0/).

1. Introduction

In implant therapy, a soft tissue seal is crucial for preventing bacterial invasion and maintaining the long-term health of peri-implant tissues [1–5]. Peri-implant soft tissue heals and develops during various stages of provisionalization, including implant surgery, second-stage surgery, subsequent wound healing, emergence profile formation, and tissue maturation [6–9]. Implant provisional restorations are typically made of polymer-based materials. Depending on the modality of cure and chemical composition, these resins can be classified as self-cured, light-cured, or heat-cured acrylic or composite resins.

The biological properties and responses of the peri-implant soft tissues can be influenced by the chemical composition of provisional restorative materials [10–13]. Some studies have reported that acrylic materials are cytotoxic mainly due to the continued release of residual monomers even after polymerization [14–19]. In chemical cured materials, almost all materials are mainly polymethyl methacrylate (PMMA)-based or bis-acrylic-based. PMMA is generated by the polymerization of MMA, and it has high moldability and tractability [20–23]. Some bis-acrylics contain fillers to prevent curing shrinkage and photoinitiators to improve handling and reduce curing time [24–27]. PMMA disks are chemically pre-polymerized under high-temperature and high-pressure conditions. These disks can be milled into the desired shape using computer-aided design/computer-aided manufacturing (CAD/CAM) systems. Prefabricated PMMA disks are assumed to produce minimal or no residual monomers or free radicals [28–31]. In light-cured materials, composites with high fluidity contain fewer fillers and lower viscosity matrix materials, such as triethylene glycol dimethacrylate (TEGDMA), while composites with lower fluidity contain more fillers and higher viscosity matrix materials, such as urethane dimethacrylate (UDMA) [32–34]. In many studies, the addition of eluates from acrylic materials to culture medium is commonly performed [17,35–39]. This method allows for maintaining consistent concentration; however, the actual components leaching from the materials tend to decrease over time. Therefore, by conducting cell culture in contact and proximity to the actual material itself, a more accurate evaluation of cellular responses to an in vivo environment is possible.

Fibroblasts play a key role in wound healing by contributing to formation of the extracellular matrix (ECM) components such as collagen, glycoproteins, and other constituents of the developed peri-implant soft tissue, creating a soft tissue seal in the connective tissue [40–42]. To maintain tissue function, cells adhere, settle, and interact with other cells and the extracellular matrix in order to maintain the structure and organization of tissues through these complex cellular adhesion mechanisms [43,44]. The influence of external factors, such as restorative materials, may vary depending on the condition of cell settlement [45]. In our previous in vitro studies, the direct contact and close proximity effects of five different provisional materials on fibroblasts and osteoblasts were assessed [46,47]. The cells were seeded simultaneously with the placement of materials; thus, the cell settlement conditions were not taken into consideration. Peri-implant soft tissue heals and develops during various stages of provisionalization, including implant surgery, second-stage surgery, subsequent wound healing, emergence profile formation, and tissue maturation. We need to investigate the cell response to materials using a systemic experimental model that considers these tissue healing processes.

Therefore, the objective of this study is to investigate the effects of eight different provisional materials, including titanium (Ti) alloy as a control, on human gingival fibroblasts at various stages of cell settlement by examining initial cell attachment, growth, and function. We hypothesized that there is a substantial difference between materials before cell settlement, but the differences decrease as cells settle.

2. Results

2.1. Growth of Fibroblasts on Test Materials

To evaluate the success of cell settlement to test plates, fibroblasts on the plates were visualized by fluorescent microscopy, and the number of attached cells was quantified

(Figure 1A). Cell nuclei were stained with DAPI, and actin filaments were stained with rhodamine. Fibroblasts on Ti alloy exhibited a spindle shape and were aligned in the same direction. The cells on milled acrylic were spindle-shaped and spread randomly. A few small fibroblasts were attached to self-curing acrylics. No cells were attached to any other materials.

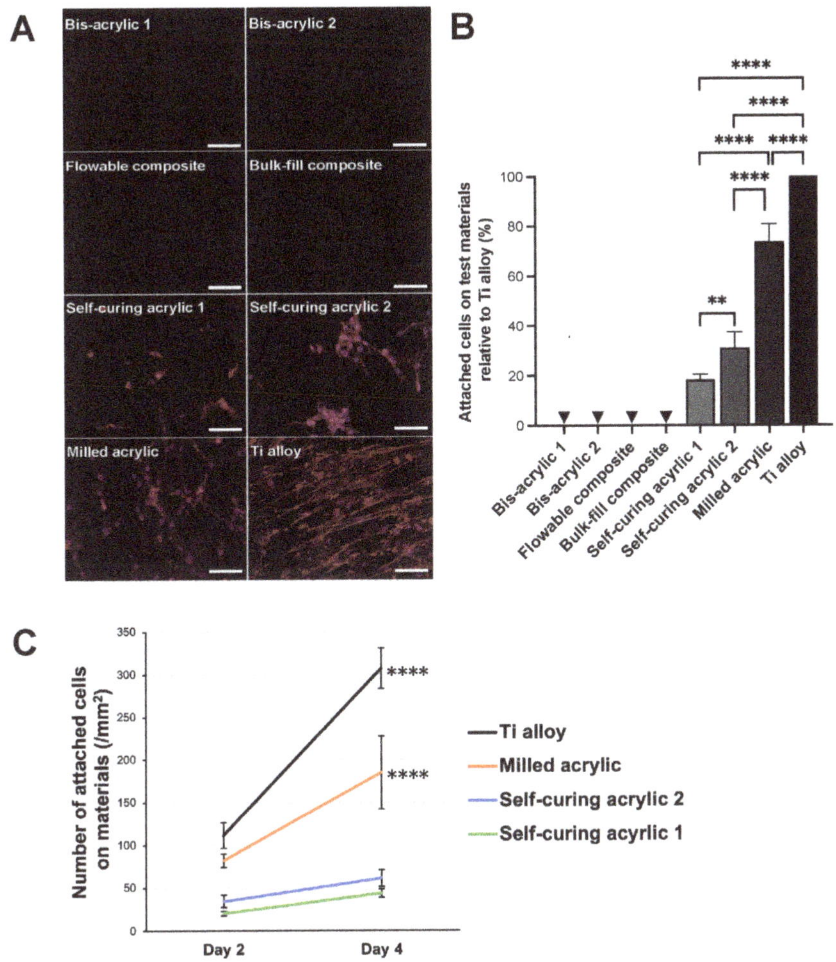

Figure 1. Fibroblasts growth on test materials. (**A**) Visualization of fibroblasts on test materials 2 days after seeding. Fibroblasts were dual stained with fluorescent dyes, DAPI for nuclei, and rhodamine–phalloidin for actin filaments. Scale bars indicate 1 mm. (**B**) Percentage of the number of cells attached to the test materials relative to the number of cells on Ti alloy was measured. Arrowheads indicate not applicable. One-way ANOVA, followed by Tukey–Kramer post hoc test, p ** < 0.01, p **** < 0.0001. (**C**) Cell propagation from day 2 to day 4. Data shown are means ± SD. Two-way ANOVA, followed by the Fisher's LSD post hoc test. p **** < 0.0001.

Cell numbers were determined by counting cell nuclei, and the percentage of cells attached to each test material was calculated relative to that of Ti alloy (Figure 1B). The percentage for milled acrylic was 73.52 ± 7.25% of that for Ti alloy. Irrespective of the same category, the percentage for self-curing acrylic 2 was approximately two times higher than that of self-curing acrylic 1 (31.00 ± 6.44% and 18.21 ± 2.26%, respectively) (p = 0.042).

To assess cell proliferation, changes in cell numbers from day 2 to day 4 were evaluated (Figure 1C). On day 4, cell numbers for Ti alloy and milled acrylic were more than two times higher than on day 2 (307.33 ± 23.62 cells/mm^2 vs. 111.43 ± 14.79 cells/mm^2, and 184.83 ± 42.67 cells/mm^2 vs. 81.93 ± 8.08 cells/mm^2, respectively) ($p < 0.0001$), while there were no significant increases for self-curing acrylic 1 and 2.

2.2. Cell Retention Ability

Next, we assessed the retention of fibroblasts once attached to the materials. Fibroblasts attached to the materials for four days were subjected to chemical detachment, and the percentage of remaining cells was calculated (Figure 2). No cells remained in bis-acrylics and composite materials because no cells adhered to these materials. There was no significant difference between Ti alloy and milled acrylic (37.47 ± 4.99% and 31.42 ± 4.67%, respectively) ($p = 0.22$). At the same time, self-cured acrylic 1 and 2 showed 60–80% lower cell retention than milled acrylic (14.26 ± 3.76% and 7.10 ± 1.37%, respectively) ($p < 0.0001$).

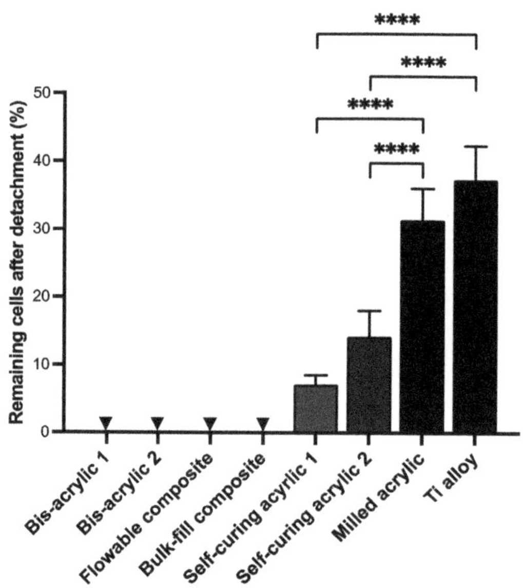

Figure 2. Cell retention on test materials. Cell retention was evaluated by calculating the percentage of remaining cells after chemical detachment. Arrowheads indicate not applicable. Data shown are means ± SD. One-way ANOVA followed by the Tukey–Kramer post hoc test. **** $p < 0.0001$.

2.3. Fibroblast Growth at Different Phases of Cell Settlement

The cells attached to the surface of the well were measured two days after the placement of a cell culture insert containing the test plate. Unlike the contact experiment, cells were observed in all groups in the proximity experiment (Figure 3A). Small cells were sparsely dispersed in the bis-acrylic and flowable composite groups until early settlement, and there were gaps between the increased cells at late settlement. The spindle-shaped cells were densely spread in the self-curing acrylic, milled acrylic, and Ti alloy group after the early settlement phase. Cell numbers were determined by counting cell nuclei, and the percentage of cells attached to the well relative to the attached cells in the Ti alloy group was calculated (Figure 3B). During the pre-settlement phase, bis-acrylic 1 and flowable composite had less than 5% of cells relative to the Ti alloy group (3.96 ± 0.40% and 4.90 ± 3.63%, respectively), while bis-acrylic 2 and bulk–fill composite had 10–20% of cells (10.72 ± 1.76% and 17.02 ± 6.50%, respectively).

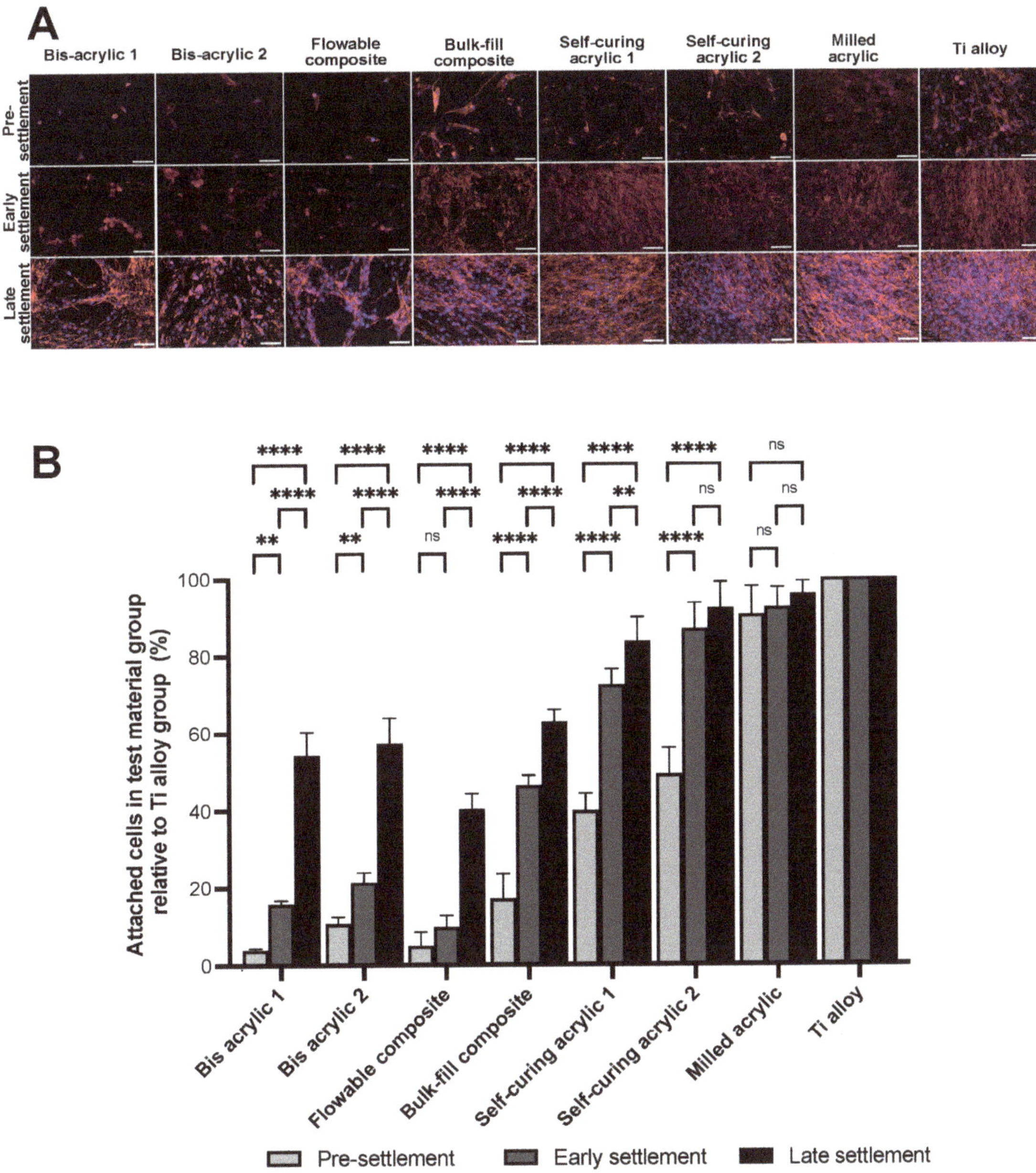

Figure 3. Fibroblast growth on the well surface two days after placement of test materials. (**A**) Visualization of fibroblasts: Fibroblasts were dual stained with fluorescent dyes, DAPI for nuclei, and rhodamin–phalloidin for actin filaments. Scale bars indicate 1 mm. (**B**) Percentage of the number of cells in test material groups relative to the number of cells in Ti alloy group was measured. Data shown are means ± SD. Two-way ANOVA, followed by the Fisher's LSD post hoc test. p ** < 0.01, p **** < 0.0001. ns, Not significant.

The number of fibroblasts attached to the well increased from pre-settlement to the early settlement phase, except for the flowable composite group. In the early settlement phase, the percentage of self-curing acrylic 2 and milled acrylic reached above 80% (86.81 ± 6.66% and 92.42 ± 5.14%, respectively), while only 10–20% of cells relative to Ti alloy group were attached in the flowable composite group and the bis-acrylic 1 and 2 groups (9.71 ± 3.07%, 15.71 ± 1.03%, and 21.32 ± 2.61%, respectively). Even in the late settlement phase, the flowable composite was less than half of Ti alloy (40.24 ± 4.01%).

Next, we evaluated cell proliferation from day 2 to day 4 (Figure 4). In the pre-settlement phase, Ti alloy, milled acrylic, and self-curing acrylic 1 and 2 groups showed significant increases, while there was no increase in the other groups. In the early settlement phase, flowable composite and bis-acrylic 1 and 2 groups showed a significant decrease. In the late settlement phases, the number of cells did not increase in flowable composite, bulk–fill composite, and bis-acrylic 1 and 2 groups, even though the cells did not reach confluence.

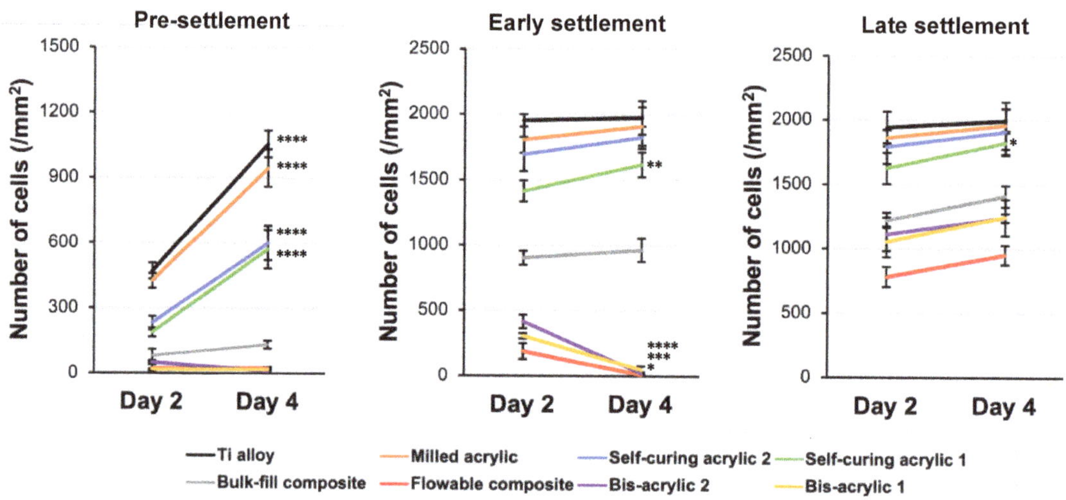

Figure 4. Cell propagation under three different conditions from day 2 to day 4. Data shown are means ± SD. Two-way ANOVA, followed by the Fisher's LSD post hoc test. p * < 0.05, p ** < 0.01, p *** < 0.001, p **** < 0.0001.

2.4. Collagen Deposition

Finally, we evaluated the collagen deposition at three different settlement phases (Figure 5). The collagen deposition on the well was evaluated two days after the placement of the test materials. In the pre-settlement phase, the amount of collagen deposition in bis-acrylic 1 and 2 and flowable composite groups was less than one-fifth of that in the Ti alloy group. From the pre-settlement to the early settlement phase, there was no increase in bis-acrylic 2 and flowable groups. In the late settlement phase, the amount of collagen deposition remained at a low level in bis-acrylic 2 and flowable composite groups, and once exposed to these materials, there was even a reduction in the amount of collagen that was already produced. The collagen deposition in the self-curing composite and the milled acrylic group were comparable to the Ti alloy group in all phases.

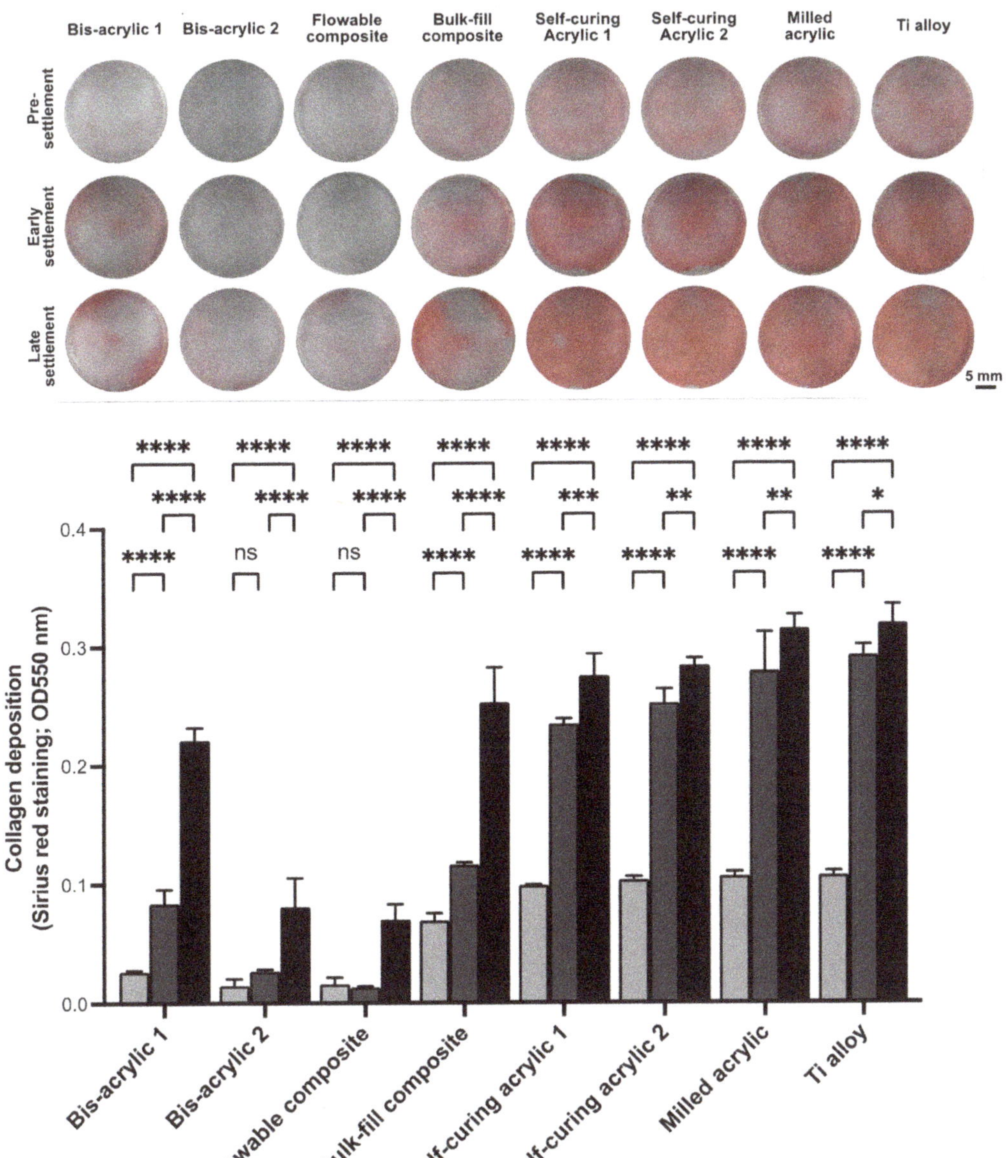

Figure 5. Imaging and quantification of collagen deposition using Sirius red staining two days after placement of test materials. Data shown are means ± SD. Two-way ANOVA, followed by the Fisher's LSD post hoc test. p * < 0.05, p ** < 0.01, p *** < 0.001, p **** < 0.0001. ns, Not significant.

3. Discussion

In this study, the effects of different implant provisional restorative materials on the attachment, growth, and function of human gingival fibroblasts were evaluated under different phases of cell settlement. The cells were evaluated in direct contact with the materials and in close proximity to materials. The proximity experiment was divided into pre-settlement, early settlement, and late-settlement phases. These phases mimicked the healing process of immediate implant placement with provisionals, secondary surgery for the placement of provisionals, and the replacement from healing abutment to provisionals, respectively. As a control, a biocompatible Ti alloy was used, and a total of eight different materials were tested. The results demonstrate that the harmful effects of materials on fibroblasts varied depending on the material compositions, even with the same curing modality. There was a significant difference in cell attachment, growth, or function between flowable and bulk–fill composite, between bis-acrylics, and between self-curing acrylics. In addition, we found that the harmful effects were mitigated when materials were in proximity to cells compared to direct contact. However, the harmful effects of bis-acrylics and flowable composite persisted even in the late settlement phase. Cytotoxicity of these materials would exceed the damage tolerance of fibroblasts. The summary of these results is shown in Figure 6. To our knowledge, this is the first study to evaluate the influence of provisional materials on cells with different phases of cell settlement.

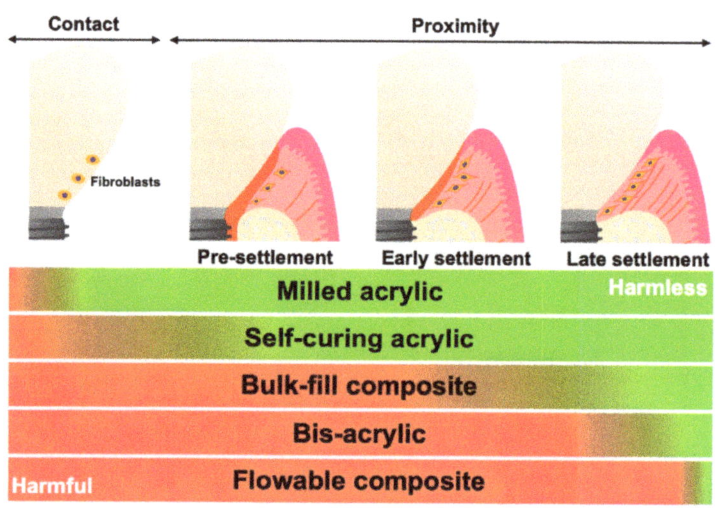

Figure 6. Schematic summary of results. Pre-settlement, early settlement, and late settlement mimicked the healing process of immediate implant placement with provisionals, secondary surgery for the placement of provisionals, and replacement from healing abutment to provisionals, respectively. The increased red indicated an increased level of harm, while the increased green indicated an increased level of harmlessness.

Some studies have shown that material compositions such as monomers, polymerization initiators, and filler particles influence their cytotoxicity [48–54]. It is known that unreacted monomers exert critical biological effects on cells [15,55,56]. Bis-GMA, a main component of bulk–fill and flowable composites, is released at high levels even 28 days after polymerization [57]. UDMA, a main component of bulk–fill composite, is also persistently released, and bis-GMA is eluted at higher concentrations than UDMA [35,58]. Results on cell attachment and proliferation being the highest in the milled acrylic group suggest that there is lower residual monomer elution from milled acrylic [28]. Our results suggest that bis-GMA and UDMA are more toxic than MMA, while UDMA is less toxic than bis-GMA.

Polymerization initiators such as benzoyl peroxide (BPO) and camphorquinone (CQ) compromise cell viability [47,59–61]. BPO, a major initiator for self-curing acrylic resin, is broken down during polymerization to release radicals that injure surrounding cells [52–54]. CQ is well known as a photoinitiator for light-curing acrylic and composite resin, and it produces a pair of free radicals through proton abstraction [62,63]. In our study, self-curing acrylic was less cytotoxic than flowable and bulk–fill composites. This suggests that BPO is likely to be less cytotoxic than CQ. Milled acrylic is made of PMMA disks molded under high temperature and pressure in an anhydrous environment, and it has superior mechanical properties to conventional heat-polymerizing acrylics [64]. In addition, milled acrylics have favorable cytocompatibility due to lower residual monomer composition than self-cured acrylics [29], which was confirmed in the present results. The number of attached cells was different within self-curing acrylic groups in the contact experiment. The major difference between self-curing acrylics is diethyl phthalate as a plasticizer. Diethyl phthalate is leached from the acrylic materials, causing cellular damage by producing reactive oxygen species [65]. Self-curing acrylic 1 contains 10–20% diethyl phthalate as a composition of the powder. Thus, self-curing acrylic 1 is thought to be more harmful than self-curing acrylic 2.

After the early settlement phase, cell proliferation was not observed in self-curing acrylic 2, milled acrylic, and Ti alloy groups. Considering their low cytotoxicity and the time from cell seeding to the material placement, it was thought that the number of cells already reached a plateau two days after the placement of the materials. On the other hand, self-curing acrylic 1 slightly impaired cell growth, so there was capacity for cell propagation through to the late settlement phase. In bis-acrylics and flowable composite groups, only a few cells could survive in pre-settlement exposure. In the early settlement phase, the deleterious effect was still at a high level and resulted in a decreased number of cells. Even in the late settlement, the negative effect persisted. It would be explained that TEGDMA has various effects on cells, not only causing apoptosis but also inhibiting cell proliferation or differentiation [66,67]. The reduced cytotoxicity of bulk–fill composite compared to flowable composite may be partially attributed to the lower concentration of TEGDMA. In addition, the wettability and roughness of the materials' surface affect the initial cellular behavior [68–73]. Cell attachment and proliferation are more favorable on a hydrophilic surface compared to a hydrophobic surface. Additionally, the smooth surface is more favorable compared to the rough surface. SEM images in a previous study showed that the surface roughness differed among test materials [46]. Milled acrylic and Ti alloy have a smooth surface; on the other hand, bis-acrylic, composite, and self-curing acrylics have a rough surface. Therefore, not only the compositions of materials but also the surface topography may affect cytocompatibility.

The soft tissue seal at the abutment and prosthetic material interface, if established, may play an important role in preventing peri-implantitis. Therefore, cell adhesion or cell retention by the material is an essential factor in determining the cytocompatibility of provisional implant materials. Of note, the milled acrylic retained a comparable number of cells to Ti alloy. Considering that cell retention was significantly compromised on self-curing acrylics, this result represents an additional benefit favoring milled acrylic. Increased cell attachment and proliferation do not necessarily result in increased cell retention. However, as vividly observed in the fluorescence microscopy images, the increased cell density, cell spreading, and cytoskeletal development thus improved intercellular adhesion and may have increased cell retention on milled acrylic. Although this study used a chemical detachment protocol, other methods of detachment, such as mechanical and vibrational detachment or their combination with chemical detachment, are an area of interest since the provisional material–fibroblast interface may be subjected to micro-movements in vivo.

In the proximity experiment, a cell culture insert containing each test plate was placed in a well where fibroblasts were pre-cultured. As expected, the damage to the fibroblasts was reduced compared to the contact experiment. In all test groups, as the number of settled cells before placement of the insert increased, the harmful effect of materials decreased. Cell proliferation from day 2 to day 4 indicated that Ti alloy, milled acrylic, self-curing acrylics, and bulk–fill composite showed an increase in the pre-settlement phase, whereas bis-acrylics and flowable composite decreased. Of note, bis-acrylics and flowable composite had almost no cells remaining. A similar trend was also observed in the early settlement phase. Therefore, the cytotoxicity of these materials might exceed the tolerance of fibroblasts without direct contact.

Regarding collagen deposition, in the early and late phases, collagen production was thought to be comparable among all groups before material placement. Thus, the amount of collagen was not influenced by the materials, although the number of cells decreased. However, the collagen deposition in bis-acrylic 2 and flowable composite was limited. Not only was there the inability for new collagen production from the reduced number of cells, but also the damaged fibroblasts were secreting matrix metalloproteinases (MMPs), resulting in decreased deposition [74–76]. Bis-acrylic 1 exhibited two times more collagen deposition compared to bis-acrylic 2, while bulk–fill composite showed three to five times more collagen deposition than flowable composite. Considering these results, the difference between materials may not only be due to the decrease mediated by fibroblast-produced collagenases but also to the inherent capacity of certain materials to degrade collagen. BisGMA and UDMA do not degrade collagen, but other proprietary components may possess collagen-degrading properties. Further studies are needed with a more comprehensive analysis to determine whether the properties of provisional materials degrade collagen or not.

Due to the chemistry of the resinous/polymerizing materials, we believe that the standardized preparation of materials, in particular the storage time, directly affects the results. The materials elute residual monomer after initial polymerization [15,18,39]. Although the amount of residual monomer reaches its peak within 24 h, the peak differs depending on the materials. Considering the number and types of materials examined, conducting the experiments immediately after materials fabrication and comparing the peaks is extremely difficult, especially because these fibroblasts are so susceptible when isolated from the host response. Therefore, we standardized the storage time of all test materials for two weeks under standardized conditions prior to proceeding with the experiments. Although the concentration decreases after the peak, there is a sustained release of the elutes that impacts the cellular interactions. We are fully aware that materials with different storage times need to be tested in future studies.

This study focused on the effect of the materials on the initial activity of fibroblasts under different conditions. The initial cellular response and reaction are critically important to subsequent initial site healing, cellular function, and long-term tissue health. Indeed, if cells are exposed to cytotoxins, they undergo rapid cell death and cannot function properly. This study reveals a great variation in the initial fibroblast activities in various environments in response to different provisional materials. However, from a clinical perspective, it is necessary to examine whether the materials cause inflammation in the peri-implant tissues. In order to assess inflammatory reactions in peri-implant tissue, in vivo studies should be conducted. In vivo studies enable the investigation of site-specific inflammatory cytokine production and the infiltration of inflammatory cells within the tissues. Such studies will contribute to a more comprehensive understanding of the clinical use of provisional materials.

Peri-implant tissue is composed of junctional epithelium and connective tissue, consisting of keratinocytes and fibroblasts, respectively. In this study, we focused on the soft tissue seal around the most apical portion of provisional restorations, in the closest proximity to the crestal bone, and established an experimental model to mimic this relationship by investigating the effects on fibroblasts. The rationale behind this focus is that if the connective tissue breaks, inflammation will rapidly spread to the alveolar bone, leading to bone resorption. However, the epithelium also plays an important role in establishing peri-implant soft tissue seals as keratinocytes adhere to titanium through hemidesmosomes. If the hemidesmosomes formed onto low-cytotoxic provisional materials, it potentially contributes to better cell retention. Further experiments using keratinocytes and the derivatives of epithelial cells were desired. Future studies other than those already mentioned may include the development of materials with high cytocompatibility properties (i.e.,

N-acetyl cysteine and tri-n-butyl borane) to improve reduction–oxidation systems and reduce radical production [53,61,77,78].

4. Materials and Methods

4.1. Material Preparation

Eight different specimens were fabricated in a rectangular plate form (6 mm × 14 mm, 2 mm thickness) for evaluation. The prepared test plates and their principal compositions are shown in Table 1. Two bis-acrylics, flowable and bulk–fill composites, and two self-curing acrylics were prepared using standardized silicone molds and according to the manufacturer's instructions (Figure 7A). Three plates of each material were prepared for each experiment. A light curing device (Coltolux LED; Coltène, Altstätten, Switzerland) was used to polymerize the light curing materials with a wavelength of 450–470 nm and an intensity of 1275 mW/cm^2 for 30 s. Milled acrylic plates were designed using CAD software (123D® Design version 2.2.14, Autodesk, Inc., San Rafael, CA, USA), and manufactured from PMMA disks with a milling machine (Versamill 5 × 200, Axsys Dental Solutions, Wixom, MI, USA) using CAM software (HyperDENT® version 9.0.2, Synergy Health, Sydney, Australia). Two weeks after preparation, all acrylic plates were washed with a steam cleaner and disinfected with 75% ethanol. Machined Ti alloy plates were also manufactured as a positive control. No surface polishing was performed to milled PMMA and Ti alloy.

Table 1. Materials used in this study.

Materials	Principal Compositions	Curing Modality
Bis-acrylic 1 (Integrity® Multi + Cure Temporary Crown and Bridge Material, Dentsply Sirona, Chariotte, NC, USA)	Acrylates and methacrylates (bis- and multifunctional) Barium boro alumino silicate glass	Dual-curing (chemical-curing and light-curing)
Bis-acrylic 2 (Visalys® Temp, Kettenbach GmbH & Co. KG, Eschenburg, Germany)	Aliphatic dimethacrylate, Poly(alkyleneglycol) diacrylate, hydroquinone monomethyl ether	Chemical-curing
Flowable composite (Aeliteflo™, BISCO Inc., Schaumburg, IL, USA)	Bis-GMA, TEGDMA	Light-curing
Bulk–fill composite (Aelite™ Aesthetic Enamel, BISCO Inc.)	Ytterbium Fluoride, Bis-GMA, UDMA Bis-EMA, TEGDMA	Light-curing
Self-curing acrylic 1 (JET Tooth Shade, Lang Dental Manufacturing Company Inc., Wheeling, IL, USA)	(liquid) MMA, N,N-Dimethyl-p-Toluidine (powder) 2-Propenoic acid, 2-methyl-, methyl ester homopolymer, Diethyl Phthalate	Chemical-curing
Self-curing acrylic 2 (UNIFAST™ Trad, GC, Tokyo, Japan)	(liquid) MMA, N,N-dimethyl-p-toluidine (powder) PMMA, Dibenzoyl peroxide	Chemical-curing
Milled acrylic (Vivid PMMA Disc, Pearson™ Dental Supply Co., Sylmar, CA, USA)	PMMA	Pre-curing (chemical-curing with high pressure and high temperature)
Ti alloy	Ti-6Al-4V (Grade 5)	-

Abbreviations: UDMA, urethane dimethacrylate; Bis-EMA, bisphenol A Ethoxylate Dimethacrylate; TEGDMA, triethylene glycol dimethacrylate; Bis-GMA, bisphenol A glycidyl methacrylate; MMA, methyl methacrylate; PMMA, poly(methyl methacrylate).

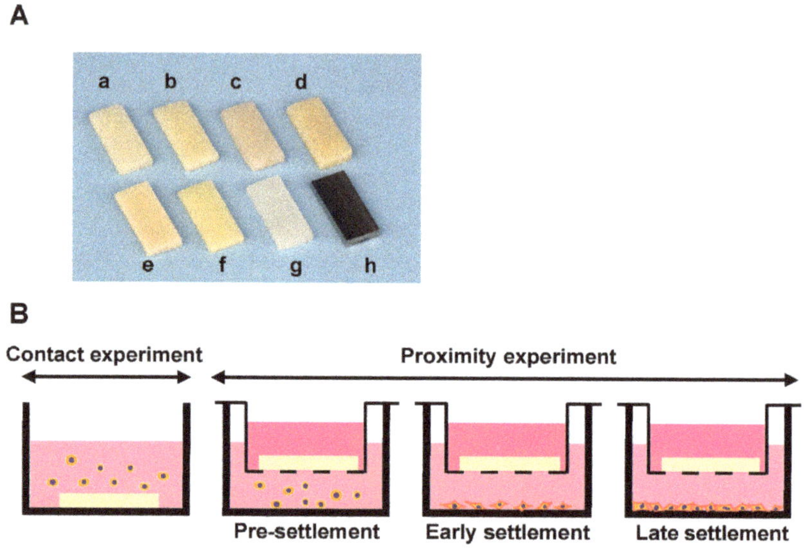

Figure 7. Test materials and experimental design for counting cells. (**A**) Rectangular plates (6 mm × 14 mm, 2 mm thick) were prepared. a, bis-acrylic 1; b, bis-acrylic 2; c, flowable composite; d, bulk–fill composite; e, self-curing acrylic 1; f, self-curing acrylic 2; g, milled acrylic; and h, Ti alloy. (**B**) Attached fibroblasts were counted to determine the effect of test materials under various cell conditions. The experiment was divided into contact experiments and proximity experiments. The contact experiment was the quantification of fibroblasts directly attached to test material. The proximity experiment was divided into three phases: pre-settlement, early settlement, and late settlement. In the proximity experiment, the quantification of fibroblasts attached to the well of the culture dish (20 mm diameter) was conducted immediately, 24 h or 72 h after the placement of a cell culture insert containing each test material.

4.2. Cell Culture and Material Placement

Human gingival fibroblasts were obtained from ScienCell Research Laboratories (Carlsbad, CA, USA) and grown in a fibroblast medium supplemented with 5% fetal bovine serum (FBS), 1% Fibroblast Growth Supplement-2, and 1% penicillin/streptomycin solution. At 80% confluence, the cells were detached using 0.05% trypsin-EDTA solution and seeded at density of 4×10^4 cells/well. To evaluate the influence of materials on cells in various conditions, seeding process was divided into two experiments: the cells in direct contact with test materials (contact experiment) and in close proximity to test materials (proximity experiment). The proximity experiment was further divided into three phases: pre-settlement, early settlement, and late settlement phase (Figure 7B). In the contact experiment, cells were seeded onto each test material placed in a well (20 mm diameter) of 12-well culture plates. In the proximity experiment, cells were seeded onto each well without materials. For evaluation of the pre-settlement phase, a cell culture insert with a 0.4 µm pore size containing the test plate was placed into a well of 12-well immediately after cell seeding. In the early settlement phase, the test plates were placed at 24 h after cell seeding. In the late settlement phase, the plates were placed at 72 h after cell seeding. The culture medium was renewed every three days. The UCLA Institutional Biosafety Committee (BUA-2-22-036-001) approved the study protocol.

4.3. Quantification of Attached and Propagated Cells with Fluorescence Microscopy

The number of attached fibroblasts was counted to determine the effect of test materials under various culture conditions. The contact experiment referred to the quantification of fibroblasts directly attached to test materials, while the proximity experiment referred to the quantification of fibroblasts attached to the well after placement of the cell culture insert containing the test material. In the contact experiment, two and four days after cell seeding, cells on the test plates were fixed in 10% formalin, permeabilized with 0.5% Triton X-100, and blocked with 1% BSA. Subsequently, the cells were dual stained with fluorescent dyes: 4′,6-diamidino-2-phenylindole (DAPI) to identify nuclei and rhodamine–phalloidin for actin filaments and observed with fluorescence microscopy (DMI6000B, Leica Microsystems, Wetzlar, Germany). In the proximity experiment, two and four days after incubating the cells with the test materials, the attached cells on the wells were fixed and stained. The number of cells was quantified by counting the cell nuclei in the taken images (Image J version 1.53, NIH, Bethesda, MD, USA). Increases in the number of cells from day 2 to day 4 were measured as cell proliferation.

4.4. Cell Retention Assay

The retention of fibroblasts attached to each test plate was evaluated by calculating the percentage of remaining cells after chemical detachment, as reported previously [79]. Four days after seeding, to remove any extra cells not attached to the samples, carefully transfer and dip the samples into a new well containing PBS. After removing the PBS completely by aspiration, 0.0125% trypsin/EDTA was added to detach the attached cells, and the samples were incubated at 37 °C for 2 min. A hematocytometer was used to count the number of detached cells. The remaining cells on the material surface were completely detached with 0.0125% trypsin-EDTA at 37 °C for 10 min, and completely detached cells were counted. The percentage of remaining cells was calculated using the following formula:

Percentage of remaining cells (%) = {(Number of completely detached cells at 10 min − Number of detached cells at 2 min)/(Number of completely detached cells at 10 min)} × 100

4.5. Collagen Deposition

Collagen deposition produced by fibroblasts in the proximity experiment was measured two days after placement of inserts containing test plates by picrosirius red staining (Polysciences Inc., Warrington, PA, USA). The cells were washed with PBS and fixed in 10% formaldehyde for 10 min. Subsequently, the collagen fibers were stained with 0.1% picrosirius red solution for 60 min at room temperature, after which 0.1 N sodium hydroxide was added for 60 min to elute the bound dye. The supernatant was measured at an absorbance of 550 nm using a microplate reader.

4.6. Statistical Analysis

Results are expressed as means ± standard deviations (SD). All experiments were performed in triplicate (n = 3). The eight materials were compared by one-way analysis of variance (ANOVA) followed by the Tukey–Kramer post hoc test. Furthermore, two-way ANOVA followed by post hoc comparisons using Fisher's LSD test was performed to evaluate the changes between test materials at different time points. p-values less than 0.05 were deemed statistically significant.

5. Conclusions

This study indicated that the harmful effects of provisional materials on gingival fibroblasts vary depending on the curing modality and material composition. Pre-settlement of cells mitigated the harmful effects, implying the susceptibility to material toxicity varies depending on the progress of wound healing and tissue condition. However, cell pre-settlement was not sufficient to fully restore the fibroblastic function to the normal level. Notably, bis-acrylics and flowable composites still exhibited considerable adverse effects even in the late-settlement phase, whereas milled and self-curing acrylics demonstrated excellent and acceptable biocompatibility when compared to other materials. These results provide valuable information for clinical practice to optimize peri-implant health and enhance material cytocompatibility in future developments of provisional materials.

Author Contributions: Conceptualization, T.M. and T.O.; data curation T.M. and J.C.; formal analysis, T.M. and K.K.; funding acquisition, T.-L.C. and T.O.; investigation, S.F. and D.C.; methodology, S.S., J.C. and T.B.; project administration, T.M.; resources, T.B., D.C. and T.-L.C.; supervision T.O.; validation, S.S. and A.P.; visualization T.M. and K.K.; writing—original draft, T.M.; writing—review and editing, S.S. and T.O. All authors have read and agreed to the published version of the manuscript.

Funding: This research received no external funding.

Institutional Review Board Statement: Not applicable.

Informed Consent Statement: Not applicable.

Data Availability Statement: The data are available upon request from the corresponding author.

Conflicts of Interest: The authors declare no conflict of interest.

References

1. Schwarz, F.; Ramanauskaite, A. It is all about peri-implant tissue health. *Periodontology 2000* **2022**, *88*, 9–12. [CrossRef] [PubMed]
2. Scarano, A.; Khater, A.G.A.; Gehrke, S.A.; Serra, P.; Francesco, I.; Di Carmine, M.; Tari, S.R.; Leo, L.; Lorusso, F. Current Status of Peri-Implant Diseases: A Clinical Review for Evidence-Based Decision Making. *J. Funct. Biomater.* **2023**, *14*, 210. [CrossRef] [PubMed]
3. Korsch, M.; Marten, S.M.; Stoll, D.; Prechtl, C.; Dotsch, A. Microbiological findings in early and late implant loss: An observational clinical case-controlled study. *BMC Oral Health* **2021**, *21*, 112. [CrossRef] [PubMed]
4. Welander, M.; Abrahamsson, I.; Berglundh, T. The mucosal barrier at implant abutments of different materials. *Clin. Oral Implant. Res.* **2008**, *19*, 635–641. [CrossRef]
5. Bordin, S.; Flemmig, T.F.; Verardi, S. Role of fibroblast populations in peri-implantitis. *Int. J. Oral Maxillofac. Implant.* **2009**, *24*, 197–204.
6. Kan, J.Y.K.; Rungcharassaeng, K.; Deflorian, M.; Weinstein, T.; Wang, H.L.; Testori, T. Immediate implant placement and provisionalization of maxillary anterior single implants. *Periodontology 2000* **2018**, *77*, 197–212. [CrossRef] [PubMed]
7. Saito, H.; Chu, S.J.; Reynolds, M.A.; Tarnow, D.P. Provisional Restorations Used in Immediate Implant Placement Provide a Platform to Promote Peri-implant Soft Tissue Healing: A Pilot Study. *Int. J. Periodontics Restor. Dent.* **2016**, *36*, 47–52. [CrossRef]
8. Zhuang, J.; Wang, Y.; Song, Y.; Qu, F.; Xu, C. The application of individualized abutment-crown integrated provisional restoration in optimizing the peri-implant soft tissue contour in the esthetic zone. *J. Esthet. Restor. Dent.* **2021**, *33*, 560–566. [CrossRef]
9. Gonzalez-Martin, O.; Lee, E.; Weisgold, A.; Veltri, M.; Su, H. Contour Management of Implant Restorations for Optimal Emergence Profiles: Guidelines for Immediate and Delayed Provisional Restorations. *Int. J. Periodontics Restor. Dent.* **2020**, *40*, 61–70. [CrossRef]
10. Grossner-Schreiber, B.; Herzog, M.; Hedderich, J.; Duck, A.; Hannig, M.; Griepentrog, M. Focal adhesion contact formation by fibroblasts cultured on surface-modified dental implants: An in vitro study. *Clin. Oral Implant. Res.* **2006**, *17*, 736–745. [CrossRef]
11. Kim, Y.S.; Ko, Y.; Kye, S.B.; Yang, S.M. Human gingival fibroblast (HGF-1) attachment and proliferation on several abutment materials with various colors. *Int. J. Oral Maxillofac. Implant.* **2014**, *29*, 969–975. [CrossRef] [PubMed]
12. Bajantri, P.; Rodrigues, S.J.; Shama Prasada, K.; Pai, U.Y.; Shetty, T.; Saldanha, S.; Mahesh, M.; Hegde, P.; Sales, A.; Mukherjee, S.; et al. Cytotoxicity of Dental Cements on Soft Tissue Associated with Dental Implants. *Int. J. Dent.* **2022**, *2022*, 4916464. [CrossRef] [PubMed]
13. Bandarra, S.; Neves, J.; Paraiso, A.; Mascarenhas, P.; Ribeiro, A.C.; Barahona, I. Biocompatibility of self-adhesive resin cement with fibroblast cells. *J. Prosthet. Dent.* **2021**, *125*, 705.e1–705.e7. [CrossRef] [PubMed]
14. Gautam, R.; Singh, R.D.; Sharma, V.P.; Siddhartha, R.; Chand, P.; Kumar, R. Biocompatibility of polymethylmethacrylate resins used in dentistry. *J. Biomed. Mater. Res. B Appl. Biomater.* **2012**, *100*, 1444–1450. [CrossRef] [PubMed]
15. Ulker, M.; Ulker, H.E.; Zortuk, M.; Bulbul, M.; Tuncdemir, A.R.; Bilgin, M.S. Effects of current provisional restoration materials on the viability of fibroblasts. *Eur. J. Dent.* **2009**, *3*, 114–119. [CrossRef] [PubMed]
16. Shim, J.S.; Kim, H.C.; Park, S.I.; Yun, H.J.; Ryu, J.J. Comparison of Various Implant Provisional Resin Materials for Cytotoxicity and Attachment to Human Gingival Fibroblasts. *Int. J. Oral Maxillofac. Implant.* **2019**, *34*, 390–396. [CrossRef] [PubMed]
17. Van Landuyt, K.L.; Nawrot, T.; Geebelen, B.; De Munck, J.; Snauwaert, J.; Yoshihara, K.; Scheers, H.; Godderis, L.; Hoet, P.; Van Meerbeek, B. How much do resin-based dental materials release? A meta-analytical approach. *Dent. Mater.* **2011**, *27*, 723–747. [CrossRef]
18. Duruk, G.; Akkuc, S.; Ugur, Y. Evaluation of residual monomer release after polymerization of different restorative materials used in pediatric dentistry. *BMC Oral Health* **2022**, *22*, 232. [CrossRef]
19. Bezgin, T.; Cimen, C.; Ozalp, N. Evaluation of Residual Monomers Eluted from Pediatric Dental Restorative Materials. *Biomed. Res. Int.* **2021**, *2021*, 6316171. [CrossRef]
20. Ogawa, T.; Tanaka, M.; Matsuya, S.; Aizawa, S.; Koyano, K. Setting characteristics of five autopolymerizing resins measured by an oscillating rheometer. *J. Prosthet. Dent.* **2001**, *85*, 170–176. [CrossRef]

21. Çelebi, N.; Yüzügüllü, B.; Canay, Ş.; Yücel, Ü. Effect of polymerization methods on the residual monomer level of acrylic resin denture base polymers. *Polym. Adv. Technol.* **2008**, *19*, 201–206. [CrossRef]
22. Nakagawa, K.; Saita, M.; Ikeda, T.; Hirota, M.; Park, W.; Lee, M.C.; Ogawa, T. Biocompatibility of 4-META/MMA-TBB resin used as a dental luting agent. *J. Prosthet. Dent.* **2015**, *114*, 114–121. [CrossRef] [PubMed]
23. Kojima, N.; Yamada, M.; Paranjpe, A.; Tsukimura, N.; Kubo, K.; Jewett, A.; Ogawa, T. Restored viability and function of dental pulp cells on poly-methylmethacrylate (PMMA)-based dental resin supplemented with N-acetyl cysteine (NAC). *Dent. Mater.* **2008**, *24*, 1686–1693. [CrossRef] [PubMed]
24. Goncalves, F.P.; Alves, G.; Guimaraes, V.O.J.; Gallito, M.A.; Oliveira, F.; Scelza, M.Z. Cytotoxicity Evaluation of Two Bis-Acryl Composite Resins Using Human Gingival Fibroblasts. *Braz. Dent. J.* **2016**, *27*, 492–496. [CrossRef] [PubMed]
25. Parikh, V.; Cheng, D.H.; Linsley, C.; Shah, K.C. Bond strength of three chairside crown reline materials to milled polymethyl methacrylate resin. *J. Prosthet. Dent.* **2021**, *125*, 544.e1–544.e8. [CrossRef] [PubMed]
26. Shibasaki, S.; Takamizawa, T.; Suzuki, T.; Nojiri, K.; Tsujimoto, A.; Barkmeier, W.W.; Latta, M.A.; Miyazaki, M. Influence of Different Curing Modes on Polymerization Behavior and Mechanical Properties of Dual-Cured Provisional Resins. *Oper. Dent.* **2017**, *42*, 526–536. [CrossRef] [PubMed]
27. Amin, B.M.; Aras, M.A.; Chitre, V. A comparative evaluation of the marginal accuracy of crowns fabricated from four commercially available provisional materials: An in vitro study. *Contemp. Clin. Dent.* **2015**, *6*, 161–165. [CrossRef]
28. Hussain, B.; Thieu, M.K.L.; Johnsen, G.F.; Reseland, J.E.; Haugen, H.J. Can CAD/CAM resin blocks be considered as substitute for conventional resins? *Dent. Mater.* **2017**, *33*, 1362–1370. [CrossRef]
29. Wei, X.; Pan, Y.; Wang, M.; Wang, Y.; Lin, H.; Jiang, L.; Lin, D.; Cheng, H. Comparative analysis of leaching residual monomer and biological effects of four types of conventional and CAD/CAM dental polymers: An in vitro study. *Clin. Oral Investig.* **2022**, *26*, 2887–2898. [CrossRef]
30. Burgers, R.; Schubert, A.; Muller, J.; Krohn, S.; Rodiger, M.; Leha, A.; Wassmann, T. Cytotoxicity of 3D-printed, milled, and conventional oral splint resins to L929 cells and human gingival fibroblasts. *Clin. Exp. Dent. Res.* **2022**, *8*, 650–657. [CrossRef]
31. Aati, S.; Chauhan, A.; Shrestha, B.; Rajan, S.M.; Aati, H.; Fawzy, A. Development of 3D printed dental resin nanocomposite with graphene nanoplatelets enhanced mechanical properties and induced drug-free antimicrobial activity. *Dent. Mater.* **2022**, *38*, 1921–1933. [CrossRef] [PubMed]
32. Lempel, E.; Czibulya, Z.; Kovacs, B.; Szalma, J.; Toth, A.; Kunsagi-Mate, S.; Varga, Z.; Boddi, K. Degree of Conversion and BisGMA, TEGDMA, UDMA Elution from Flowable Bulk Fill Composites. *Int. J. Mol. Sci.* **2016**, *17*, 732. [CrossRef] [PubMed]
33. Lin, C.H.; Lin, Y.M.; Lai, Y.L.; Lee, S.Y. Mechanical properties, accuracy, and cytotoxicity of UV-polymerized 3D printing resins composed of Bis-EMA, UDMA, and TEGDMA. *J. Prosthet. Dent.* **2020**, *123*, 349–354. [CrossRef] [PubMed]
34. Leyva Del Rio, D.; Johnston, W.M. Effect of monomer composition and filler fraction on surface microhardness and depth of cure of experimental resin composites. *Eur. J. Oral Sci.* **2023**, *131*, e12933. [CrossRef] [PubMed]
35. Gul, P.; Miloglu, F.D.; Akgul, N.; Kadioglu, Y. Effect of Different Extraction Media on Quantification of the Released Monomers from Dental Composite. *Asian J. Chem.* **2013**, *25*, 2994–3000. [CrossRef]
36. Hampe, T.; Wiessner, A.; Frauendorf, H.; Alhussein, M.; Karlovsky, P.; Burgers, R.; Krohn, S. A comparative in vitro study on monomer release from bisphenol A-free and conventional temporary crown and bridge materials. *Eur. J. Oral Sci.* **2021**, *129*, e12826. [CrossRef] [PubMed]
37. Landenberger, P.; Baumann, L.; Gerhardt-Szep, S.; Ruttermann, S. The effect of new anti-adhesive and antibacterial dental resin filling materials on gingival fibroblasts. *Dent. Mater.* **2021**, *37*, 1416–1424. [CrossRef]
38. Tsitrou, E.; Kelogrigoris, S.; Koulaouzidou, E.; Antoniades-Halvatjoglou, M.; Koliniotou-Koumpia, E.; van Noort, R. Effect of extraction media and storage time on the elution of monomers from four contemporary resin composite materials. *Toxicol. Int.* **2014**, *21*, 89–95. [CrossRef]
39. Alshali, R.Z.; Salim, N.A.; Sung, R.; Satterthwaite, J.D.; Silikas, N. Analysis of long-term monomer elution from bulk-fill and conventional resin-composites using high performance liquid chromatography. *Dent. Mater.* **2015**, *31*, 1587–1598. [CrossRef]
40. Bainbridge, P. Wound healing and the role of fibroblasts. *J. Wound Care* **2013**, *22*, 407–412.
41. Nevins, M.; Nevins, M.L.; Camelo, M.; Boyesen, J.L.; Kim, D.M. Human histologic evidence of a connective tissue attachment to a dental implant. *Int. J. Periodontics Restor. Dent.* **2008**, *28*, 111–121.
42. Roy, M.; Corti, A.; Dominici, S.; Pompella, A.; Cerea, M.; Chelucci, E.; Dorocka-Bobkowska, B.; Daniele, S. Biocompatibility of Subperiosteal Dental Implants: Effects of Differently Treated Titanium Surfaces on the Expression of ECM-Related Genes in Gingival Fibroblasts. *J. Funct Biomater.* **2023**, *14*, 59. [CrossRef] [PubMed]
43. Garrod, D.R. Cell to cell and cell to matrix adhesion. *BMJ* **1993**, *306*, 703–705. [CrossRef] [PubMed]
44. Taghva, O.; Amini Sedeh, S.; Ejeian, F.; Amini, S. Comparison of Adhesion and Proliferation of Human Gingival Fibroblasts on Acellular Dermal Matrix with and without Low Level Diode Laser Irradiation, an in vitro Study. *J. Dent.* **2022**, *23*, 106–112.
45. Moon, Y.H.; Yoon, M.K.; Moon, J.S.; Kang, J.H.; Kim, S.H.; Yang, H.S.; Kim, M.S. Focal adhesion linker proteins expression of fibroblast related to adhesion in response to different transmucosal abutment surfaces. *J. Adv. Prosthodont.* **2013**, *5*, 341–350. [CrossRef] [PubMed]
46. Matsuura, T.; Komatsu, K.; Chao, D.; Lin, Y.C.; Oberoi, N.; McCulloch, K.; Cheng, J.; Orellana, D.; Ogawa, T. Cell Type-Specific Effects of Implant Provisional Restoration Materials on the Growth and Function of Human Fibroblasts and Osteoblasts. *Biomimetics* **2022**, *7*, 243. [CrossRef] [PubMed]

47. Matsuura, T.; Komatsu, K.; Choi, K.; Suzumura, T.; Cheng, J.; Chang, T.L.; Chao, D.; Ogawa, T. Conditional Mitigation of Dental-Composite Material-Induced Cytotoxicity by Increasing the Cure Time. *J. Funct. Biomater.* **2023**, *14*, 119. [CrossRef]
48. Jorge, J.H.; Giampaolo, E.T.; Machado, A.L.; Vergani, C.E. Cytotoxicity of denture base acrylic resins: A literature review. *J. Prosthet. Dent.* **2003**, *90*, 190–193. [CrossRef]
49. Schweikl, H.; Spagnuolo, G.; Schmalz, G. Genetic and cellular toxicology of dental resin monomers. *J. Dent. Res.* **2006**, *85*, 870–877. [CrossRef]
50. Kraus, D.; Wolfgarten, M.; Enkling, N.; Helfgen, E.H.; Frentzen, M.; Probstmeier, R.; Winter, J.; Stark, H. In-vitro cytocompatibility of dental resin monomers on osteoblast-like cells. *J. Dent.* **2017**, *65*, 76–82. [CrossRef]
51. Haugen, H.J.; Marovic, D.; Par, M.; Thieu, M.K.L.; Reseland, J.E.; Johnsen, G.F. Bulk Fill Composites Have Similar Performance to Conventional Dental Composites. *Int. J. Mol. Sci.* **2020**, *21*, 5136. [CrossRef] [PubMed]
52. Komatsu, K.; Hamajima, K.; Ozawa, R.; Kitajima, H.; Matsuura, T.; Ogawa, T. Novel Tuning of PMMA Orthopedic Bone Cement Using TBB Initiator: Effect of Bone Cement Extracts on Bioactivity of Osteoblasts and Osteoclasts. *Cells* **2022**, *11*, 3999. [CrossRef] [PubMed]
53. Sugita, Y.; Okubo, T.; Saita, M.; Ishijima, M.; Torii, Y.; Tanaka, M.; Iwasaki, C.; Sekiya, T.; Tabuchi, M.; Mohammadzadeh Rezaei, N.; et al. Novel Osteogenic Behaviors around Hydrophilic and Radical-Free 4-META/MMA-TBB: Implications of an Osseointegrating Bone Cement. *Int. J. Mol. Sci.* **2020**, *21*, 2405. [CrossRef] [PubMed]
54. Tsukimura, N.; Yamada, M.; Aita, H.; Hori, N.; Yoshino, F.; Chang-Il Lee, M.; Kimoto, K.; Jewett, A.; Ogawa, T. N-acetyl cysteine (NAC)-mediated detoxification and functionalization of poly(methyl methacrylate) bone cement. *Biomaterials* **2009**, *30*, 3378–3389. [CrossRef] [PubMed]
55. Emmler, J.; Seiss, M.; Kreppel, H.; Reichl, F.X.; Hickel, R.; Kehe, K. Cytotoxicity of the dental composite component TEGDMA and selected metabolic by-products in human pulmonary cells. *Dent. Mater.* **2008**, *24*, 1670–1675. [CrossRef] [PubMed]
56. Lagocka, R.; Jakubowska, K.; Chlubek, D.; Buczkowska-Radlinska, J. Elution study of unreacted TEGDMA from bulk-fill composite (SDR Dentsply) using HPLC. *Adv. Med. Sci.* **2015**, *60*, 191–198. [CrossRef] [PubMed]
57. Polydorou, O.; Trittler, R.; Hellwig, E.; Kummerer, K. Elution of monomers from two conventional dental composite materials. *Dent. Mater.* **2007**, *23*, 1535–1541. [CrossRef] [PubMed]
58. Sideridou, I.; Tserki, V.; Papanastasiou, G. Effect of chemical structure on degree of conversion in light-cured dimethacrylate-based dental resins. *Biomaterials* **2002**, *23*, 1819–1829. [CrossRef]
59. Att, W.; Yamada, M.; Kojima, N.; Ogawa, T. N-Acetyl cysteine prevents suppression of oral fibroblast function on poly(methylmethacrylate) resin. *Acta Biomater.* **2009**, *5*, 391–398. [CrossRef]
60. Chang, M.C.; Lin, L.D.; Wu, M.T.; Chan, C.P.; Chang, H.H.; Lee, M.S.; Sun, T.Y.; Jeng, P.Y.; Yeung, S.Y.; Lin, H.J.; et al. Effects of Camphorquinone on Cytotoxicity, Cell Cycle Regulation and Prostaglandin E2 Production of Dental Pulp Cells: Role of ROS, ATM/Chk2, MEK/ERK and Hemeoxygenase-1. *PLoS ONE* **2015**, *10*, e0143663. [CrossRef]
61. Matsuura, T.; Komatsu, K.; Ogawa, T. N-Acetyl Cysteine-Mediated Improvements in Dental Restorative Material Biocompatibility. *Int. J. Mol. Sci.* **2022**, *23*, 15869. [CrossRef] [PubMed]
62. Volk, J.; Ziemann, C.; Leyhausen, G.; Geurtsen, W. Non-irradiated campherquinone induces DNA damage in human gingival fibroblasts. *Dent. Mater.* **2009**, *25*, 1556–1563. [CrossRef] [PubMed]
63. Stansbury, J.W. Curing dental resins and composites by photopolymerization. *J. Esthet. Dent.* **2000**, *12*, 300–308. [CrossRef]
64. Iwaki, M.; Kanazawa, M.; Arakida, T.; Minakuchi, S. Mechanical properties of a polymethyl methacrylate block for CAD/CAM dentures. *J. Oral Sci.* **2020**, *62*, 420–422. [CrossRef] [PubMed]
65. Yin, Q.; Pi, X.; Ren, G.; Liu, Z.; Liu, H.; Wang, M.; Hu, C.; Zhao, H.; Li, D.; Yin, J. A novel bispecific antibody alleviates bleomycin-induced systemic sclerosis injury. *Int. Immunopharmacol.* **2020**, *85*, 106644. [CrossRef] [PubMed]
66. Lefeuvre, M.; Amjaad, W.; Goldberg, M.; Stanislawski, L. TEGDMA induces mitochondrial damage and oxidative stress in human gingival fibroblasts. *Biomaterials* **2005**, *26*, 5130–5137. [CrossRef]
67. Stanislawski, L.; Lefeuvre, M.; Bourd, K.; Soheili-Majd, E.; Goldberg, M.; Perianin, A. TEGDMA-induced toxicity in human fibroblasts is associated with early and drastic glutathione depletion with subsequent production of oxygen reactive species. *J. Biomed. Mater. Res. A* **2003**, *66*, 476–482. [CrossRef]
68. Webb, K.; Hlady, V.; Tresco, P.A. Relative importance of surface wettability and charged functional groups on NIH 3T3 fibroblast attachment, spreading, and cytoskeletal organization. *J. Biomed. Mater. Res.* **1998**, *41*, 422–430. [CrossRef]
69. Huang, H.H.; Ho, C.T.; Lee, T.H.; Lee, T.L.; Liao, K.K.; Chen, F.L. Effect of surface roughness of ground titanium on initial cell adhesion. *Biomol. Eng.* **2004**, *21*, 93–97. [CrossRef]
70. Zareidoost, A.; Yousefpour, M.; Ghaseme, B.; Amanzadeh, A. The relationship of surface roughness and cell response of chemical surface modification of titanium. *J. Mater. Sci. Mater. Med.* **2012**, *23*, 1479–1488. [CrossRef]
71. Pires, L.A.; de Meira, C.R.; Tokuhara, C.K.; de Oliveira, F.A.; Dainezi, V.B.; Zardin Graeff, M.S.; Fortulan, C.A.; de Oliveira, R.C.; Puppin-Rontani, R.M.; Borges, A.F.S. Wettability and pre-osteoblastic behavior evaluations of a dense bovine hydroxyapatite ceramics. *J. Oral Sci.* **2020**, *62*, 259–264. [CrossRef] [PubMed]
72. Suzumura, T.; Matsuura, T.; Komatsu, K.; Sugita, Y.; Maeda, H.; Ogawa, T. Vacuum Ultraviolet (VUV) Light Photofunctionalization to Induce Human Oral Fibroblast Transmigration on Zirconia. *Cells* **2023**, *12*, 2542. [CrossRef] [PubMed]
73. Majhy, B.; Priyadarshini, P.; Sen, A.K. Effect of surface energy and roughness on cell adhesion and growth—Facile surface modification for enhanced cell culture. *RSC Adv.* **2021**, *11*, 15467–15476. [CrossRef] [PubMed]

74. Page-McCaw, A.; Ewald, A.J.; Werb, Z. Matrix metalloproteinases and the regulation of tissue remodelling. *Nat. Rev. Mol. Cell Biol.* **2007**, *8*, 221–233. [CrossRef]
75. Williams, R.C.; Skelton, A.J.; Todryk, S.M.; Rowan, A.D.; Preshaw, P.M.; Taylor, J.J. Leptin and Pro-Inflammatory Stimuli Synergistically Upregulate MMP-1 and MMP-3 Secretion in Human Gingival Fibroblasts. *PLoS ONE* **2016**, *11*, e0148024. [CrossRef]
76. Beklen, A.; Ainola, M.; Hukkanen, M.; Gurgan, C.; Sorsa, T.; Konttinen, Y.T. MMPs, IL-1, and TNF are regulated by IL-17 in periodontitis. *J. Dent. Res.* **2007**, *86*, 347–351. [CrossRef]
77. Zhang, Y.; Xiao, J.F.; Yang, H.F.; Jiao, Y.; Cao, W.W.; Shi, H.M.; Cun, J.F.; Tay, F.R.; Ping, J.; Xiao, Y.H. N-Acetyl Cysteine as a Novel Polymethyl Methacrylate Resin Component: Protection against Cell Apoptosis and Genotoxicity. *Oxidative Med. Cell Longev.* **2019**, *2019*, 1301736. [CrossRef]
78. Nishimiya, H.; Yamada, M.; Ueda, T.; Sakurai, K. N-acetyl cysteine alleviates inflammatory reaction of oral epithelial cells to poly (methyl methacrylate) extract. *Acta Odontol. Scand.* **2015**, *73*, 616–625. [CrossRef]
79. Ikeda, T.; Ueno, T.; Saruta, J.; Hirota, M.; Park, W.; Ogawa, T. Ultraviolet Treatment of Titanium to Enhance Adhesion and Retention of Oral Mucosa Connective Tissue and Fibroblasts. *Int. J. Mol. Sci.* **2021**, *22*, 12396. [CrossRef]

Disclaimer/Publisher's Note: The statements, opinions and data contained in all publications are solely those of the individual author(s) and contributor(s) and not of MDPI and/or the editor(s). MDPI and/or the editor(s) disclaim responsibility for any injury to people or property resulting from any ideas, methods, instructions or products referred to in the content.

International Journal of *Molecular Sciences*

MDPI

Review

Considerations about Cytotoxicity of Resin-Based Composite Dental Materials: A Systematic Review

Kacper Wiertelak-Makała [1], Izabela Szymczak-Pajor [2], Kinga Bociong [3,*] and Agnieszka Śliwińska [2,*]

[1] Student Scientific Society of Civilization Diseases, Medical University of Lodz, 251 Pomorska Str., 92-213 Lodz, Poland
[2] Department of Nucleic Acid Biochemistry, Medical University of Lodz, 251 Pomorska Str., 92-213 Lodz, Poland
[3] Department of General Dentistry, Medical University of Lodz, 251 Pomorska Str., 92-213 Lodz, Poland
* Correspondence: kinga.bociong@umed.lodz.pl (K.B.); agnieszka.sliwinska@umed.lodz.pl (A.Ś.)

Abstract: The dental material industry is rapidly developing resin-based composites (RBCs), which find widespread use in a variety of clinical settings. As such, their biocompatibility has gained increasing interest. This literature review presents a summary of research into the cytotoxicity of methacrylate-based composites published from 2017 to 2023. Subject to analysis were 14 in vitro studies on human and murine cell lines. Cytotoxicity in the included studies was measured via MTT assay, LDH assay, and WST-1 assay. The QUIN Risk of Bias Tool was performed to validate the included studies. Included studies (based entirely on the results of in vitro studies) provide evidence of dose- and time-dependent cytotoxicity of dental resin-based composites. Oxidative stress and the depletion of cellular glutathione (GSH) were suggested as reasons for cytotoxicity. Induction of apoptosis by RBCs was indicated. While composites remain the golden standard of dental restorative materials, their potential cytotoxicity cannot be ignored due to direct long-term exposure. Further in vitro investigations and clinical trials are required to understand the molecular mechanism of cytotoxicity and produce novel materials with improved safety profiles.

Keywords: dental materials; resin-based composite; biocompatibility; cytotoxicity; oxidative stress; apoptosis

check for updates

Citation: Wiertelak-Makała, K.; Szymczak-Pajor, I.; Bociong, K.; Śliwińska, A. Considerations about Cytotoxicity of Resin-Based Composite Dental Materials: A Systematic Review. *Int. J. Mol. Sci.* **2024**, *25*, 152. https://doi.org/10.3390/ijms25010152

Academic Editor: Mary Anne Melo

Received: 16 November 2023
Revised: 15 December 2023
Accepted: 19 December 2023
Published: 21 December 2023

Copyright: © 2023 by the authors. Licensee MDPI, Basel, Switzerland. This article is an open access article distributed under the terms and conditions of the Creative Commons Attribution (CC BY) license (https://creativecommons.org/licenses/by/4.0/).

1. Introduction

1.1. Background

Dental caries, the most common oral health disorder, is a societal disease that affects patients of every age and social group worldwide [1,2]. Its multifactorial mechanism is explained by the loss of minerals from susceptible enamel, cementum, and dentin due to an acidic environment induced by cariogenic microorganisms when physiological remineralization processes are insufficient to restore lost minerals [1,2]. Cariogenic bacteria play an important role in the development of caries. As such, it can be defined as an infectious, dental plaque-dependent and biofilm-mediated disease [1,2]. Unbalanced mineral loss disrupts the structure of enamel, cementum, and dentin and leads to the development of carious lesions, which seriously impact masticatory function, cause pain, and have a negative aesthetic impact [3]. As such, carious lesions require clinical intervention: the removal of carious tissue and its replacement with either a direct filling or, in extensive tissue loss, prosthetic restoration [1,2]. Hard tissue loss caused by other factors, such as trauma, requires a similar restoration process [3].

Methacrylate-based composite materials have been used as filling materials in dentistry since the 1960s [4]. Since then, there have been many developments in the field of dental material science, resulting in the production of a new class of materials with versatile applications. Nowadays, resin-based composite restorative materials (RBCs) are widespread as the first choice of restorative material, largely replacing amalgam as the gold standard in conservative dentistry [5–7]. This change has been further accelerated by the Minamata Convention on Mercury, which advised a phase-down in the application of dental amalgam due to environmental concerns regarding the impact of mercury pollution [3,8]. The interdisciplinary interactions of dentistry, material science, and molecular biology have led to dynamic and productive developments in the field in recent years. The paradigm shift to minimally invasive dentistry in clinical practice and the increasing demand for aesthetic, tooth-colored restorative materials contributed to the popularity of RBCs [5,6,8–13]. The major advantages of RBCs as the gold standard of restorative material are that they are adhesive to the tooth structure, restore the structural integrity of the tooth, and have excellent aesthetics as well as clinically favorable parameters of mechanical performance, durability, and resistance to degradation [3,6,8,13].

Currently, RBCs are rapidly developing, with 3D-printable resins or bioactive materials with antibacterial and remineralizing properties being tested and introduced into the dental market [1,2,13–16]. Novel composites seek to not merely replace but fully replicate lost tissues [17]. Challenges such as toxicity, high polymerization shrinkage (and, in consequence, contraction stress), secondary carries, wear resistance, chipping and bulk fracture, and degradation in time of resin composite-based restorations are still being explored [3,6,7,18–20]. The degradation of adhesive materials and their bonds with the dentin is a complex, multifactorial process resulting from the hydrolysis of the resin as well as masticatory and hydraulic stress [21]. Collagen fiber hydrolysis and the enzymatic activity of host-derived matrix metalloproteinases and cysteine cathepsins further contribute to the degradation of the hybrid layer interface of restorative material and tooth structure [21]. Resin-based composites are applied in combination with adhesive systems following a wide range of clinical protocols, such as two-step etch and rinse, three-step etch and rinse, one-step self-etch, two-step self-etch, and universal adhesives, further modifying the bond strength of composite restorations [20]. These factors allow for degraded or unpolymerized monomers to diffuse through saliva and the dentinal tubules and interact directly with the pulp and soft tissues of the oral cavity, leading to biological interactions described in the further parts of this manuscript. These challenges are being addressed via the modification of RBC composition in manufacturing and new clinical application protocols.

The objective of this review was to evaluate the biocompatibility of dental resin-based composite materials and their mechanism of toxicity in in vitro models. The following question was formed to guide the research: "In what way do resin-based composite materials used in dentistry induce cytotoxicity on the cells of oral cavity tissues"?

1.2. Basic Characteristics of Resin-Based Composites

Resins are a wide class of viscous materials capable of hardening. They are required to meet appropriate criteria in order to be used in dental restorations. Their mechanical properties, wear resistance, and optical properties are expected to be similar to dental tissue [22]. They should present a high color stability. Dental materials should be adhesive to the tissues and easy to work with [22].

Since resin-based composites are intended to replace carious tissues, they must possess mechanical properties required to withstand stress without damage. Compressive strength, bending strength and elastic modulus are the main experimental criteria tested while selecting composite materials [22].

Resin-based composites include a diverse group of complex polymers [23]. Their change from viscous liquid to hardened solid state is caused by a polymerization reaction triggered by external energy in the form of heat, chemical, or radiant energy conditional on the presence of initiators [5]. Most monomers used in RBCs are linear molecules with a methacrylate group at each end [10]. Their conversion to complex polymers occurs via vinyl-free radical chain growth polymerization divided into three stages: initiation, propagation, and termination [5,10]. There are two modes of initiation: chemically activated and light-activated [3,19]. Activated photo-initiators generate free radicals, which convert double C=C bonds of methacrylate monomers into single C-C bonds, triggering a chain reaction. It should be noted that polymerization is never fully complete, and the maximum conversion of monomers in the cured composite is measured as the Degree of Conversion of resin [3,4,10,11,15,24]. Reduction in the filler–matrix ratio, as well as using nano-filled composites rather than particle-filled materials, tend to increase the Degree of Conversion [22]. Curing depth is another important consideration. The time of irradiation, effective wavelength, and light intensity of the curing lamp, and its distance from the material's surface all influence the curing depth [22]. Due to incomplete conversion, as well as aging, mechanical weardown, hydrolysis, and enzymatic degradation over time, the elution of uncured, leachable monomers—as well as initiators and other additives—into the liquids of the oral cavity occurs [3,4,7,11,12,24–26].

RBCs consist of an organic resin matrix (typically 15–50% of its weight), inorganic filler particles (reinforcing phase), a coupling agent, and a photo-initiator [5,8,19,22,27]. The chemical composition and function of each phase are summarized in Table 1. Most common resin monomers chemical structure is presented in Figure 1.

Table 1. Function and composition of resin-based composite phases based on [5,10,27,28].

Phase	Function	Composition
Resin matrix	Polymerization	BisGMA, TEGDMA, UDMA, HEMA, BisEMA, EGDMA
Filler particles	Improvement in mechanical and wear properties, aesthetic qualities, and reduction in polymerization shrinkage	Soft and hard glass: borosilicate, quartz, aluminum silicate, lithium aluminum silicate, ytterbium fluoride, barium, strontium, zirconium and zinc glass
Coupling agent	Combining resin matrix and filler particles, reduced water sorption	Silanes, zirconates, titanates
Photo-initiator	Initiation of resin matrix polymerization by providing free radicals upon exposure to external energy	CQ, PQ, TPO

BisGMA—bisphenol A-glycidyl methacrylate; TEGDMA—triethylene glycol dimethacrylate; UDMA—urethane dimethacrylate; HEMA—2-hydroxyethyl methacrylate; BisEMA—ethoxylatedbisphenol-A dimethacrylate; EGDMA—ethylene glycol dimethacrylate; CQ—camphorquinone; PQ—phenanthrenequinone; TPO—trimethylbenzoyl-diphenylphosphine oxide.

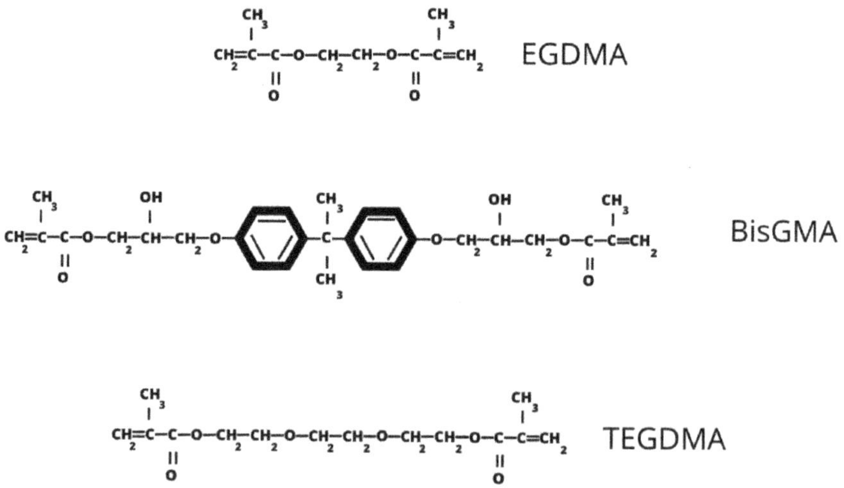

Figure 1. Most common resin monomers: EGDMA—ethylene glycol dimethacrylate; BisGMA—bisphenol A-glycidyl methacrylate; TEGDMA—triethylene glycol dimethacrylate.

1.3. Biological Properties of Resin-Based Composites

Biocompatibility is a vital characteristic of biomaterials, defining organisms' reactions to them [29]. It can be defined as the ability of a biomaterial to produce a suitable host reaction when applied according to standard procedure [12,25]. To be classified as biocompatible, biomaterials must avoid toxic, harmful, and otherwise physiologically undesirable reactions, such as cytotoxicity, induction of oxidative stress, induction of inflammatory response, mutagenicity, and immunological rejection [24,25,30,31]. Biocompatible materials should also not interfere with the healing process [30]. Toxicity is defined as the damage induced by the biomaterial to the organism [12]. Mercury toxicity controversy has been a major rationale behind the replacement of amalgam with RBCs as the golden standard of restorative material in dentistry [5]. A comprehensive understanding of the biocompatibility and potential toxicity of RBCs is of vital importance since dental biomaterials are in long-term direct and indirect contact with the surrounding tissues of the oral cavity. As such, ideal biomaterials should be chemically stable, biocompatible, and tasteless [11,12,22,25]. In the oral cavity, RBCs are exposed to saliva, food, bacteria, and by-products of their metabolism, changes in pH and temperature, as well as mechanical weardown [12]. In commercially available RBCs, there is some extent of degradation and solution leading to the elution of toxic components, making local exposure via both direct contact of dentin, pulp, and gingival cells as well as indirect exposure of gingiva via saliva possible and worth consideration [4,12,25,32]. Concerns have been raised regarding the allergies and hypersensitivities in patients as well as cytotoxic, mutagenic, and estrogenic effects on cells exposed to residual monomers [4]. Biocompatibility is important not only for the patients but also for the medical professionals [30]. Adverse skin and mucosal reactions have been reported in about 12% of patients and 27% of dentists utilizing RBCs [12,25]. Resin is responsible for such adverse reactions, which are mostly induced by uncured RBCs.

So far, dental materials were intended to simply replicate missing tissues. However, in recent years, developments have been made to introduce a new generation of bioactive materials. Remineralizing and antibacterial properties or the ability to inhibit biofilm formation or neutralize acids are researched [33].

Cytotoxicity, the main point of focus of this manuscript, can be defined as toxicity caused by exogenous substances in living cells. Uncured composites exert much greater cytotoxicity due to the higher content of free monomers able to induce oxidative stress [29]. The consideration of cytotoxicity is of paramount importance, as resin-based composites are in prolonged close contact with the pulp and gingival.

2. Methods

2.1. Research Strategy

A PICO framework was used to guide the literature search. The following question, as mentioned above, was posed: "In what way do resin-based composite materials used in dental practice induce cytotoxicity on the oral cavity tissues"? The target population (P) was human and animal cells naturally found in the oral cavity tissues (studied in vitro), and the intervention (I) was exposure to commercially available resin-based materials (adhesive systems, composite resins, and resin luting agents) or to experimental resins or to free resin monomers, the comparison (C) was untreated cells or positive controls, and the outcome (O) was a change in the viability of the treated cells.

In order to research the current literature, the library of the Medical University of Lodz, Google Scholar, and PUBMED search engines were used to identify relevant papers. Subject to screening were original studies: in vitro studies. The search for literature was performed up to 25 May 2023. The following keyword combinations were employed: "cytotoxicity of dental materials" OR "cytotoxicity of resin composite" OR "biocompatibility of dental materials" OR "biocompatibility of resin composite" OR "oxidative stress of dental materials" OR "oxidative stress of resin composite" OR "genotoxicity of dental materials" OR "genotoxicity of resin composite" OR "mutagenicity of dental materials" OR "mutagenicity of resin composite". Only English language literature was analyzed.

2.2. Study Selection

Abstracts of publications returned from the search engines were selected using inclusion and exclusion criteria summarized in Table 2. Duplicates were removed manually. A hand search was performed in the reference lists of valuable studies to identify additional relevant papers. For all publications that met the inclusion criteria, the electronic version of the paper was retrieved in full and analyzed.

Table 2. The inclusion and exclusion criteria.

Inclusion Criteria	Exclusion Criteria
Publication in English Published since 2017 In vitro study Containing the following keyword combinations: "cytotoxicity of dental materials" OR "cytotoxicity of resin composite" OR "biocompatibility of dental materials" OR "biocompatibility of resin composite" OR "oxidative stress of dental materials" OR "oxidative stress of resin composite" OR "genotoxicity of dental materials" OR "genotoxicity of resin composite" OR "mutagenicity of dental materials" OR "mutagenicity of resin composite"	Publication not in English Published before 2017 Review articles, abstracts, book chapters, live animal studies

2.3. Risk of Bias

The quality of selected in vitro studies required further validation. The Quality Assessment Tool for In Vitro Studies (the QUIN Tool) developed by Sheth et al. was used in the process [34]. The QUIN Tool takes into consideration a list of 12 criteria: clearly stated aims/objectives; detailed explanation of sample size calculation; detailed explanation of sampling technique; details of the comparison group; detailed explanation of methodology; operator details; randomization; method of measurement of outcome; outcome assessor details; blinding; statistical analysis; and presentation of results. Each criterion is then assigned a score: score = 2 for "adequately specified", score = 1 for "inadequately specified", or score = 0 for "not specified". Criteria that are recognized as not applicable are not included in the final score. The final score is obtained using the following formula:

$$\text{Final score} = \frac{\text{Total score} \times 100}{2 \times \text{number of applicable criteria}}$$

The obtained scores are used to assign the risk of bias: >70% for low risk of bias, 50–70% for medium risk of bias, and <50% for high risk of bias. Studies assigned a high risk of bias were rejected.

3. Results

3.1. Study Selection

There were 55 studies identified in the databases. Of these studies, 27 were excluded from screening. The remaining 28 papers were retrieved and assessed for eligibility. Of these, 14 studies were selected for further analysis based on the inclusion and exclusion criteria presented in Table 2. PRISMA flow chart of the study selection process is presented in Figure 2 [35]. A brief summary of the included studies is presented in Table 3.

Table 3. The Quality Assessment Tool for In Vitro Studies (the QUIN Tool) risk of bias.

Study	Final Score (%)	Risk of Bias
Neves et al., 2019 [36]	65	Medium
Cengiz et al., 2022 [37]	80	Low
Carrillo-Cotto et al., 2020 [38]	80	Low
Beltrami et al., 2021 [39]	70	Medium
Kavuncu et al., 2020 [40]	70	Medium

Table 3. *Cont.*

Study	Final Score (%)	Risk of Bias
Sulek et al., 2022 [41]	80	Low
Schneider et al., 2019 [42]	55	Medium
Lovász et al., 2021 [43]	55	Medium
Agnes et al., 2017 [44]	80	Low
Lovász et al., 2021 [45]	80	Low
Sun et al., 2018 [46]	75	Low
Yang et al., 2022 [47]	85	Low
Lee et al., 2022 [48]	75	Low
Chang et al., 2020 [49]	75	Low
Wawrzynkiewicz et al., 2021 [50]	85	Low
Wawrzynkiewicz et al., 2020 [51]	85	Low

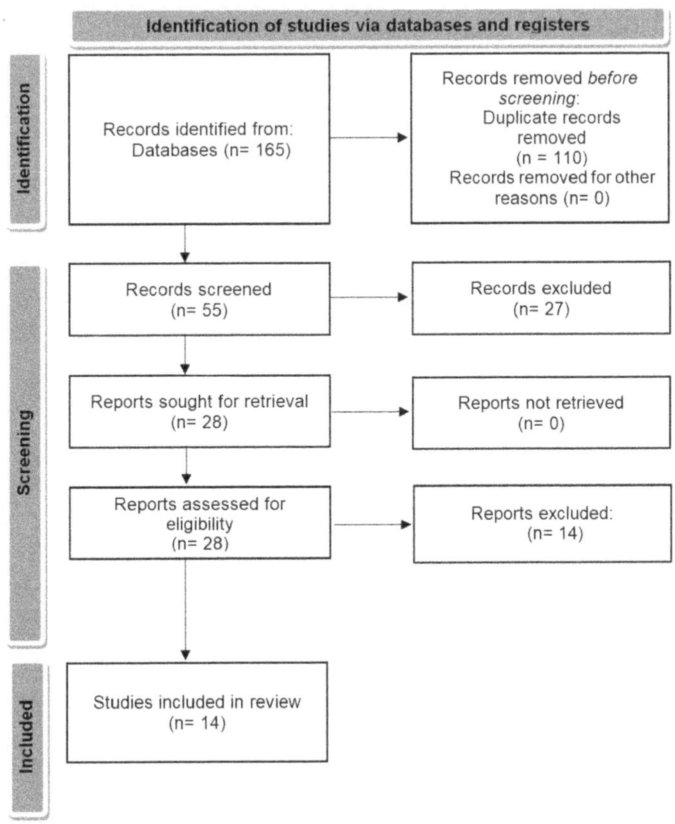

Figure 2. PRISMA flow chart of the study selection process.

3.2. *Risk of Bias*

The risk of bias for selected studies was assessed using the Quality Assessment Tool for In Vitro Studies (the QUIN Tool). Out of 14 studies, there were 9 that were assigned a low risk of bias and 5 with a medium risk of bias.

3.3. *Cytotoxicity Results*

It should be noted that studies included in this review had differing methodologies. One major difference was the material of choice in the study. For the purpose of this review, included studies were divided with regard to the method of exposure. Group one analyzed free resin monomers suspended in a medium. Group two focused on eluates obtained from commercially available composites after polymerization.

Four types of cell viability assays were used to assess the cytotoxicity of tested materials: MTT, XTT, WST-1, and LDH. MTT assay is a colorimetric assay that utilizes 3-(4,5-dimethylthiazol-2-yl)-2,5-diphenyl-2H-tetrazolium bromide, a monotetrazolium salt. Its reduction results in the formation of a violet water-insoluble formazan [52]. XTT assay is a related procedure that uses as its reagent 2,3-bis (2-methoxy-4-nitro-5-sulfophenyl)-5-[(phenylamino) carbonyl]-2H-tetrazolium hydroxide [53]. WST-1 is another tetrazolium salt assay that utilizes 2-(4-iodophenyl)-3-(4-nitrophenyl)-5-(2,4-disulfophenyl)-2H-tetrazolium [54]. LDH assay is based on the activity of L-lactate dehydrogenase [55]. All assays are commonly used in testing cytotoxicity and can be treated as the gold standard.

Neves et al. revealed in MTT assay that BisGMA, UDMA, and TEGDMA induced dose-dependent cytotoxicity on human peripheral blood mononuclear cells with toxic effects present at high concentrations: BisGMA at 0.06–1 mM induced a 44–95% decrease in mitochondrial activity, UDMA at 0.05–2 mM caused 50–93% decrease, and TEGDMA at 2.5–10 mM induced a 26–93% decrease [36]. Schneider et al. took on a different approach, measuring the LDH release of human dental pulp cells as an indicator of toxicity of their exposure to dental composite resin monomers: BisGMA, UDMA, and TEGDMA [42]. Their findings indicated BisGMA and UDMA as highly toxic, with BisGMA-induced toxicity beginning at the concentration of 30 μm and UDMA starting at 100 μm. TEGDMA failed to cause toxicity at any concentration tested [42]. Lovász et al. used a water-soluble tetrazolium salts (WST-1) colorimetric assay to measure the viability of human dental pulp cells exposed to TEGDMA monomers [43]. They noted a significant reduction in viability after exposure to 1.5 and 3 mM of TEGDMA for 24 h. Lower concentrations at this time did not produce significant effects [43]. In another study by Lovász et al., human dental pulp cells exposed to 0.75, 1.5, and 3 mM TEGDMA for 5 days were analyzed via WST-1 assay [45]. After 24 h, there was a decrease in viability from 1.237 WST values in the control group to 0.970, 0.814, and 0.518 WST values for 0.75, 1.5, and 3 mM, respectively. Second-day results presented a decrease from 1.961 to 1.290, 0.472, and 0.056 WST values in the same group order. Similarly, the fifth-day results were a decrease from 2.259 to 0.893, 0.105, and 0.089 WST values [45]. Sun et al. studied HEMA cytotoxicity on human dental pulp cells incubated for 72 h with 50, 100, 200, 400, 800, 1000, 1500, 2000, 2500, and 3000 μg/mL HEMA [46]. Cell viability levels were consistently reduced. Yang et al. investigated the cytotoxicity of TEGDMA monomers on murine macrophages via MTT assay [47]. The cells were incubated with or without TEGDMA for 24 h. TEGDMA was found to induce approximately 75% cytotoxicity in comparison to the control group [47]. Lee et al. conducted a similar study on murine macrophages incubated with 0, 0.5, 1, 5, and 10 mM HEMA for 24 h, testing the cytotoxicity with an MTT assay [48]. The results presented a dose-dependent cytotoxicity, with a decrease in viability by almost 20% for 1 mM, 40% for 5 mM, and 55% for 10 mM [48]. Chang et al. studied murine macrophages incubated with UDMA at 0, 0.1, 1, 10, and 100 μM for 24 h, measuring the cytotoxicity via LDH (lactate dehydrogenase) assay [49]. They observed 12% cytotoxicity for 1 μM and as high as 40% cytotoxicity for 10 μM [49].

Cengiz et al. recorded a reduction in MTT absorbance in murine fibroblasts exposed to Signum and Adoro composites over 2 weeks, with significant differences for both materials at each incubation time [37]. Carrillo-Cotto analyzed MTT assay in human spontaneously transformed aneuploid immortal keratinocytes exposed to adhesive systems OptiBond FL, Clearfil SE Bond and Adper Single Bond Universal, conventional composite resin Filtek Z350 XT, flowable composite resin Filtek Flow Z350 XT, self-adhesive composite resin Dyad Flow and luting agents Variolink II and RelyXU200 in combinations simulating their clinical use with or without the presence of dentin [38]. They observed that cytotoxicity varied with the combination of materials tested and that for all tested materials, most toxic effects were observed in the first 24 h. Dentin was found to increase cell viability. Beltrami et al. used an MTT assay to determine the cytotoxicity of immortalized human gingival fibroblasts exposed to different nano-hybrid composite resins [39]. It was found that after 72 h of incubation, the cell viability was significantly lower than after 48 h of incubation for all tested materials except for Enamel Plus HRi and G-aenial. After 48 h, Omnichroma, Omnichroma Blocker, Admira Fusion x-tra, and Enamel Plus HRi Bio Function Enamel showed the lowest grade of cytotoxicity, with cell viability above 80% [39]. Enamel Plus HRi and G-aenial showed, respectively, severe and moderate toxicity after 48 h. After 72 h of incubation, Omnichroma and Omnichroma Blocker showed mild cytotoxicity with a significant decrease in cell viability rates as compared to levels after 48 h. Admira Fusion x-tra and Enamel Plus HRi Bio Function Enamel showed a significant reduction to moderate cytotoxicity after 72 h. G-aenial Flo X and Enamel Plus HRi Bio Function Bio Dentine showed similar results after 48 h and 72 h. Both materials showed a lower cell viability rate after 72 h as compared to 48 h incubation [39]. Kavuncu et al. tested three composite materials, Admira Fusion, Charisma Topaz, and Estelite Quick Sigma, on human gingival fibroblasts and periodontal ligament fibroblasts [40]. In comparison to the control group, Admira Fusion, and Estelite Quick Sigma presented no cytotoxic effects on gingival fibroblasts after 24 h incubation. Only Charisma Topaz was cytotoxic to gingival fibroblasts at this time. After 1 week of incubation, the viability of gingival fibroblasts exposed to Charisma Topaz was significantly lower than the control, while the other materials were similar to the control. In contrast, Charisma Topaz did not produce significant cytotoxicity on periodontal ligament fibroblasts after 24 h of incubation and was the only cytotoxic material in 1 week of exposure [40]. Sulek et al. tested the cytotoxicity of Charisma, Estelite Sigma Quick, and Filtek Z250 on human gingival fibroblasts via MTT assay [41]. All tested materials significantly decreased cell viability while freshly cured. Charisma generated a 58% decrease in cell viability, and Filtek Z250 and Estelite Sigma Quick resulted in a 31% and 22% decrease, respectively. In delayed toxicity testing by LDH assay, pre-incubated Charisma toxicity was significantly lower, Estelite Sigma Quick resulted in similar toxicity, and Filtek Z250 was significantly more toxic [41]. Agnes et al. examined human dental pulp cells exposed to Flow Line and Durafill resin composites combined with mineral trioxide aggregate (MTA) and Dycal capping materials using LDH assay [44]. Their findings showed that 24 h exposure to Durafill and Flow Line resulted in cell death rates of approximately 35–40% and 35%, respectively. It should be noted that exposure to the combination of Durafill with Dycal did not alter toxicity significantly, and Flow Line with Dycal actually decreased cell death rates to around 15%. Meanwhile, the combination of MTA with Durafill enhanced Durafill cytotoxicity up to approximately 85%, and the cytotoxicity of Flow Line did not alter significantly [44]. Wawrzynkiewicz et al. used XTT assay (2H-tetrazolium-5-Carboxanilide colorimetric assay) to assess the cytotoxicity of dental adhesives—All-Bond Universal, CLEARFIL Universal Bond Quick, G-Premio BOND and Single Bond Universal—on human monocytes/macrophage peripheral blood cells [50]. Their results indicated significant differences in the cytotoxicity of tested eluates. Only G-Premio BOND was found to significantly decrease the cell viability by around 22% in comparison with the control. In another study by Wawrzynkiewicz et al. on human monocytes/macrophage peripheral blood cells, only OptiBond Universal induced a significant decrease in cell viability by around 25% [51].

The results presented above indicate a dose- and time-dependent cytotoxicity of resin-based composite. As could be predicted, increasing the dose of the material and time of exposure enhance cytotoxicity. It should be noted, however, that this review is based entirely on the results of in vitro studies. As such, it is up for further consideration to what extent the conditions of included studies are representative of the actual environment of the oral cavity. The resin monomer concentration is likely to be lower in vivo. Additionally, concentration fluctuates throughout the day due to many changing factors in the oral cavity. These arguments should be well considered while interpreting the cytotoxicity results of in vitro studies. Moreover, the type and origin of cells used in the included studies should be considered while drawing comparisons. These considerations will be further reflected upon in Section 4 (Discussion) of this review.

3.4. Summary of Study Characteristics

A brief summary of the included studies is presented in Table 4.

Table 4. Summary of study characteristics.

Study	Cell Viability Assay	Other Methods	Cell Line	Resin-Based Materials
Neves et al., 2019 [36]	MTT assay	APC Annexin V apoptosis detection kit, ELISA assay	Human peripheral blood mononuclear cells (hPBMC)	Monomers: BisGMA, TEGDMA, UDMA
Schneider et al., 2019 [42]	LDH assay	MCB assay, DCF assay	Human dental pulp cells (hDPC)	Monomers: BisGMA, TEGDMA, UDMA
Lovász et al., 2021 [43]	WST-1 assay	EncCheckGelatinolytic Collagenolitic activity assay, Western blotting	Human dental pulp cells (hDPC)	Monomers: TEGDMA
Lovász et al., 2021 [45]	WST-1 assay	Hemocytometer, fluorescence microscopy, Western blotting	Human dental pulp cells (hDPC)	Monomers: TEGDMA
Sun et al., 2018 [46]	MTT assay	RT-PCR, gelatinezymography, transwell migration assay, Western blotting	Human dental pulp cells (hDPC)	Monomers: HEMA
Yang et al., 2022 [47]	MTT assay	FITC Annexin V, comet assay	Murine macrophages	Monomers: TEGDMA
Lee et al., 2022 [48]	MTT assay	FITC Annexin V, micronucleus assay, comet assay	Murine macrophages	Monomers: HEMA
Chang et al., 2020 [49]	LDH assay	FITC Annexin V, MN assay, comet assay	Murine macrophage	Monomers: UDMA
Cengiz et al., 2022 [37]	MTT assay	None	Murine fibroblasts (mF)	Eluates of polymerized specimens: Signum (S), Adoro (A)
Carillo-Cotto et al., 2020 [38]	MTT assay	Fourier-transform infrared spectroscopy	Human keratinocytes (hK)	Eluates of polymerized specimens: OptiBond FL (OB), Clearfil SE Bond (CB), Adper Single Bond Universal (AS), Filtek Z350 XT (FZ3), Filtek Flow Z350 XT (FFZ3), Dyad Flow (DF), Variolink II (VII), RelyXU200 (RX)
Beltrami et al., 2021 [39]	MTT assay	None	Human gingival fibroblasts (hGF)	Eluates of polymerized specimens: Omnichroma (OC), Omnichroma Blocker (OCB), Admira Fusion x-tra (AFX), Enamel Plus Hri Bio Function Enamel (EPE), Enamel Plus Hri (EP), G-aenial (GA), G-aenial Flo X (GAF), Enamel Plus Hri Bio Function Bio Dentine (EPD)
Kavuncu et al., 2020 [40]	MTT assay	None	Human gingival fibroblasts (hGF), human periodontal ligament fibroblasts (hPLF)	Polymerized samples placed directly in cell culture medium: Admira Fusion (AF), Charisma Topaz (CT), Estelite Sigma Quick (ESQ)
Sulek et al., 2022 [41]	MTT assay, LDH assay	Flow cytometry, FITC Annexin V, Western blotting	Human gingival fibroblasts (hGF)	Eluates of polymerized specimens: Charisma (CH), Estelite Sigma Quick (ESQ), Filtek Z550 (FZ5)
Agnes et al., 2017 [44]	LDH assay	DCF assay	Human dental pulp cells (hDPC)	Polymerized samples placed directly in cell culture medium: Flow Line (FL), Durafill VS (DF)
Wawrzynkiewicz et al., 2021 [50]	XTT assay	Comet assay, flow cytometry, FITC Annexin V	Human monocytes/macrophage peripheral blood cells	Eluates of polymerized specimens: All-Bond Universal, CLEARFIL Universal Bond Quick, G-Premio BOND, Single Bond Universal
Wawrzynkiewicz et al., [51]	XTT assay	Comet assay, flow cytometry, FITC Annexin V	Human monocytes/macrophage peripheral blood cells	Eluates of polymerized specimens: OptiBond Universal, Prime&Bond Universal, AdheseUniversal

MTT—3-(4,5-dimethylthiazol-2-yl)-2,5-diphenyl-2H-tetrazolium bromide; LDH—L-lactate dehydrogenase; WST-1—2-(4-iodophenyl)-3-(4-nitrophenyl)-5-(2,4-disulfophenyl)-2H-tetrazolium; XTT—2,3-bis (2-methoxy-4-nitro-5-sulfophenyl)-5-[(phenylamino) carbonyl]-2H-tetrazolium hydroxide; BisGMA—bisphenol A-glycidyl methacrylate; TEGDMA—triethylene glycol dimethacrylate; UDMA—urethane dimethacrylate; HEMA—2-hydroxyethyl methacrylate.

4. Discussion

4.1. Mechanism of Toxicity

Resin-based materials are in prolonged, intimate contact with the oral cavity. This includes both direct contact between the filling and restored dentin, enamel, or adjacent gingiva, as well as indirect contact with the dentin and pulp via dentinal tubules or soft tissues via saliva. As such, cytotoxicity should be thoroughly studied and its mechanism well understood so that novel materials with improved safety profile can be developed.

Incomplete polymerization plays a major role in RBCs cytotoxicity, allowing for the elution of toxic methacrylate and dimethacrylate monomers: BisGMA, TEGDMA, UDMA, HEMA, BisEMA [11,15,41]. Most authors agree that the extent of the toxic effect is dependent on several factors, e.g., the chemical composition of the material, dose, and time of exposure, as well as internal factors, e.g., cell line, cell lining parameters, cellular membrane integrity, cell volume, and cytoplasm volume, the refractive index of the cell, the propensity to cleave DNA, and related nuclear condensation [12,36–40,43,45,46].

Smaller monomers with hydrophilic properties (e.g., HEMA and TEGDMA) have been found to penetrate cellular membranes, inhibit glutathione and lipids synthesis, and induce damage to mitochondria and DNA [12,25,38,42,45]. This damage to DNA is best understood as the loss of integrity and fragmentation of DNA strands [48]. Mitochondrial dysfunction can be explained by the generation of reactive oxygen species and depolarization of mitochondria, which leads to caspases-mediated apoptosis [48,49]. Mitochondria play a vital role in cellular metabolism and homeostasis, and their dysfunction initiates cell death [49]. Toxicity of BisGMA can be connected to this monomer being a derivative of Bisphenol A (BPA), which is known to be involved with genital developmental problems, immune function, thyroid function, and neurodevelopment in children [15,25]. Synthetic pathways of BisGMA production that do not require the use of bisphenol A have been introduced, but BPA presence has been found in urine and saliva of patients after dental treatment [19].

Monomers such as TEGDMA have been found to increase the generation of mitogen-activated protein kinases (MAPKs), which may interfere with biomineralization and initiate apoptosis [12,47].

Suggested molecular mechanisms of toxicity after exposure to RBCs include the depletion of glutathione (GSH), a tripeptide responsible for maintaining the redox balance of the cell [12,25,30,38]. GSH can be consumed to detoxify the monomers even in non-lethal concentrations, which induce little changes in viability [30]. Schneider et al. tested the effects of BisGMA, UDMA, and TEGDMA monomers on cystine uptake in dental pulp cells since it has been demonstrated to alter cellular glutathione levels [42]. After 2 h, BisGMA and UDMA decreased cystine uptake, and TEGDMA increased it, with significant changes observed at the monomer concentrations of 300 μM [42]. After 48 h, a significant decrease in cystine uptake was observed for BisGMA and UDMA at concentrations of above 30 μM and 100 μM, respectively. Interestingly, lower UDMA concentrations caused increased uptake at 48 h. TEGDMA induced increased cystine uptake [42]. An incoherent relation between cystine uptake and cellular GSH level was observed, contrary to the expectation that since cystine is necessary for GSH production, its changes would mirror one another. At 2 h, there was a noticeable decrease in cellular GSH for all monomers independently of the concentration. At 48 h, there was a dramatic decrease in cellular GSH levels for BisGMA and UDMA in concentrations of 30 μM and above. At the same time, TEGDMA caused a slight increase at 100 μM and a decrease at 300 μM [42]. Depletion of GSH levels upsets the redox balance of the cell. Reactive oxygen species are produced in increased quantities, inducing redox imbalance resulting in damage to DNA and cell death [12,30]. This redox imbalance activates adaptive cell mechanisms, up- and down-regulating enzyme systems [30].

Lovász et al. found via Western blotting that TEGDMA exposure induced strongly increased levels of metalloproteinases MMP-2, MMP-8, and MMP-9 in dental pulp cells at the concentration of 0.2 mM. These metalloproteinases are protein hydrolases produced in the pulp and found in dentin. They can be divided into two groups. MMP-2 and MMP-9 are classified as gelatinases involved in tissue remodeling and tertiary dentin formation [43]. MMP-8 is classified as a collagenase involved in the organization of the pre-mineralized collagen fiber network in the dentin [43]. A lower concentration of 0.1 mM increased MMP-2 expression alone, and a higher 0.75 mM concentration increased MMP-2 significantly and MMP-8 slightly [43]. HEMA has been found to inhibit MMP-2 and MMP-9 expression by Sun et al. [46].

Sulek et al. analyzed changes to the expression of two epigenetic and biochemical biomarkers of toxicity: miR-9 and heat shock protein 70 (HSP70) in human gingival fibroblasts. mi-R9 is a stress-related micro-RNA with indicative potential for the transition of healthy fibroblasts into cancer cells. This micro-RNA binds to 3′-untranslated regions of the target mRNA, regulating the transcription of several genes relevant to cell physiology and pathology, and its increased level is typical for epithelial cells undergoing epithelial-mesenchymal transition, indicating cancer transformation [41]. HSP70 is an important chaperone and stress marker protein involved in protection mechanisms against thermal, chemical, and oxidative stress, responsible for refolding damaged proteins and inhibiting apoptosis [41]. Sulekat et al. observed that miR-9 was significantly increased in all groups—there was a 3.4-fold increase in cells incubated with Filtek Z550 and less than that 2-fold increase for Charisma and Estelite Quick Sigma [41]. As for HSP70, only Charisma induced a significant 7-fold increase in its expression [41]. In other studies, HEMA and TEGDMA were observed to decrease heat shock protein expression in human monocytes [25].

The production of reactive oxygen species is another important consideration [25,45]. While reactive oxygen species are physiologically instrumental in immunoregulation, and antimicrobial and antiviral functions, their excessive production may lead to cellular damage, DNA damage, lipid peroxidation, and inflammatory response [47,48]. Since all RBCs contain toxic monomers and free-radical-based photo-initiators, they can be expected to produce oxidative stress. As mentioned above, depletion of glutathione levels upsets the redox balance of the cell, and reactive oxygen species are produced in increased quantities, inducing redox imbalance [12,30]. Sulek et al. analyzed oxidative stress in human gingival fibroblast cells grown for 3 and 6 h with Charisma, Estelite Quick Sigma, and Filtek Z550. In all samples, there was a significant increase in the mean DCF fluorescence [41]. Filtek Z550 induced the highest increase at 12-fold. A slightly lower 10-fold and 9-fold increase was observed in fluorescence intensity for Charisma and Estelite Quick Sigma, respectively [41]. Schneider et al. demonstrated a significant increase in free radicals in high BisGMA and UDMA monomer concentrations, while TEGDMA had no influence on free radicals in dental pulp cells [42]. The accumulation of reactive oxygen species induces oxidative stress, damaging DNA, lipids, and proteins. Genotoxicity and induction of apoptosis by accumulation of reactive oxygen species are strictly associated with cytotoxicity.

Some authors have demonstrated that RBCs influence inflammatory response in soft tissues adjacent to the restoration [39]. Incubation of macrophages with TEGDMA initiates several proinflammatory mechanisms, such as the upregulation of cyclooxygenase-2 and inducible nitric oxide synthase [47]. Neves et al. investigated the influence of BisGMA, TEGDMA, and UDMA on pro-inflammatory cytokine IL-1β and TNF-α release on peripheral blood mononuclear cells incubated with TC_{20} concentration of those monomers [36]. Those cells typically exhibit an increased cytokine production when exposed to *Porphyromonas gingivalis*. BisGMA did not influence the expression of cytokines in any way. However, TEGDMA had an inhibitory influence on the secretion of both IL-1β and TNF-α, and UDMA decreased IL-1β alone in cells exposed to *P. gingivalis* [36]. These findings suggest that monomers could interfere with the local immune inflammatory response. However, no negative impact was observed in cells that were not stimulated with *P. gingivalis*, suggesting that low concentrations of resin monomers assure relative safety [36]. It should be noted that the outcomes could be influenced by a decreased number of viable cells capable of cytokine production.

In animal studies, methacrylates increased the number of micro-nucleated cells in bone marrow, indicating mutagenicity [25]. Sulek et al. found that Charisma produced a 5-fold increase in hypodiploid cell numbers in comparison with the control group, signifying high DNA damage in human gingival fibroblasts [41]. In the same study, Estelite Sigma Quick and Filtek Z550 induced a 2- and 3-fold increase in damaged cells, respectively [41].

4.2. Impact on Cell Viability

Most studies conclude that RBCs induce cytotoxicity in in vitro cell models and that the impact on cell viability is dose- and time-dependent and changes depending on the chemical composition of the product and is specific for each cell line [36–38,40,49].

Neves et al. found that three tested monomers—BisGMA, UDMA, and TEGDMA—caused dose-dependent cytotoxicity on peripheral blood mononuclear cells, as evidenced by inhibited mitochondrial metabolic activity [36]. They established the following order of toxicity, based on TC_{50} (concentration that caused a 50% decrease in cell viability) and TC_{20} (concentration that caused a 20% decrease in cell viability) values: BisGMA > UDMA > TEGDMA with TC_{50} values of 69.0 mM, 505.0 mM, and 3161.0 mM; and TC_{20} values of 50.5 μM, 167.0 μM, and 2150.0 μM, respectively [36]. Schneider et al. established that BisGMA and UDMA caused cytotoxicity in concentrations beginning at 30 μm and 100 μm, respectively [42]. Most studies support that the order of cytotoxicity of monomers is BisGMA > UDMA > TEGDMA > HEMA [41,42].

Beltrami et al. investigated the biocompatibility of several RBCs on human gingival fibroblasts. MTT assay revealed that after 72 h of incubation, cell viability was significantly lower than after 48 h of incubation for all tested materials except for Enamel Plus HRi and G-aenial [39]. Both Enamel Plus HRi and G-aenial did, however, show severe and moderate cytotoxicity after 48 h, respectively. After 48 h, Omnichroma, Omnichroma Blocker, Admira Fusion x-tra, and Enamel Plus HRi Bio Function Enamel presented a low degree of cytotoxicity, with cell viability rates above 80% [39]. After 72 h, Omnichroma and Omnichroma Blocker induced mild cytotoxicity, with a significant decrease in cell viability rates in comparison to 48 h incubation [39]. Admira Fusion x-tra and Enamel Plus HRi Bio Function Enamel showed a reduction to moderate cytotoxicity rates, while G-aenial Flo X and Enamel Plus HRi Bio Function Bio Dentine showed similar results after 48 h and 72 h [39]. Kavuncu et al. investigated nanohybrid ormocerAdmira Fusion, Charisma Topaz, and resin-based EsteliteQuick Sigma [40]. Using MTT assay, they found that Admira Fusion and EsteliteQuick Sigma induced no cytotoxicity on human gingival fibroblasts after 24 h, while Charisma Topaz was cytotoxic [40]. Similarly, after one week of exposure, only Charisma Topaz was significantly cytotoxic. In the same study, for human periodontal ligament fibroblasts minor differences were observed in comparison to human gingival fibroblasts. Charisma Topaz was not significantly cytotoxic after 24 h when compared to the control group [40]. In one week, only Charisma Topaz was significantly cytotoxic, and cell viability was lower than in the control group [40]. For all materials, cytotoxicity was higher after one week than 24 h [40]. Sulek et al. investigated interactions of RBCs Charisma, Estelite Sigma Quick, and Filtek Z550 with human gingival fibroblasts. MTT assay revealed that all freshly cured materials caused a significant decrease in cell viability with the highest toxicity (a 58% decrease in viable cell numbers) induced by Charisma, while Estelite Sigma Quick and Filtek Z550 caused 22% and 31% declines, respectively [41]. Charisma was slightly less cytotoxic in the pre-cured application, while Filtek Z550 increased its toxicity considerably, and Estelite Sigma Quick induced similar toxicity in this state [41]. All three materials induced damage to the cell membrane. LDH assay used to assess time-dependent cytotoxicity saw increased LDH activity in cells incubated with Charisma after 1 h [41]. After 24 h, cell membrane damage increased to more than 40% of total enzyme activity, and the cytotoxicity plot was hyperbolic [41]. The trend of LDH release was similar for Estelite Sigma Quick, but the maximal values after 24 h were lower, reaching 28% of total enzyme activity [41]. Meanwhile, Filtek Z550 produced 37% of total enzyme activity after 24 h and, interestingly, the time-dependent cytotoxicity plot was a linear correlation [41].

4.3. Impact on Cell Cycle and Mechanism of Cell Death

The production of reactive oxygen species and depletion of cellular glutathione reserves described before is the likely reason for the induction of cell death in pulp cells, gingival fibroblasts, odontoblasts, and other cells exposed to monomers [25]. Evidence regarding the specific pathway of cell death induced by RBCs is scarce and inconclusive. This knowledge gap needs to be addressed, as it is relevant to the identification of efficient prevention strategies.

There are two pathways of cell death—apoptosis and necrosis. Cytotoxicity and genotoxicity via violation of DNA integrity are closely connected to apoptosis, which is a programmed cell death [47]. This pathway is regulated by proteolytic cysteinyl enzymes caspases. Caspase-3 acts as an executioner caspase which initiates DNA strand breaking and fragmentation [47–49]. This caspase can be activated via intrinsic or extrinsic pathways. Intrinsic pathway is connected to caspase-9 activity, which is triggered by mitochondrial disruption. The extrinsic pathway, meanwhile, is mediated by caspase-8 via death receptor activation [47]. The accumulation of intracellular reactive oxygen species is responsible for the activation of caspase chain [47,48].

Neves et al. found that incubation with BisGMA and *Porphyromonas gingivalis* resulted in a significant increase in the percentage of necrotic monocytes when compared with culture exposed to BisGMA alone [36]. As for TEGDMA, there was a noticeable increase in apoptotic cells in monocytes incubated with TEGDMA and *P. gingivalis* compared to exposure to *P. gingivalis* alone, while no such differences were observed for necrotic cells, thus suggesting that most of TEGDMA-induced cell death was via apoptosis [36]. UDMA was similarly found to induce cell death due to apoptosis [36]. Lovász et al. investigated the pathway of apoptotic cell death induced by TEGDMA in dental pulp cells [45]. They observed that TEGDMA exposure caused an increase in caspases and apoptosis-induced factor (AIF) production. Caspase-3 was significantly increased after exposure to 1.5 mM and 3 mM TEGDMA, caspase-8 in 0.1 and 0.2 mM, and caspase-9 in 0.75, 1.5 and 3 mM. Caspase-12 was significantly increased at concentrations above 0.75 mM. The production of AIF—a mitochondrial polypeptide responsible for chromatin condensation and DNA degradation—was induced by 0.2, 0.75, and 1.5 mM TEGDMA [45]. Similar induction of caspase-3, caspase-8, and caspase-9 activity in mouse macrophage after TEGDMA exposure at the concentration of 3 μM was found by Yang et al. [47]. These results indicate that TEGDMA-induced apoptosis, as evidenced by the increase in apoptosis-specific caspases. This information is of clinical relevance, as apoptosis does not involve an inflammatory process, in contrast to necrosis. There is no satisfactory evidence, however, to determine the specific pathway of apoptosis. Chang et al. studied the level of apoptosis and necrosis induced by UDMA in macrophages and proved that low concentrations induced early apoptosis, while at high concentrations, late apoptosis and necrosis were induced [49]. In another study by Wawrzynkiewicz et al. on human monocytes/macrophage peripheral blood cells, only OptiBond Universal induced a significant increase in apoptosis, as approximately 45% of cells were found at the early or late stage of apoptosis [51].

In the study by Sulek et al., the pathway of cell death in human gingival fibroblasts after 24 h incubation with RBCs was investigated using Annexin V/propidium iodide [41]. Estelite Sigma Quick produced an inconclusive pattern of changes with 48% necrotic cells and 28% of cells with mixed apoptotic and necrotic cells [41]. Filtek Z550 and Charisma produced mostly nonspecific necrotic changes, with 73% and 75% of cells stained as necrotic, respectively [41]. In cells incubated with Filtek Z550 a substantial number of cells—16%—developed both necrotic and apoptotic features, however. Wawrzynkiewicz et al. observed via FITC Annexin V apoptosis kit that G-Premio Bond induced a significant increase in the apoptosis of human peripheral blood cells, with approximately 39% of cells at the early or late stage of apoptosis [50].

Free monomers have been found to inhibit cell proliferation [25]. Sulek et al. demonstrated that Charisma, Estelite Sigma Quick, and Filtek Z550 caused significant antiproliferative effects [41]. Estelite Sigma Quick inhibited cell proliferation by about 60%, Filtek Z550 by 35%, and Charisma by 17% [41]. Interestingly, only Charisma additionally decreased the number of resting cells by 40% [41]. Sun et al. observed a time- and dose-dependent anti-proliferative influence of HEMA in human dental pulp cells [46]. HEMA was also found by Lee et al. to induce dose-dependent cytotoxicity via apoptosis and not necrosis in murine macrophages [48]. The intrinsic pathway of apoptosis was activated further in this study.

Mitochondrial dysfunction and toxic influence on DNA were suggested as possible mechanisms of apoptosis induction [36]. TEGDMA-induced apoptosis has been explained by the generation of reactive oxygen species, phosphorylation of mitogen-activated protein kinase, and downstream transcription factors [47].

4.4. Prevention Strategies

Prevention strategies are of vital importance in clinical settings as RBCs remain widespread. Selecting the most biocompatible monomers is an important direction in the development of more suitable composite resins. This approach has its limitations, however, as other factors, such as the degree of conversion or the species of photo-initiator, can influence cytotoxicity, accounting for differences in cytotoxicity levels in materials with the same monomer composition. The quantity and composition of eluate should both be taken into consideration separately, as there is no direct relation between the amount of released monomers and the composition of the biomaterial [30]. For example, while many RBCs contain a mixture of BisGMA and TEGDMA, the more hydrophilic TEGDMA is likely to be eluted in higher quantities [30].

Dentin has been reported to have a physiological protective effect [30]. Carrillo-Cotto et al. tried to replicate the conditions of the oral cavity by investigating the cytotoxicity of RBCs on keratinocytes in the presence of dentin and without [38]. Interestingly, they found out that after 1 day of incubation, the cell viability of the control group was ten times higher in the presence of dentin. This behavior was observed for all groups where dentin was present [38]. Dentin mechanically lowers the concentration of toxic substances, serving as a diffusion barrier [30]. It is also possible for dentin to absorb the unbound monomers and exert a buffering effect [30,38]. Dentin was observed to stimulate proliferation and increase cell viability. The responsible mechanism may be explained by the presence of growth factors in this tissue and its release into the medium [38]. Based on these findings, it should be investigated whether the presence of growth factors in dentin is capable of counteracting the cytotoxic impact of leached monomers from RBCs in vivo. Indirect contact tests with dentin should be used to produce an understanding of hard tissue interactions with diffusing monomers [11].

Antioxidants such as rutin or melatonin have been proposed with some success as a prevention method to reduce the genotoxicity of RBCs [25,47]. Cengiz et al. investigated the possible protective influence of melatonin in saliva, as melatonin is reported to have properties such as DNA protection, reducing inflammation, antioxidative characteristics, and potentially being an antiapoptotic agent [37,56]. However, the MTT assay revealed no significant differences in cytotoxicity between murine fibroblasts incubated in artificial saliva with the addition of melatonin and those without [37]. Yang et al. investigated the protective effect of rutin—a bioflavonoid antioxidant—on TEGDMA-induced cytotoxicity in murine macrophages [47]. Rutin, also known as quercetin-3-rhamnosyl glucoside, is a natural flavanol glycoside known to reduce inflammatory response and genotoxicity [47]. Pretreatment with rutin was found to decrease TEGDMA cytotoxicity in a concentration-dependent manner, with significant reduction beginning at 30 μM [47]. Genotoxicity, apoptosis, necrosis, and reduction in reactive oxygen species generation decreased similarly in the same study.

Another important factor determining the extent of cytotoxicity is the degree of polymerization, measured as the Degree of Conversion, which is dependent on curing time, light source, viscosity and thickness of the RBCs layer, type and mixture of photo-initiators as well as type and proportion of monomers and filler [11,39,40]. Incomplete polymerization contributes to the increased release of unreacted monomers and therefore supports cytotoxicity [11,12,38–40,43]. The monomer composition of the material is not the only factor influencing its biocompatibility. Filler content should also be considered. It has been found that nano-hybrid ormocers exert less cytotoxicity and release less unbound monomers [12,39,40]. A higher filler content minimizes the organic resin component, thus improving biocompatibility [39]. Beltrami et al. have found that Omnichroma, Omnichroma Blocker, Admira Fusion x-tra, and Enamel Plus HRi Bio Function Enamel—in all of which classic monomers are absent—induced lower cytotoxicity in human gingival fibroblasts when compared with conventional materials. Similarly, Kavuncu et al. investigated nanohybrid ormocerAdmira Fusion, nanohybrid Charisma Topaz, and resin-based Estelite Quick Sigma in human gingival fibroblasts and periodontal ligament fibroblasts. Their findings indicated Charisma Topaz as the most toxic material of the three after both 24 h and 1 week exposure [40]. This suggests that ormocer group composites can be considered the most biocompatible materials in clinical cases, especially in close contact with gingiva and periodontium.

Pulp capping refers to the process of placing capping material, e.g., calcium hydroxide, over the exposed pulp in order to preserve its vitality and protect against toxic substances and physical stimuli. Novel capping materials include mineral trioxide aggregate (MTA). Restorative material is then placed directly over the capping material. RBCs are known to impair healing processes and the formation of a reparative dentin barrier while in direct contact with the exposed pulp [30]. Agnes et al. investigated possible combined toxic effects of RBCs and capping materials in human dental pulp cells [44]. They found that calcium hydroxide Dycal demonstrated dose-dependent toxicity, while MTA remained bioinert. With cultures exposed to Dycal and RBCs Durafill or Flow Line for 24 h, cell death rates reached 30–40%. Meanwhile, MTA increased the cytotoxicity of Durafill but had no impact on the Flow Line. Flow Line and Durafill were also found to induce some oxidative stress, but the results varied between groups. Interestingly, MTA enhanced oxidative stress induced by Durafill while Dycal reduced the oxidative stress of Flow Line. This study suggests the need for further investigations into the interactions between pulp capping materials and restorative composites. Another issue in pulp capping is to research the ability of eluted monomers able to diffuse through the capping material.

4.5. Analysis of Methodologies

RBCs should undergo several steps of biocompatibility and toxicity tests. The more standardized, systematic approach toward methodology should be developed to enable qualitative and quantitative analysis and synthesis of results with clinical relevance. Presently, however, there is little cohesion in methodologies among studies in the field [11]. Methodological standardization should be of primary concern to provide easier comparison of studies and reduce conflicting results.

The oral cavity is a complex environment with many diverse niches and interactions between host, restorative material, and microorganisms that are hard to replicate in in vitro studies. Limited volume, flow of saliva and dentinal fluid, and chemical interactions with dentine are some of the factors that contribute to the complexity of replicating this environment.

In vitro cytotoxicity tests are the first step in the evaluation of biocompatibility. In biocompatibility studies, there are three modes of in vitro methods, as outlined in Table 5. Direct contact tests are the least complex and therefore widely used. However, scientists argue that direct contact tests are actually of limited clinical relevance, as most cells are not in direct contact with the biomaterial on site [25]. As such, an indirect contact test seems to be a preferable alternative. Of these methods, an indirect contact test with a dentin layer can yield clinically relevant results when testing RBCs' impact on the pulp tissue.

Table 5. In vitro methods in biocompatibility studies [12,25,40].

Method	Characteristics
Direct contact test	Direct contact between the material and cell culture, typically in mono-layer.
Indirect contact test	Separation of the material and cell culture with an intermediate layer, e.g., agar gel, Millipore filter, dentin layer.
Extract test	Application of eluates from the material to cell culture.

Many studies on the subject tend to report the influence of resin composite materials at TC_{50} (concentration that caused a 50% decrease in cell viability) concentrations. However, some authors argue that such high concentrations of eluted monomers are not likely to be released from fillings in the site. As such, a lower concentration of methacrylates, such as TC_{20}, could yield more relevant results [36]. The determination of TC_{50} concentration is also dependent on the type of host cells investigated [36]. This is contested by other researchers who argue that limited volume, especially in niches such as the pulp chamber, can allow for the accumulation of higher concentrations of eluted monomers [42]. Monomers such as TEGDMA and HEMA can diffuse through dentinal tubules, reaching the pulp chamber in toxic concentrations [11]. The highest intrapulpal concentration has been reported as 4 mM [43]. The quantity of monomers released from polymerized RBCs is dependent on the ratio of the material sample surface to cell culture volume [30,39,40]. In clinical practice, the mean surface area of mesial-occlusal-distal fillings was calculated as 95 mm^2, in cervical fillings as 12 mm^2, and in veneers as 86 mm^2 [40].

In biocompatibility studies, some continuous cell lines, e.g., mouse fibroblasts, are commonly used due to their easy production and control of the culture. However, it should be noted that using the primary cells of the target tissue, e.g., pulp cells and gingival or periodontal ligament fibroblasts, results in studies of higher quality, producing more meaningful outcomes to consider in clinical applications [11]. Thus, primary cells should be preferred, regardless of the difficulties they might cause in the laboratory, e.g., slower growth or shorter life span of the cells. It is important to remember that monolayer cell cultures exhibit higher sensitivity to toxins than three-dimensional (3D) cultures, which is an important limitation of most in vitro studies. Substitution of monolayer cell culture with micro-tissue 3D models allows improved replication of the tissue microenvironment by enabling cellular communication and increased cell–cell and extracellular matrix–cell interactions [11,30].

Longer experimental periods could provide further insight into the chronic effects of the continuous long-term elution of unbound monomers on cell cultures, as such chronic exposure should not be discarded in human health risk considerations [39,40].

Interpretation of the underlying molecular mechanisms of toxicity is complicated, and it is hard to establish causality—biochemical changes are measured after toxicity has already occurred and may be either the cause of the cell death or its result. Appropriate selection of viability assay is another important consideration. The colorimetric MTT assay is widely described in the literature. It delivers objective results in a short time and is based on the assessment of cell metabolic activity. MTT is also recommended by international standards, including ISO 10993 [11]. However, its results can be affected by cell numbers and provide evidence of cell death only, bringing no contribution to the understanding of cytotoxicity pathways [11,41]. LDH release assay is frequently used as well and delivers information on the measure of dead cells. Toxicologists have established new paradigms that go beyond the measure of cell death and deliver additional information on the mechanism of toxicity based on the methodologies of proteomics, genomics, and pathway analyses, but these paradigms are only beginning to influence dental materials research [11,30]. Such methods include scanning electron microscopy to observe cell morphology, investigation of necrosis and apoptosis pathways, and flow cytometry to evaluate the effects on cell cycle [11].

Without a comprehensive understanding of the mechanisms of toxicity, materials with improved biocompatibility cannot be developed.

5. Summary

In this manuscript, the authors analyzed in what way resin-based composite materials used in dental practice induce cytotoxicity in the oral cavity tissues. Biocompatibility of dental materials remains an important factor in the development of novel RBCs. In recent years, there has been significant development in the field. RBCs have been markedly improved in regard to their aesthetic value, adhesive abilities, and mechanical properties. Researchers and manufacturers strive to further improve these materials, focusing on reducing their polymerization shrinkage, preventing secondary caries, strengthening adhesion, and further improving durability and resistance to degradation or endowing materials with bioactive, antibacterial properties. Safety profile considerations should be taken into account in the early stages of these developments. Analyzed papers point out that most resin-based composites induce dose- and time-dependent cytotoxic effects, causing oxidative stress and depletion of cellular glutathione reserves as well as disrupting enzymatic activity. Interestingly, they were also found in several studies to induce apoptosis. Viability loss, oxidative stress, depletion of glutathione, and induction of apoptosis in cells exposed to composite materials is well documented, and other studies support the conclusions of this manuscript [57]. The results of this review indicate that there is still a need for further research and improvement in this area. According to the current state of knowledge, RBCs meet legal regulations and standards, such as the ones outlined in ISO 14971, ISO 10993, and ISO 7405. However, the acquisition of knowledge concerning biocompatibility can provide a better understanding of material–host interaction and lead to the development of novel materials with improved safety profiles or the introduction of clearer safety measures in clinical practice. The benefits of such developments for the patients, clinicians, public and environmental health are obviously overwhelming, even if hard to quantify. Possible solutions which can improve the biocompatibility of dental resin-based composite materials include improving their degree of conversion and resistance to weardown, thus reducing the number of free monomers. Monomers liberated from these materials should be carefully identified and quantified as they determine the extent of cytotoxicity. Research indicates that newer monomers tend to present higher biocompatibility than older types, such as BisGMA. Direct and indirect pulp capping is recommended to protect the pulp from exposure and enable reparative dentin formation. The addition of antioxidants is being considered to reduce oxidative stress. Nanomaterials are reported to present great improvement in material properties, including biocompatibility [30]. There is also a need for the development of more standardized methods of in vitro tests for use in dental material science, which will allow for more accurate conclusions and comparisons and eliminate conflicting outcomes. Biocompatibility research in this area should provide more insight into the molecular mechanisms of cytotoxicity, which are not yet fully understood. A better understanding of these mechanisms may drive forward the development of more biocompatible materials. It should be noted that in vitro settings are limited in their ability to represent the intricately complex and changing environment of the oral cavity. The components of saliva and dentinal fluid, the presence of dentin and the quality of monomers capable of diffusing through this tissue to the pulp, the impact of physiological and pathological bacteria, changes in pH, and mechanical factors all influence the release of resin monomers from composite restorations and, as such, have impact on the adverse effects of resin-based composite materials in the patient.

Author Contributions: Conceptualization, K.W.-M., I.S.-P. and A.Ś.; methodology, K.W.-M., I.S.-P. and K.B.; visualization, K.W.-M.; writing—original draft preparation, K.W.-M. and I.S.-P.; writing—review and editing, A.Ś. and K.B.; funding acquisition, A.Ś. All authors have read and agreed to the published version of the manuscript.

Funding: This paper was supported by a grant from the Medical University of Lodz (No. 503/1-159-01/503-21-001).

Conflicts of Interest: The authors declare no conflict of interest.

References

1. Bin-Jardan, L.I.; Almadani, D.I.; Almutairi, L.S.; Almoabid, H.A.; Alessa, M.A.; Almulhim, K.S.; AlSheikh, R.N.; Al-Dulaijan, Y.A.; Ibrahim, M.S.; Al-Zain, A.O.; et al. Inorganic Compounds as Remineralizing Fillers in Dental Restorative Materials: Narrative Review. *Int. J. Mol. Sci.* **2023**, *24*, 8295. [CrossRef] [PubMed]
2. Al-hijazi, A.Y.; Hasan, N.; Nasr, B.K.; Jasim Al-Khafaji, H.H.; Al-Khafaji, B.; Abdah Alanssari, B.F.; Jalil, A.T. Recent Advances in the Use of Inorganic Nanomaterials as Anti Caries Agents. *Heliyon* **2023**, *9*, e15326. [CrossRef] [PubMed]
3. Guo, X.; Yu, Y.; Gao, S.; Zhang, Z.; Zhao, H. Biodegradation of Dental Resin-Based Composite—A Potential Factor Affecting the Bonding Effect: A Narrative Review. *Biomedicines* **2022**, *10*, 2313. [CrossRef] [PubMed]
4. Berghaus, E.; Klocke, T.; Maletz, R.; Petersen, S. Degree of Conversion and Residual Monomer Elution of 3D-Printed, Milled and Self-Cured Resin-Based Composite Materials for Temporary Dental Crowns and Bridges. *J. Mater. Sci. Mater. Med.* **2023**, *34*, 23. [CrossRef] [PubMed]
5. Riva, Y.R.; Rahman, S.F. Dental Composite Resin: A Review. *AIP Conf. Proc.* **2019**, *2193*, 020011.
6. Demarco, F.F.; Cenci, M.S.; Montagner, A.F.; De Lima, V.P.; Correa, M.B.; Moraes, R.R.; Opdam, N.J.M. Longevity of Composite Restorations Is Definitely Not Only about Materials. *Dent. Mater.* **2023**, *39*, 1–12. [CrossRef] [PubMed]
7. Gitalis, R.; Zhou, L.; Marashdeh, M.Q.; Sun, C.; Glogauer, M.; Finer, Y. Human Neutrophils Degrade Methacrylate Resin Composites and Tooth Dentin. *Acta Biomater.* **2019**, *88*, 325–331. [CrossRef] [PubMed]
8. Smith, L.; Ali, M.; Agrissais, M.; Mulligan, S.; Koh, L.; Martin, N. A Comparative Life Cycle Assessment of Dental Restorative Materials. *Dent. Mater.* **2023**, *39*, 13–24. [CrossRef]
9. Arbildo-Vega, H.I.; Lapinska, B.; Panda, S.; Khan, A.S.; Lukomska-Szymanska, M. Clinical Effectiveness of Bulk-Fill and Conventional Resin Composite Restorations: Systematic Review and Meta-Analysis. *Polymers* **2020**, *12*, 1786. [CrossRef]
10. Pratap, B.; Gupta, R.K.; Bhardwaj, B.; Nag, M. Resin Based Restorative Dental Materials: Characteristics and Future Perspectives. *Jpn. Dent. Sci. Rev.* **2019**, *55*, 126–138. [CrossRef]
11. Caldas, I.P.; Alves, G.G.; Barbosa, I.B.; Scelza, P.; de Noronha, F.; Scelza, M.Z. In Vitro Cytotoxicity of Dental Adhesives: A Systematic Review. *Dent. Mater.* **2019**, *35*, 195–205. [CrossRef] [PubMed]
12. Bapat, R.A.; Parolia, A.; Chaubal, T.; Dharamadhikari, S.; Abdulla, A.M.; Sakkir, N.; Arora, S.; Bapat, P.; Sindi, A.M.; Kesharwani, P. Recent Update on Potential Cytotoxicity, Biocompatibility and Preventive Measures of Biomaterials Used in Dentistry. *Biomater. Sci.* **2021**, *9*, 3244–3283. [CrossRef] [PubMed]
13. Wuersching, S.N.; Högg, C.; Kohl, L.; Reichl, F.-X.; Hickel, R.; Kollmuss, M. Leaching Components and Initial Biocompatibility of Novel Bioactive Restorative Materials. *Dent. Mater.* **2023**, *39*, 293–304. [CrossRef] [PubMed]
14. Lankes, V.; Reymus, M.; Mayinger, F.; Coldea, A.; Liebermann, A.; Hoffmann, M.; Stawarczyk, B. Three-Dimensional Printed Resin: Impact of Different Cleaning Protocols on Degree of Conversion and Tensile Bond Strength to a Composite Resin Using Various Adhesive Systems. *Materials* **2023**, *16*, 3580. [CrossRef] [PubMed]
15. Jung, Y.S.; Ro, S.T.; Kang, S.W.; Lee, H.; Lee, J.S.; Chae, Y.K.; Lee, K.E.; Lee, H.-S.; Kwack, K.H.; Kim, S.K.; et al. Bisphenol A Release from Commercially Available 3-Dimensionally Printed Resins and Human Cell Apoptosis to Bisphenol A: An in-Vitro Study. *J. Clin. Pediatr. Dent.* **2023**, *47*, 89–95. [CrossRef] [PubMed]
16. Vulović, S.; Nikolić-Jakoba, N.; Radunović, M.; Petrović, S.; Popovac, A.; Todorović, M.; Milić-Lemić, A. Biofilm Formation on the Surfaces of CAD/CAM Dental Polymers. *Polymers* **2023**, *15*, 2140. [CrossRef] [PubMed]
17. Tiu, J.; Belli, R.; Lohbauer, U. A Step toward Bio-Inspired Dental Composites. *Biomater. Investig. Dent.* **2023**, *10*, 1–7. [CrossRef]
18. Khan, A.A.; Zafar, M.S.; Fareed, M.A.; AlMufareh, N.A.; Alshehri, F.; AlSunbul, H.; Lassila, L.; Garoushi, S.; Vallittu, P.K. Fiber-Reinforced Composites in Dentistry—An Insight into Adhesion Aspects of the Material and the Restored Tooth Construct. *Dent. Mater.* **2023**, *39*, 141–151. [CrossRef]
19. He, J.; Lassila, L.; Garoushi, S.; Vallittu, P. Tailoring the Monomers to Overcome the Shortcomings of Current Dental Resin Composites—Review. *Biomater. Investig. Dent.* **2023**, *10*, 2191621. [CrossRef]
20. Hardan, L.; Devoto, W.; Bourgi, R.; Cuevas-Suárez, C.E.; Lukomska-Szymanska, M.; Fernández-Barrera, M.Á.; Cornejo-Ríos, E.; Monteiro, P.; Zarow, M.; Jakubowicz, N.; et al. Immediate Dentin Sealing for Adhesive Cementation of Indirect Restorations: A Systematic Review and Meta-Analysis. *Gels* **2022**, *8*, 175. [CrossRef]
21. Frassetto, A.; Breschi, L.; Turco, G.; Marchesi, G.; Di Lenarda, R.; Tay, F.R.; Pashley, D.H.; Cadenaro, M. Mechanisms of Degradation of the Hybrid Layer in Adhesive Dentistry and Therapeutic Agents to Improve Bond Durability—A Literature Review. *Dent. Mater.* **2016**, *32*, e41–e53. [CrossRef] [PubMed]
22. Zhang, X.; Zhang, Q.; Meng, X.; Ye, Y.; Feng, D.; Xue, J.; Wang, H.; Huang, H.; Wang, M.; Wang, J. Rheological and Mechanical Properties of Resin-Based Materials Applied in Dental Restorations. *Polymers* **2021**, *13*, 2975. [CrossRef] [PubMed]
23. Cervino, G.; Cicciù, M.; Herford, A.S.; Germanà, A.; Fiorillo, L. Biological and Chemo-Physical Features of Denture Resins. *Materials* **2020**, *13*, 3350. [CrossRef] [PubMed]

24. Mulligan, S.; Hatton, P.V.; Martin, N. Resin-Based Composite Materials: Elution and Pollution. *Br. Dent. J.* **2022**, *232*, 644–652. [CrossRef] [PubMed]
25. Shahi, S.; Özcan, M.; Maleki Dizaj, S.; Sharifi, S.; Al-Haj Husain, N.; Eftekhari, A.; Ahmadian, E. A Review on Potential Toxicity of Dental Material and Screening Their Biocompatibility. *Toxicol. Mech. Methods* **2019**, *29*, 368–377. [CrossRef] [PubMed]
26. Pałka, K.; Miazga-Karska, M.; Pawłat, J.; Kleczewska, J.; Przekora, A. The Effect of Liquid Rubber Addition on the Physicochemical Properties, Cytotoxicity, and Ability to Inhibit Biofilm Formation of Dental Composites. *Materials* **2021**, *14*, 1704. [CrossRef] [PubMed]
27. Szczesio-Wlodarczyk, A.; Domarecka, M.; Kopacz, K.; Sokolowski, J.; Bociong, K. An Evaluation of the Properties of Urethane Dimethacrylate-Based Dental Resins. *Materials* **2021**, *14*, 2727. [CrossRef]
28. Kowalska, A.; Sokolowski, J.; Bociong, K. The Photoinitiators Used in Resin Based Dental Composite—A Review and Future Perspectives. *Polymers* **2021**, *13*, 470. [CrossRef]
29. Mulla, S.A.; Kondkari, S.A.; Patil, A.; Jain, A.; Mali, S.; Jaiswal, H.C.; Jakhar, A.; Ansari, Z.M.; Agarwal, S.; Yadav, P. A Look Into the Cytotoxicity of Composite Fillings: Friend or Foe? *Cureus* **2023**, *15*, e46327. [CrossRef]
30. Schmalz, G.; Galler, K.M. Biocompatibility of Biomaterials—Lessons Learned and Considerations for the Design of Novel Materials. *Dent. Mater.* **2017**, *33*, 382–393. [CrossRef]
31. Arab-Nozari, M.; Zamani, E.; Evazalipour, M.; Soleimani, B.; Jorbonian, A.; Nahvi, A. Histocompatibility of Light-Curing Composites Used in Pediatric Dentistry in Human Oral Fibroblast Cells. *J. Dent.* **2023**, *24*, 112–117. [CrossRef]
32. Folwaczny, M.; Ahantab, R.; Kessler, A.; Ern, C.; Frasheri, I. Cytotoxicity of 3D Printed Resin Materials for Temporary Restorations on Human Periodontal Ligament (PDL-hTERT) Cells. *Dent. Mater.* **2023**, *39*, 529–537. [CrossRef] [PubMed]
33. Zhang, K.; Zhang, N.; Weir, M.D.; Reynolds, M.A.; Bai, Y.; Xu, H.H.K. Bioactive Dental Composites and Bonding Agents Having Remineralizing and Antibacterial Characteristics. *Dent. Clin. N. Am.* **2017**, *61*, 669–687. [CrossRef] [PubMed]
34. Sheth, V.H.; Shah, N.P.; Jain, R.; Bhanushali, N.; Bhatnagar, V. Development and Validation of a Risk-of-Bias Tool for Assessing in Vitro Studies Conducted in Dentistry: The QUIN. *J. Prosthet. Dent.* **2022**. *ahead of print*. [CrossRef] [PubMed]
35. Page, M.J.; McKenzie, J.E.; Bossuyt, P.M.; Boutron, I.; Hoffmann, T.C.; Mulrow, C.D.; Shamseer, L.; Tetzlaff, J.M.; Akl, E.A.; Brennan, S.E.; et al. The PRISMA 2020 Statement: An Updated Guideline for Reporting Systematic Reviews. *BMJ* **2021**, *372*, n71. [CrossRef] [PubMed]
36. Neves, S.O.; Magalhães, L.M.D.; Corrêa, J.D.; Dutra, W.O.; Gollob, K.J.; Silva, T.A.; Horta, M.C.R.; Souza, P.E.A. Composite-Derived Monomers Affect Cell Viability and Cytokine Expression in Human Leukocytes Stimulated with Porphyromonas Gingivalis. *J. Appl. Oral. Sci.* **2019**, *27*, e20180529. [CrossRef] [PubMed]
37. Cengiz, S.; Velioğlu, N.; Cengiz, M.İ.; Çakmak Özlü, F.; Akbal, A.U.; Çoban, A.Y.; Özcan, M. Cytotoxicity of Acrylic Resins, Particulate Filler Composite Resin and Thermoplastic Material in Artificial Saliva with and without Melatonin. *Materials* **2022**, *15*, 1457. [CrossRef] [PubMed]
38. Carrillo-Cotto, R.; Etges, A.; Jardim, P.S.; Torre, E.; Kaizer, M.R.; Ferrúa, C.P.; Nedel, F.; Cuevas-Suárez, C.E.; Moraes, R.R. Cytotoxicity of Contemporary Resin-based Dental Materials in Contact with Dentin. *Eur. J. Oral Sci.* **2020**, *128*, 436–443. [CrossRef]
39. Beltrami, R.; Colombo, M.; Rizzo, K.; Di Cristofaro, A.; Poggio, C.; Pietrocola, G. Cytotoxicity of Different Composite Resins on Human Gingival Fibroblast Cell Lines. *Biomimetics* **2021**, *6*, 26. [CrossRef]
40. Kavuncu, G.; Yilmaz, A.M.; Karademir Yilmaz, B.; Yilmaz Atali, P.; Altunok, E.C.; Kuru, L.; Agrali, O.B. Cytotoxicity of Different Nano Composite Resins on Human Gingival and Periodontal Ligament Fibroblast Cell Lines: An In Vitro Study. *Biomedicines* **2020**, *8*, 48. [CrossRef]
41. Sulek, J.; Luczaj-Cepowicz, E.; Marczuk-Kolada, G.; Rosłan, M.; Holownia, A. Cytotoxicity of Methacrylate Dental Resins to Human Gingival Fibroblasts. *J. Funct. Biomater.* **2022**, *13*, 56. [CrossRef] [PubMed]
42. Schneider, T.R.; Hakami-Tafreshi, R.; Tomasino-Perez, A.; Tayebi, L.; Lobner, D. Effects of Dental Composite Resin Monomers on Dental Pulp Cells. *Dent. Mater. J.* **2019**, *38*, 579–583. [CrossRef] [PubMed]
43. Lovász, B.V.; Lempel, E.; Szalma, J.; Sétáló, G.; Vecsernyés, M.; Berta, G. Influence of TEGDMA Monomer on MMP-2, MMP-8, and MMP-9 Production and Collagenase Activity in Pulp Cells. *Clin. Oral Investig.* **2021**, *25*, 2269–2279. [CrossRef] [PubMed]
44. Agnes, A.; Long, A.; Best, S.; Lobner, D. Pulp Capping Materials Alter the Toxicity and Oxidative Stress Induced by Composite Resins in Dental Pulp Culture. *Eur. Endod. J.* **2017**, *2*, 11. [CrossRef] [PubMed]
45. Lovász, B.V.; Berta, G.; Lempel, E.; Sétáló, G.; Vecsernyés, M.; Szalma, J. TEGDMA (Triethylene Glycol Dimethacrylate) Induces Both Caspase-Dependent and Caspase-Independent Apoptotic Pathways in Pulp Cells. *Polymers* **2021**, *13*, 699. [CrossRef] [PubMed]
46. Sun, S.; Wang, G.-L.; Huang, Y.; Diwu, H.-L.; Luo, Y.-C.; Su, J.; Xiao, Y.-H. The Effects of 2-Hydroxyethyl Methacrylate on Matrix Metalloproteinases 2 and 9 in Human Pulp Cells and Odontoblast-like Cells in Vitro. *Int. Endod. J.* **2018**, *51*, e157–e166. [CrossRef] [PubMed]
47. Yang, L.-C.; Chang, Y.-C.; Yeh, K.-L.; Huang, F.-M.; Su, N.-Y.; Kuan, Y.-H. Protective Effect of Rutin on Triethylene Glycol Dimethacrylate-Induced Toxicity through the Inhibition of Caspase Activation and Reactive Oxygen Species Generation in Macrophages. *Int. J. Mol. Sci.* **2022**, *23*, 11773. [CrossRef]

48. Lee, C.-Y.; Ho, Y.-C.; Lee, S.-S.; Li, Y.-C.; Lai, M.-Y.; Kuan, Y.-H. Cytotoxicity and Apoptotic Mechanism of 2-Hydroxyethyl Methacrylate via Genotoxicity and the Mitochondrial-Dependent Intrinsic Caspase Pathway and Intracellular Reactive Oxygen Species Accumulation in Macrophages. *Polymers* **2022**, *14*, 3378. [CrossRef]
49. Chang, C.-Y.; Chiang, C.-Y.; Chiang, Y.-W.; Lee, M.-W.; Lee, C.-Y.; Chen, H.-Y.; Lin, H.-W.; Kuan, Y.-H. Toxic Effects of Urethane Dimethacrylate on Macrophages Through Caspase Activation, Mitochondrial Dysfunction, and Reactive Oxygen Species Generation. *Polymers* **2020**, *12*, 1398. [CrossRef]
50. Wawrzynkiewicz, A.; Rozpedek-Kaminska, W.; Galita, G.; Lukomska-Szymanska, M.; Lapinska, B.; Sokolowski, J.; Majsterek, I. The Toxicity of Universal Dental Adhesives: An In Vitro Study. *Polymers* **2021**, *13*, 2653. [CrossRef]
51. Wawrzynkiewicz, A.; Rozpedek-Kaminska, W.; Galita, G.; Lukomska-Szymanska, M.; Lapinska, B.; Sokolowski, J.; Majsterek, I. The Cytotoxicity and Genotoxicity of Three Dental Universal Adhesives—An In Vitro Study. *Int. J. Mol. Sci.* **2020**, *21*, 3950. [CrossRef] [PubMed]
52. Ghasemi, M.; Turnbull, T.; Sebastian, S.; Kempson, I. The MTT Assay: Utility, Limitations, Pitfalls, and Interpretation in Bulk and Single-Cell Analysis. *Int. J. Mol. Sci.* **2021**, *22*, 12827. [CrossRef] [PubMed]
53. Kuhn, D.M.; Balkis, M.; Chandra, J.; Mukherjee, P.K.; Ghannoum, M.A. Uses and Limitations of the XTT Assay in Studies of *Candida* Growth and Metabolism. *J. Clin. Microbiol.* **2003**, *41*, 506–508. [CrossRef] [PubMed]
54. Scarcello, E.; Lambremont, A.; Vanbever, R.; Jacques, P.J.; Lison, D. Mind Your Assays: Misleading Cytotoxicity with the WST-1 Assay in the Presence of Manganese. *PLoS ONE* **2020**, *15*, e0231634. [CrossRef] [PubMed]
55. Kaja, S.; Payne, A.J.; Naumchuk, Y.; Koulen, P. Quantification of Lactate Dehydrogenase for Cell Viability Testing Using Cell Lines and Primary Cultured Astrocytes. *Curr. Protoc. Toxicol.* **2017**, *72*, 2.26.1–2.26.10. [CrossRef] [PubMed]
56. Korkmaz, A.; Reiter, R.J.; Topal, T.; Manchester, L.C.; Oter, S.; Tan, D.-X. Melatonin: An Established Antioxidant Worthy of Use in Clinical Trials. *Mol. Med.* **2009**, *15*, 43–50. [CrossRef] [PubMed]
57. Samuelsen, J.T.; Dahl, J.E. Biological Aspects of Modern Dental Composites. *Biomater. Investig. Dent.* **2023**, *10*, 2223223. [CrossRef]

Disclaimer/Publisher's Note: The statements, opinions and data contained in all publications are solely those of the individual author(s) and contributor(s) and not of MDPI and/or the editor(s). MDPI and/or the editor(s) disclaim responsibility for any injury to people or property resulting from any ideas, methods, instructions or products referred to in the content.

International Journal of *Molecular Sciences*

MDPI

Article

Influence of Surface Contaminants and Hydrocarbon Pellicle on the Results of Wettability Measurements of Titanium

Daisuke Kido [1,2], Keiji Komatsu [1], Toshikatsu Suzumura [1], Takanori Matsuura [1], James Cheng [1], Jeong Kim [1], Wonhee Park [1] and Takahiro Ogawa [1,*]

1 Weintraub Center for Reconstructive Biotechnology, Division of Regenerative and Reconstructive Sciences, UCLA School of Dentistry, Los Angeles, CA 90095-1668, USA

2 Department of Oral Diagnosis and General Dentistry, Graduate School of Medical and Dental Sciences, Tokyo Medical and Dental University (TMDU), 1-5-45 Yushima, Bunkyo-ku, Tokyo 113-8510, Japan

* Correspondence: togawa@dentistry.ucla.edu; Tel.: +1-310-825-5889

Abstract: Hydrophilicity/hydrophobicity—or wettability—is a key surface characterization metric for titanium used in dental and orthopedic implants. However, the effects of hydrophilicity/hydrophobicity on biological capability remain uncertain, and the relationships between surface wettability and other surface parameters, such as topography and chemistry, are poorly understood. The objective of this study was to identify determinants of surface wettability of titanium and establish the reliability and validity of the assessment. Wettability was evaluated as the contact angle of ddH_2O. The age of titanium specimens significantly affected the contact angle, with acid-etched, microrough titanium surfaces becoming superhydrophilic immediately after surface processing, hydrophobic after 7 days, and hydrorepellent after 90 days. Similar age-related loss of hydrophilicity was also confirmed on sandblasted supra-micron rough surfaces so, regardless of surface topography, titanium surfaces eventually become hydrophobic or hydrorepellent with time. On age-standardized titanium, surface roughness increased the contact angle and hydrophobicity. UV treatment of titanium regenerated the superhydrophilicity regardless of age or surface roughness, with rougher surfaces becoming more superhydrophilic than machined surfaces after UV treatment. Conditioning titanium surfaces by autoclaving increased the hydrophobicity of already-hydrophobic surfaces, whereas conditioning with 70% alcohol and hydrating with water or saline attenuated pre-existing hydrophobicity. Conversely, when titanium surfaces were superhydrophilic like UV-treated ones, autoclaving and alcohol cleaning turned the surfaces hydrorepellent and hydrophobic, respectively. UV treatment recovered hydrophilicity without exception. In conclusion, surface roughness accentuates existing wettability and can either increase or decrease the contact angle. Titanium must be age-standardized when evaluating surface wettability. Surface conditioning techniques significantly but unpredictably affect existing wettability. These implied that titanium wettability is significantly influenced by the hydrocarbon pellicle and other contaminants inevitably accumulated. UV treatment may be an effective strategy to standardize wettability by making all titanium surfaces superhydrophilic, thereby allowing the characterization of individual surface topography and chemistry parameters in future studies.

Keywords: titanium implants; wettability; osseointegration; bone integration; UV photofunctionalization

check for updates

Citation: Kido, D.; Komatsu, K.; Suzumura, T.; Matsuura, T.; Cheng, J.; Kim, J.; Park, W.; Ogawa, T. Influence of Surface Contaminants and Hydrocarbon Pellicle on the Results of Wettability Measurements of Titanium. *Int. J. Mol. Sci.* **2023**, *24*, 14688. https://doi.org/10.3390/ijms241914688

Academic Editor: Luigi Canullo

Received: 11 August 2023
Revised: 19 September 2023
Accepted: 25 September 2023
Published: 28 September 2023

Copyright: © 2023 by the authors. Licensee MDPI, Basel, Switzerland. This article is an open access article distributed under the terms and conditions of the Creative Commons Attribution (CC BY) license (https://creativecommons.org/licenses/by/4.0/).

1. Introduction

The hydrophilic or hydrophobic state (or wettability) of titanium is a major surface characterization metric in studies of titanium and implant materials [1–14], and it is commonly assessed by measuring the contact angle of water [15–23]. Although the definition varies and is used differently in different fields, representative definitions are superhydrophilic $0° < \theta < 10°$, hydrophilic $10° < \theta < 30°$, hydrophobic $30° < \theta < 90°$, and hydrorepellent $\theta > 90°$ [23–25]. In clinical and biological studies of dental and orthopedic titanium implants, the hydrophilicity/hydrophobicity of implant surfaces is considered a critical factor that influences the biological capability–specifically osseointegration–of implants [26–38].

However, the exact relationship between surface wettability and osseointegration remains controversial, for several possible reasons. First, the hydrophilicity/hydrophobicity of most titanium surfaces is relatively constant, ranging from 60° to 120° [16,39–43], i.e., hydrophobic or hydrorepellent, making it difficult to establish strong correlations with biological parameters. Second, it is technically challenging to produce a continuous range of hydrophilicity/hydrophobicity on experimental specimens; UV treatment of titanium surfaces makes them superhydrophilic (contact angle of 0° or <10° for most titanium surfaces, regardless of surface topography) [17,44–46], and most titanium surfaces are superhydrophilic or hydrophilic immediately after surface processing [15,42,44], so titanium surfaces tend to lie at the extreme ends of the wettability spectrum. Lastly and more importantly, the reliability and validity of contact angle measurement have not been established, i.e., the surface factors and measurement conditions that influence wettability have not been fully defined [47–49]. Systematic analyses of contact angles that consider surface topography, surface conditioning, measurement protocol, titanium age, and their synergy are urgently required.

Therefore, the objective of this study was to determine the reliability and validity of hydrophilicity/hydrophobicity measurements of titanium surfaces by examining the independent and combined effects of surface topography, titanium age, measurement protocol, and surface conditioning on the contact angle of water. Establishing the critical determinants of hydrophilicity/hydrophobicity is expected to improve the design and interpretation of future studies of titanium and implant materials.

2. Results

2.1. Surface Characteristics of Titanium Specimens

SEM images of machined titanium surfaces showed no defined morphology except for the scratches and traces from machine milling (Figure 1A). Sandblasted surfaces showed relatively larger-scale roughness (range 10–20 mm) and random morphology. All acid-etched titanium surfaces showed microscale roughness consisting of peaks and pits in high-magnification images (Figure 1B). Depending on the duration of sandblasting, there was progressive formation of supra-micron larger scale roughness, with the full SB + AE surface showing the densest crater-like supra-micron concavities.

Quantitative roughness analysis showed the lowest average roughness (Sa) for machined surfaces and the highest Sa for full SB + AE surfaces (Figure 1C). A combination of sandblasting and acid-etching (SB + AE surfaces) effectively increased the Sa. Peak-to-valley roughness (Sz) results were similar, with the full SB + AE surfaces being the highest and the machined surfaces being the lowest (Figure 1D). The exception was that the full SB surfaces showed an equivalent Sz to those on SB + AE surfaces.

2.2. Effect of Aging on Hydrophilicity/Hydrophobicity of Titanium with Different Topographies

We next examined the surface wettability of titanium surfaces over time since surface processing by measuring the contact angle of 3 μL ddH_2O on acid-etched and sandblasted surfaces. Both acid-etched and sandblasted surfaces were superhydrophilic immediately after surface processing, with a contact angle of 0° (Figure 2A,B). The contact angle remained at 0° 3 days later for both surfaces. However, after 7 days, the contact angle of acid-etched surfaces was over 50°, whereas sandblasted surfaces still had a contact angle of 0°. The contact angle increased over time for both surfaces, rapidly so for acid-etched surfaces. The acid-etched and sandblasted surfaces were hydrorepellent and hydrophobic, respectively, after 90 days.

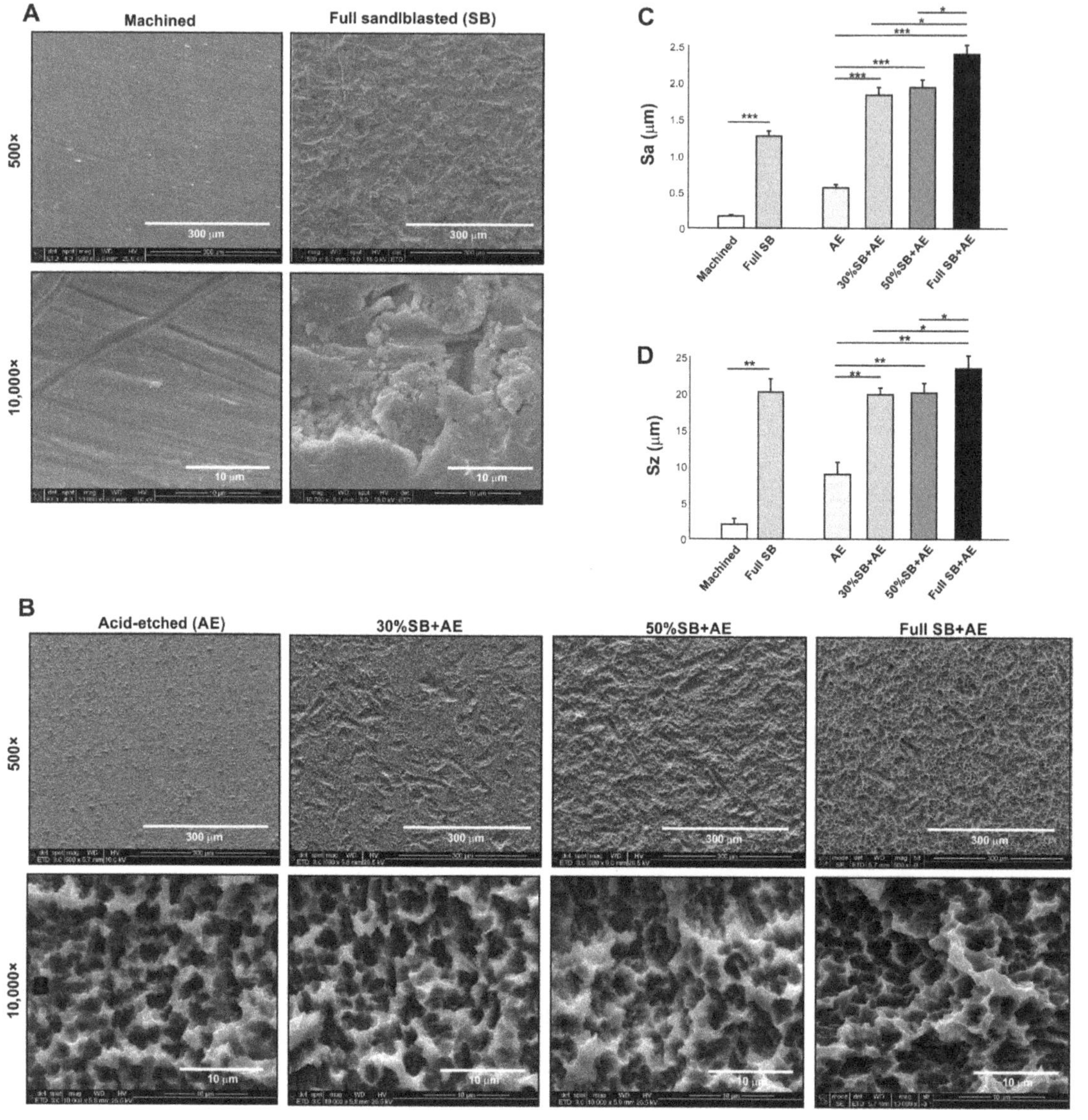

Figure 1. Surface characterization of titanium specimens used in this study. Scanning electron microscopy (SEM) images (**A**,**B**), average roughness (Sa) (**C**), and peak-to-valley roughness (Sz) (**D**) of the specimen surfaces: * $p < 0.05$, ** $p < 0.01$, *** $p < 0.001$.

We also measured the area of water spread, which was more sensitive to titanium age than the contact angle (Figure 2A,B). The area of spread was larger for acid-etched surfaces than sandblasted surfaces immediately after surface processing and started to shrink on day 3. The rate of decrease in spread area was faster for acid-etched surfaces. Although the contact angle on sandblasted surfaces remained at 0° until day 7, the area of spread steadily decreased over this time. Thus, there was age-related degradation of hydrophilicity, and the rate was topography-specific.

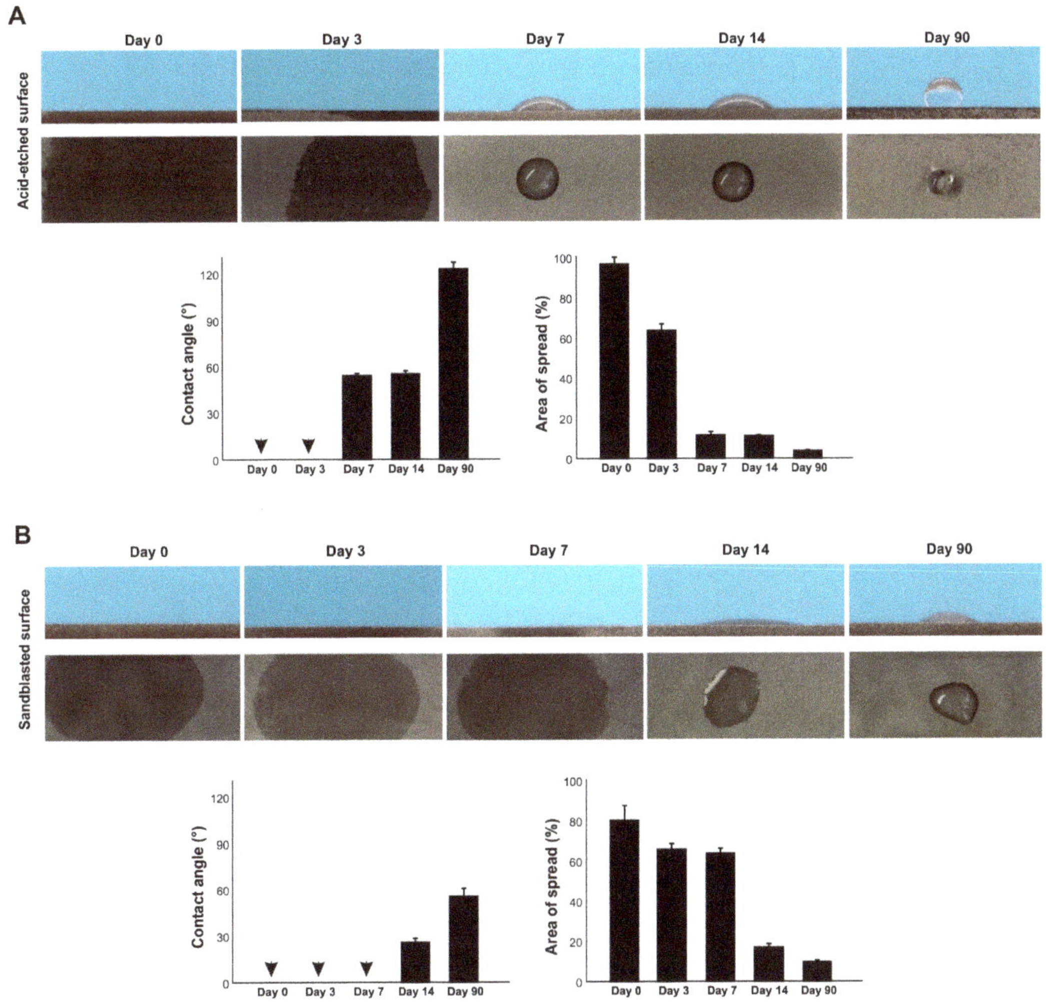

Figure 2. The effect of age on the hydrophilicity/hydrophobicity of titanium. Hydrophilicity/hydrophobicity of differently aged specimens was evaluated by the contact angle and the area of spread of 3 μL ddH_2O placed on titanium specimens. Side- and top-view photographs of a water droplet and histograms are presented. (**A**) Acid-etched titanium specimens. (**B**) Sandblasted titanium specimens. Arrowheads indicate 0°.

2.3. *Inter- and Intra-Specimen Reliability and Intra-Droplet Reliability of Contact Angle Measurements*

We next examined the inter- and intra-specimen stability of contact angle measurements using full SB + AE surfaces aged to 90 days. There was no significant difference in contact angle between three different specimens (Figure 3A) nor at three areas within an individual specimen (Figure 3B). The contact angle was the same on the right and left sides of the water droplets (Figure 3C). Therefore, at least for age-standardized titanium at 90 days, contact angle measurements showed intra- and inter-specimen reliability.

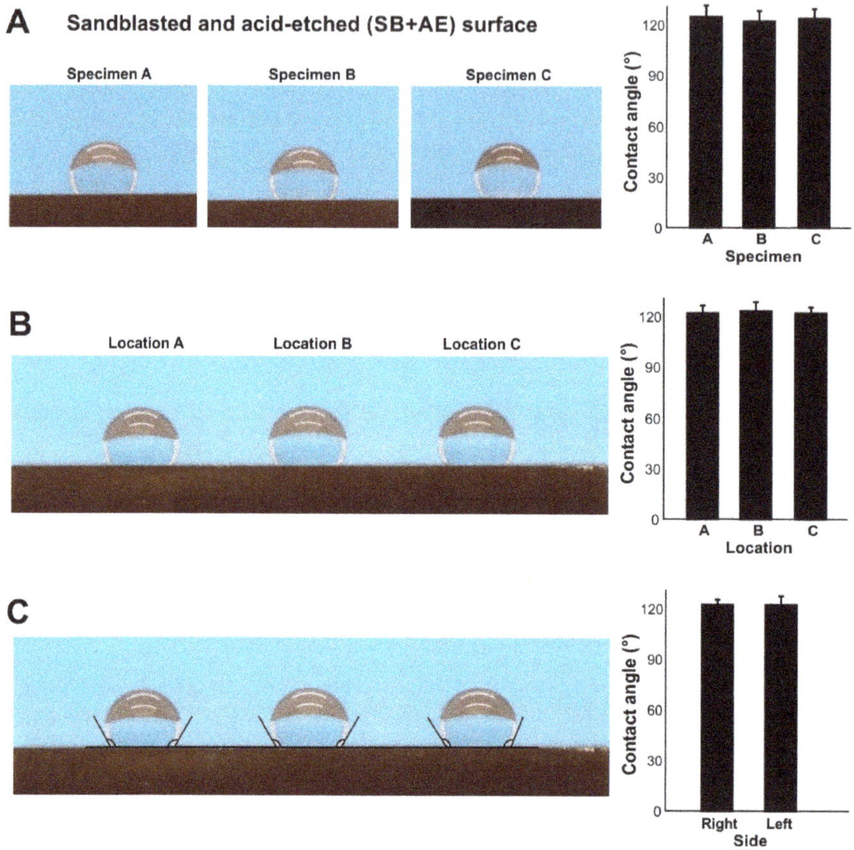

Figure 3. The reliability of contact angle measurement. To standardize the age of titanium, all specimens were stored for 90 days. (**A**) Inter-specimen reliability evaluated on sandblasted and acid-etched (SE + AE) titanium specimens. Side-view photographs of 3 μL of ddH_2O placed on the three different specimens and the calculated contact angles. (**B**) Intra-specimen reliability. Three drops of 3 μL ddH_2O placed in three different zones on the same specimen. (**C**) Intra-droplet reliability. The contact angle was measured at both sides of 3 μL ddH_2O droplets.

2.4. Effect of Water Volume on Contact Angle

We next examined the effect of different water volumes on contact angle (Figure 4A,B) on two different surfaces: full SB + AE and machined surfaces. On the full SB + AE surfaces, the contact angle increased from 1 μL to 3 μL but was then unchanged up to 20 μL (Figure 4A). On the machined surface, the contact angle was unchanged up to 5 μL but decreased with larger volumes (Figure 4A), revealing a surface topography-specific bias in measurement according to water volume.

2.5. Effect of Surface Roughness on Contact Angle

We compared the contact angle on acid-etched titanium surfaces of different roughness by altering the sandblasting time (Figure 5). The contact angle was significantly greater on the AE surface than on the machined surface and further increased on the 30% SB + AE surface (Figure 5A,B). However, further increases in sandblasting time did not further increase the contact angle, which plateaued. The regression analysis of the contact angle and Sa showed a significant exponential correlation, indicating a positive trend between surface roughness and the contact angle (Figure 5C).

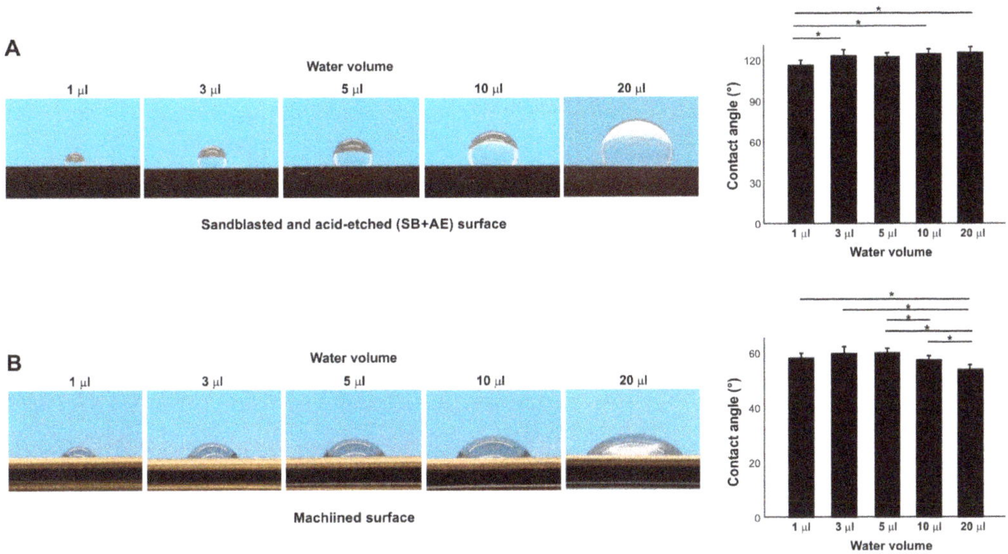

Figure 4. The effect of water volume on contact angle measurement on sandblasted and acid-etched (SB + AE) (**A**) and machined (**B**) surfaces. The specimens were used 90 days after surface processing: * $p < 0.05$.

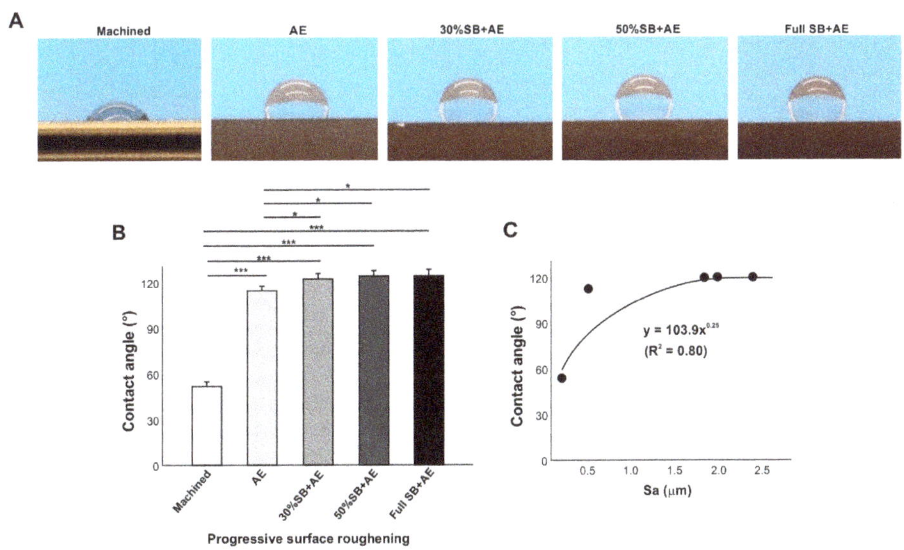

Figure 5. The effect of surface roughness on contact angle. Machined surfaces and acid-etched surfaces pre-sandblasted for various times were tested. The specimens were used 90 days after surface processing. Side-view photographs of 3 μL ddH_2O (**A**) and the calculated contact angles (**B**). The plot of contact angle against the average surface roughness (Sa) of each of the variously roughened titanium surfaces, with an exponential correlation curve (**C**): * $p < 0.05$, *** $p < 0.001$.

2.6. Effect of Surface Roughness on UV-Treated Surfaces

We next examined the effect of surface roughness on UV-treated titanium surfaces using the same specimens as in Figure 6 treated with VUV light for one minute (Figure 6). The contact angle of the machined smooth surface decreased from ~60° to less than 10°, indicating that the surface had become superhydrophilic (Figure 6A,B). Regardless of roughness, i.e., different durations of sandblasting, all acid-etched surfaces became superhydrophilic after UV treatment, with a contact angle of 0°, so there was no significant correlation in regression analysis and the trend in Figure 6 was rather opposite (Figure 6C).

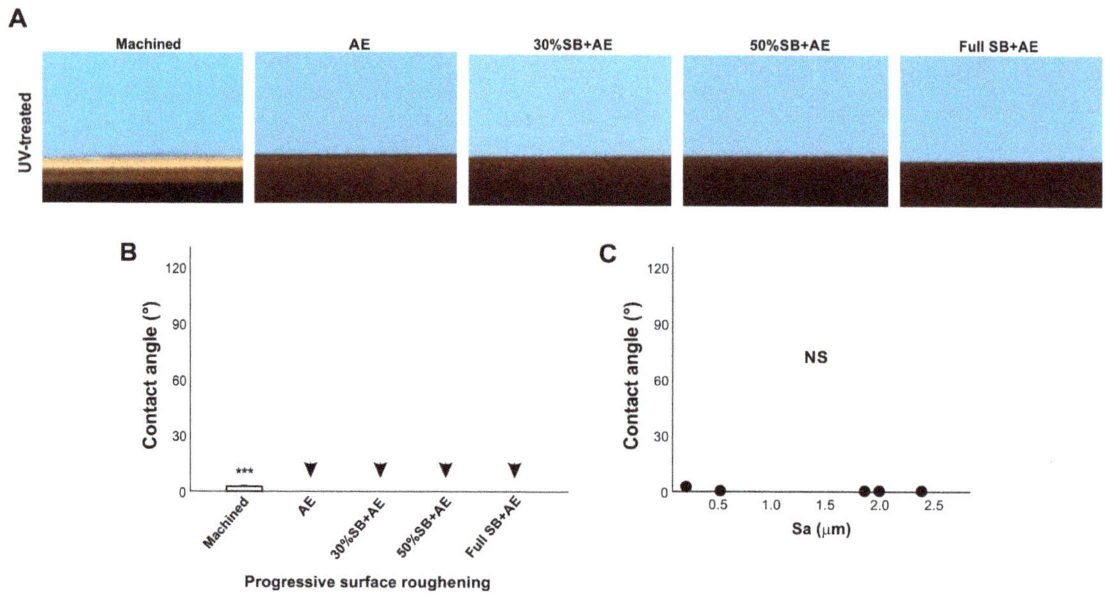

Figure 6. The effect of surface roughness on contact angle on UV-treated titanium surfaces. Machined surfaces and acid-etched surfaces pre-sandblasted for various times were stored for 90 days and treated with UV light prior to testing. Side-view photographs of 3 μL ddH_2O (**A**) and the calculated contact angles (**B**). The plot of the contact angle against the average surface roughness (Sa) of each of the various roughened titanium surfaces (**C**). There were no statistically significant correlations between parameters. Arrowheads indicate 0°: *** $p < 0.001$. NS: not significant.

2.7. Effect of Surface Conditioning on Contact Angle

We next examined the effect of various surface conditioning techniques on the contact angle. To model disinfectants used clinically and experimentally, we examined autoclaving and cleansing with 70% alcohol, while to model clinically applicable hydration, we soaked titanium specimens in ddH_2O or saline. Standard 90-day-old SB + AE specimens were tested. Compared with the baseline hydrorepellent state, autoclaving further increased the contact angle and significantly increased hydrophobicity, whereas alcohol cleansing decreased the contact angle (Figure 7A). Hydrating the specimens significantly decreased the contact angle, with saline having a greater effect than ddH_2O. However, the contact angle remained >30° and outside the range of hydrophilicity.

We next carried out the same experiment using UV-treated specimens (Figure 7B). UV-treated SB + AE surfaces were superhydrophilic with a 0° contact angle, but both disinfecting techniques reversed the wettability from superhydrophilic to hydrophobic, with autoclaving having a greater effect than alcohol cleaning and turning the specimens hydrorepellent. Hydration also reduced UV-generated superhydrophilicity, with the contact angle increasing more with ddH_2O than with saline.

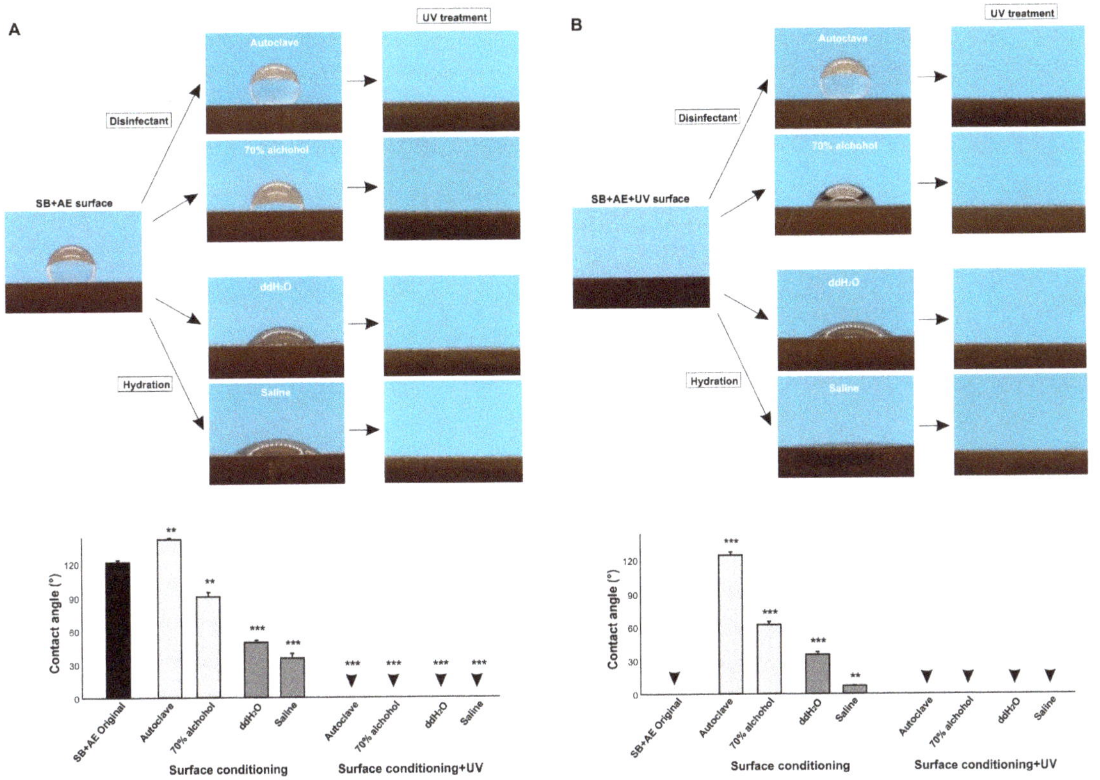

Figure 7. The effect of surface conditioning on the contact angle. The 90-day-old sandblasted and acid-etched (SB + AE) surfaces (**A**) and those with UV treatment (**B**) were tested. The specimens were conditioned using four different techniques. Side-view photographs of 3 μL ddH_2O and the calculated contact angles are shown. Arrowheads indicate 0°: ** $p < 0.01$, *** $p < 0.001$.

2.8. Generation and Recovery of Superhydrophilicity via UV Treatment

Given that surface conditioning increased hydrophobicity, we next determined if UV treatment overcame this effect. Regardless of pre-UV treatment, treatment with UV light made all surfaces superhydrophilic without exception (Figure 7A,B), indicating that superhydrophilicity can be newly generated even after surface conditioning and that superhydrophilicity can be recovered even on surfaces compromised by surface conditioning.

Finally, we determined if UV treatment can regenerate superhydrophilicity on differently aged surfaces (Figure 8A,B). Regardless of acid-etching or sandblasting, 14-day-old and 90-day-old surfaces became superhydrophilic with a 0° contact angle after UV treatment. The water spread area was even greater after UV treatment than on day 0 titanium specimens, indicating that the degree of superhydrophilicity increased further by UV treatment compared with the state of new surfaces.

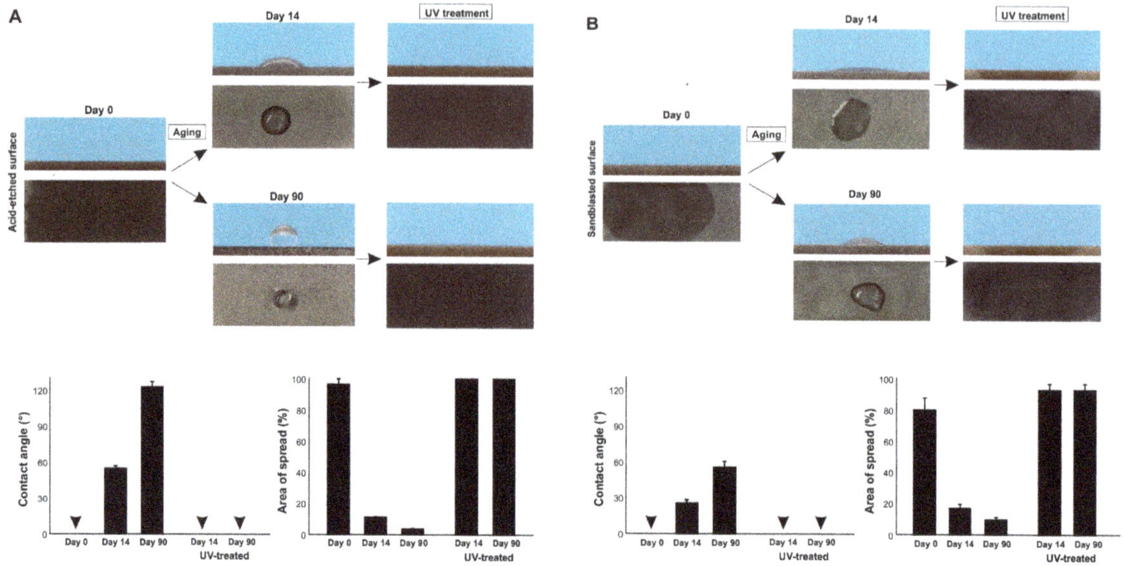

Figure 8. Restoration of superhydrophilicity via UV treatment. Differently aged titanium surfaces after acid-etching (**A**) and sandblasting (**B**) were treated with VUV light for one minute. Side- and top-view photographs of 3 μL ddH_2O and the calculated contact angles and the area of water spread are shown. Arrowheads indicate 0°.

3. Discussion

We established that some factors could have profound effects on the wettability of titanium, in some cases from superhydrophilic to hydrorepellent or vice versa. Highlights of the present results are summarized in a diagram (Figure 9). The most significant factor was the age of the titanium, with the contact angle significantly increasing (markedly reduced hydrophilicity) as the specimen aged after surface processing. This age-driven loss of hydrophilicity was previously reported for acid-etched and machined titanium surfaces [9,41,42], and is known as the "biological aging" of titanium [9]. Here we found that sandblasted surfaces also undergo biological aging, although the aging was significantly slower on sandblasted surfaces than on acid-etched surfaces. Sandblasted surfaces maintained superhydrophilicity up to day seven and still did not show hydro-repellency even after 90 days. These results suggest a combinational effect of surface aging and topography in determining titanium wettability and that age or topography alone cannot predict the contact angle. Similar aging rate modulation was observed on titanium dioxide-coated and nano-structured titanium [44]. Another study reported faster aging, i.e., degradation of hydrophilicity on rougher titanium surfaces when titanium specimens were submerged in saliva [50]. There seem to be ways to delay aging but no way to prevent it [9,51]. Furthermore, regardless of surface topography, titanium surfaces were superhydrophilic immediately after surface processing. This was surprising, since titanium surfaces are traditionally considered hydrophobic [52–55], but our data suggest that this is probably because the titanium specimens and implant products used in previous reports were sufficiently old. The physicochemical explanation for age-related loss of hydrophilicity is the pellicle of hydrocarbon molecules that naturally and inevitably form on titanium surfaces over time [9,16,41,56,57]. A few studies reported hydrophilic titanium surfaces after particular surface modification without mentioning the age of those specimens [58,59], which needs careful interpretation to preclude the possibility that they were hydrophilic just because they are new after processing. An attempt was made to prevent the pellicle accumulation by packaging titanium implants in the saline solution while handing under nitrogen inactive gas [43]. However, carbon elements were detected on these implants [43,60]. Soaking titanium in the saline solution or water did not prevent pellicle accumulation, resulting in the degradation of the bioactivity of the titanium [61]. Implant products, as medical devices, are commonly packaged and sealed individually. However, a wide variety of surface carbon ranging from 25 to 76% is detected on these products [62,63] and the surfaces are hydrophobic [38,64].

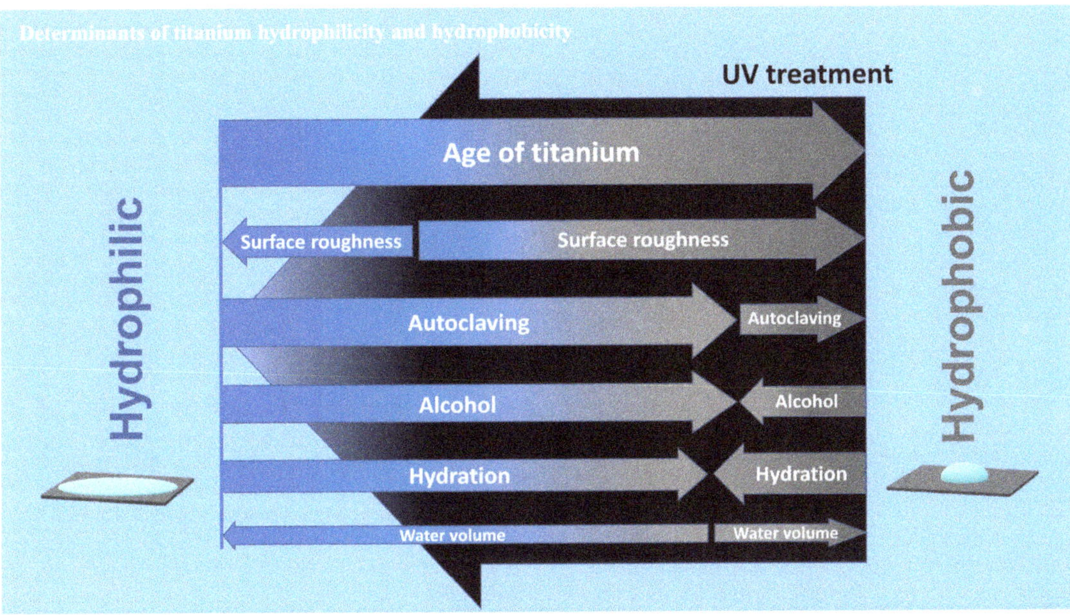

Figure 9. A diagram showing the determinants of hydrophilic/hydrophobic property or wettability of titanium surfaces. The widths of arrows represent the degree of impact of each factor on the water contact angle. Note that the impact toward hydrophilic or hydrophobic trend may not be mono-directional for some factors depending on the existing wettability. UV treatment turns titanium surfaces superhydrophilic, regardless of the degree or involvement of all factors tested.

Few studies have explored the effect of surface roughness on the wettability of titanium. We found that the rougher the surface, the more hydrophobic it became (Figure 5), with the machined surface the less hydrophobic. However, the effect of surface roughness was not linear and there was a significant exponential correlation, indicating a limit to the effect of surface roughness. The contact angle plateaued at ~120° when the average surface roughness was 1.88 mm on 30% SB + AE surfaces, and further roughening did not negatively impact wettability. A study compared the contact angle on titanium surfaces that were acid-etched at various temperatures [55]. It was shown that the contact angle was higher on titanium specimens etched at higher temperature. Since high-temperature acid-etching creates rougher surfaces, the result supported the general trend found in the present study that the rougher the surface, the more hydrophobic.

Of note, the positive correlation between surface roughness and hydrophobicity did not apply to UV-treated titanium surfaces. UV treatment converted all surfaces to superhydrophilic. Indeed, UV treatment is known to turn titanium surfaces hydrophilic by removing the hydrocarbon pellicle [10,24,27–29,61]. After UV treatment, the effect of surface roughness was nearly completely negated, and there was no correlation between the average surface roughness and the contact angle. Rougher surfaces, when aged, plausibly contain more hydrocarbon due to their larger surface area. This could be an explanation for why rougher surfaces are more hydrophobic. Conversely, when the rougher surfaces are free from hydrocarbon after UV treatment, they are superhydrophilic regardless of the degree of roughness. Although all surfaces tested were superhydrophilic after UV treatment, the contact angle was highest on machined surfaces, opposite to the principle that the rougher the surface, the more hydrophobic it becomes. Thus, UV treatment creates a special and distinct phenomenon of surface wettability. We propose that, regardless of UV treatment, surface roughness accentuates the wettability determined by the existing physicochemical properties of the titanium. For instance, regular "old" titanium surfaces are hydrophobic due to the hydrocarbon pellicle and become more hydrophobic when the surfaces are rougher. Conversely, UV-treated surfaces are hydrophilic due to their pellicle-free surface and become more hydrophilic when the surfaces are rougher.

Of the four surface conditioning techniques tested, autoclaving and alcohol cleaning, which are standard ways to disinfect specimens and devices in routine clinical and experimental settings, significantly affected the wettability of aged titanium surfaces. The two techniques had opposite effects, with autoclaving enhancing hydrophobicity and alcohol cleaning attenuating hydrophobicity, a finding that requires the assessment of other physicochemical factors that might be altered by these two techniques, including the highly sensitive elemental and isotopic analysis via X-ray photoelectron spectroscopy (XPS) [65,66] and laser ablation–inductively coupled plasma–mass spectrometry (LA-ICP-MS) [67–69]. Variations of the contact angle presumably by different cleaning techniques is reported by other studies [49,70] but no principle or mechanism was established. Autoclaving might promote carbon accumulation on titanium [71]. The autoclave used in this study was typical for clinical and research applications, but there are minor variations in steam impurities, pH, drying, remnant detergents, and other unknown factors among the devices [72]. It seems extremely difficult to control the conditions and prevent accidental contamination. Further research is needed to generalize the effects of these disinfecting techniques and identify the determinants for altered wettability. Hydration was expected to alleviate the hydrophobicity of aged titanium surfaces due to the deposition of water molecules or hydroxyl groups on the surfaces [2,73], which was supported by the results. The greater effect seen with saline compared with ddH_2O implied that the ionic interactions between saline and titanium promoted a hydrophilic state. All of the results were very different when the four conditioning techniques were applied to pre-existing (UV-treated) superhydrophilic titanium surfaces, with the contact angle remarkably increasing from 0° regardless of the conditioning technique in most cases. Superhydrophilicity was completely abolished by autoclaving, alcohol cleaning, and soaking in ddH_2O, with the greatest negative impact seen with autoclaving. Soaking in saline maintained superhydrophilicity, probably due to the positive effects of the ionic interactions. However, even with saline, UV-induced superhydrophilicity was significantly compromised.

Inter- and intra-specimen reproducibility analysis suggested that, provided that the surface processing undergoes quality control and the age of specimens is standardized, measuring the contact angle of water is a reliable means to evaluate the hydrophilicity or hydrophobicity of titanium surfaces. The effect of water volume was significant and specific; plausibly due to gravity, a greater volume of water increased the contact angle on hydrorepellent surfaces but decreased it on hydrophobic surfaces. This finding has important implications, in that absolute contact angles cannot be compared between different studies unless the water volume is standardized. Indeed, the reduced reliability of contact angle measurement was reported with the use of water of 30 mm or more [49]. On the other hand, a very small volume may introduce possible technical error. To minimize the effect of gravity and technical error, this study used a 3 μL protocol.

Our results also highlight issues in the evaluation of surface wettability across many biomedical, engineering, and clinical domains, since wettability has been used extensively in these studies as a key surface characterization metric to interpret the results and infer mechanisms associated with surface topography and chemistry. For instance, with respect to titanium age, our findings make it clear that comparing the hydrophilicity or hydrophobicity between different experimental specimens is meaningless unless they are standardized for age, and even then surface processing and topography must be considered since the rate of aging significantly differs between acid-etched and sandblasted surfaces. Similarly, surface conditioning impacted the results; regardless of experimental or clinical use, titanium specimens and devices require final conditioning prior to use, and different studies use different disinfecting techniques and rinsing protocols, the specifics of which are often omitted from the published experimental protocol. Future studies should detail the conditioning protocol.

The biological impact of the high surface energy of titanium and other biomaterials was extensively reported. High-energy titanium surfaces, represented by superhydrophilicity, recruit more osteoblasts and promote osseointegration [9,13,16,41,42,44]. The high-energy titanium surfaces created by UV treatment or UV photofunctionalization are also studied extensively for their ability to enhance osseointegration and soft tissue responses at the cell [12,74], animal [74–76], and clinical [28,77] levels. The high surface energy is created by the UV-mediated removal of the hydrocarbon pellicle via three different mechanisms (1) photochemical decomposition (ozone-mediated or non-mediated cleaning); (2) photophysical decomposition (direct bond dissociation by UV energy); and (3) photocatalytic decomposition (UV-titanium interaction) [78,79].The present study proposes another benefit of UV treatment, namely to standardize the wettability of titanium, because UV treatment converted all titanium specimens with different surface topographies, ages, and conditionings to superhydrophilic, allowing future studies to focus on the effect of other specific surface parameters such as topography and chemistry. As discussed above, the accumulation of the hydrocarbon pellicle and other chemical contaminants is nearly inevitable and there is no method to prevent the time-related loss of hydrophilicity. As shown by the rejuvenation of the aged surface, UV treatment produced even higher superhydrophilicity than new surfaces. Furthermore, with the use of high-energy VUV light, superhydrophilic standardization can be accomplished in a minute, minimally impacting the experimental protocol [78,79]. This study did not study every surface type used in the field, and further studies of other surfaces with nano- and Meso-level topography and other biological and chemical modifications and coatings are now warranted.

4. Materials and Methods

4.1. Titanium Specimens and Surface Processing

Titanium specimens in rectangular plate form (14 × 6 × 2 mm) were prepared by machine-milling grade IV commercially pure titanium. To create surface topography, specimens were modified by sandblasting, acid-etching, or a combination of both. Sandblasting (SB) was carried out with Al_2O_3 particles (70 mesh) for either 1.65 (30% SB), 2.75 (50% SB), or 5.5 (full SB) seconds. Acid-etching was performed by processing specimens in HCl and H_2SO_4 at 95 °C for 6 min and 30 s. The specimens were prepared and provided by DIO Implant (Busan, Republic of Korea) and individually packaged and sealed in a quartz ampoule. For aging studies, specimens were stored in dark, ambient conditions in a sealed package at 25 °C up to 90 days.

4.2. Surface Conditioning and UV Treatment

Four different surface conditioning techniques were used: (i) surface disinfecting by autoclaving at 2.0 bar at 121 °C for 20 min, followed by 10 min drying in a sealed sterilization pouch (Fisher Scientific, Pittsburgh, PA); (ii) cleaning by immersing specimens in 70% ethanol for 24 h followed by rinsing in ddH_2O for 20 min and drying; hydrating by immersing specimens in (iii) ddH_2O or (iv) saline solution for 24 h and then drying. UV treatment was performed at room temperature using a vacuum UV (VUV) light (172 nm vacuum UV, 60 mW/cm^2) (DIO Implant, Busan, Republic of Korea) for 1 min [71,72].

4.3. Surface Characterization and Wettability Testing

The surface morphology of titanium specimens was qualitatively examined by scanning electron microscopy (SEM; Nova 230 Nano SEM, FEI, Hillsboro, OR, USA). In addition, roughness was quantified using an optical profile microscope (MeX, Alicona Imaging GmbH, Raaba, Graz, Austria) to measure the average roughness (Sa) and peak-to-valley roughness (Sz). The hydrophilicity/hydrophobicity or wettability of specimen surfaces was evaluated by measuring the contact angle of 3 μL of ddH_2O in most experiments. To examine the effect of water volume, the contact angle was also measured with 1, 5, 10, and 20 μL ddH_2O.

4.4. Statistical Analysis

Contact angle measurements were performed on three independent titanium specimens in each group in each experiment. The effects of titanium age, different droplets, surface roughness, water volume, surface conditioning, and UV treatment were compared by one-way ANOVA. Bonferroni's test was used as a post hoc multiple comparison test where appropriate. p-values < 0.05 were considered statistically significant. Regression analysis was applied to determine associations between the average surface roughness and the contact angle.

5. Conclusions

This study identified important factors that determine the hydrophilicity/hydrophobicity (or wettability) of titanium surfaces. The effect of titanium age was critical and decreased the contact angle from superhydrophilic to hydrorepellent depending on age and topography. Different surface conditioning techniques also altered the wettability, and the alterations were diverse and specific to the existing wettability state; for instance, UV-treated superhydrophilic surfaces became hydrorepellent after autoclaving. These implied that titanium wettability is significantly influenced by the hydrocarbon pellicle and other contaminants inevitably accumulated. Our study highlights that hydrophilicity/hydrophobicity assessment and interpretation require careful consideration of measurement approach, handling, and conditioning. UV treatment may be an effective, novel strategy to standardize the wettability of titanium by making all surfaces superhydrophilic and allowing assessment of other individual factors including but not limited to surface topography and chemistry.

Author Contributions: Conceptualization, D.K. and T.O.; methodology, D.K., T.S. and J.C.; validation, T.S., J.K. and W.P.; formal analysis, T.M.; investigation, T.S., K.K. and J.C.; resources, T.O.; data curation, D.K.; writing—original draft preparation, D.K. and K.K.; writing—review and editing, T.M. and T.O.; visualization, D.K. and T.M.; supervision, T.O.; project administration, T.O.; funding acquisition, T.O. All authors have read and agreed to the published version of the manuscript.

Funding: This research was supported by a research fund from DIO IMPLANT Co., Ltd. (Busan, Korea).

Institutional Review Board Statement: Not applicable.

Informed Consent Statement: Not applicable.

Data Availability Statement: The data presented in this study are available upon request from the corresponding author.

Conflicts of Interest: The authors declare no conflict of interest.

References

1. Zhao, G.; Schwartz, Z.; Wieland, M.; Rupp, F.; Geis-Gerstorfer, J.; Cochran, D.L.; Boyan, B.D. High surface energy enhances cell response to titanium substrate microstructure. *J. Biomed. Mater. Res. A* **2005**, *74*, 49–58. [CrossRef]
2. Rupp, F.; Gittens, R.A.; Scheideler, L.; Marmur, A.; Boyan, B.D.; Schwartz, Z.; Geis-Gerstorfer, J. A review on the wettability of dental implant surfaces I: Theoretical and experimental aspects. *Acta Biomater.* **2014**, *10*, 2894–2906. [CrossRef] [PubMed]
3. Rupp, F.; Liang, L.; Geis-Gerstorfer, J.; Scheideler, L.; Huttig, F. Surface characteristics of dental implants: A review. *Dent. Mater.* **2018**, *34*, 40–57. [CrossRef] [PubMed]
4. Lai, H.C.; Zhuang, L.F.; Liu, X.; Wieland, M.; Zhang, Z.Y.; Zhang, Z.Y. The influence of surface energy on early adherent events of osteoblast on titanium substrates. *J. Biomed. Mater. Res. A* **2010**, *93*, 289–296. [CrossRef]
5. Boonrawd, W.; Awad, K.R.; Varanasi, V.; Meletis, E.I. Wettability and in-vitro study of titanium surface profiling prepared by electrolytic plasma processing. *Surf. Coat. Technol.* **2021**, *414*, 127119. [CrossRef] [PubMed]

6. Hu, Y.; Duan, J.; Yang, X.; Zhang, C.; Fu, W. Wettability and biological responses of titanium surface's biomimetic hexagonal microstructure. *J. Biomater. Appl.* **2023**, *37*, 1112–1123. [CrossRef] [PubMed]
7. Park, K.H.; Koak, J.Y.; Kim, S.K.; Heo, S.J. Wettability and cellular response of UV light irradiated anodized titanium surface. *J. Adv. Prosthodont.* **2011**, *3*, 63–68. [CrossRef] [PubMed]
8. Wassmann, T.; Kreis, S.; Behr, M.; Buergers, R. The influence of surface texture and wettability on initial bacterial adhesion on titanium and zirconium oxide dental implants. *Int. J. Implant. Dent.* **2017**, *3*, 32. [CrossRef]
9. Lee, J.H.; Ogawa, T. The biological aging of titanium implants. *Implant. Dent.* **2012**, *21*, 415–421. [CrossRef]
10. Ikeda, T.; Okubo, T.; Saruta, J.; Hirota, M.; Kitajima, H.; Yanagisawa, N.; Ogawa, T. Osteoblast Attachment Compromised by High and Low Temperature of Titanium and Its Restoration by UV Photofunctionalization. *Materials* **2021**, *14*, 5493. [CrossRef]
11. Tabuchi, M.; Hamajima, K.; Tanaka, M.; Sekiya, T.; Hirota, M.; Ogawa, T. UV Light-Generated Superhydrophilicity of a Titanium Surface Enhances the Transfer, Diffusion and Adsorption of Osteogenic Factors from a Collagen Sponge. *Int. J. Mol. Sci.* **2021**, *22*, 6811. [CrossRef] [PubMed]
12. Hirota, M.; Hori, N.; Sugita, Y.; Ikeda, T.; Park, W.; Saruta, J.; Ogawa, T. A Novel Cell Delivery System Exploiting Synergy between Fresh Titanium and Fibronectin. *Cells* **2022**, *11*, 2158. [CrossRef] [PubMed]
13. Hicklin, S.P.; Janner, S.F.; Schnider, N.; Chappuis, V.; Buser, D.; Bragger, U. Early Loading of Titanium Dental Implants with an Intraoperatively Conditioned Hydrophilic Implant Surface: 3-Year Results of a Prospective Case Series Study. *Int. J. Oral Maxillofac. Implant.* **2020**, *35*, 1013–1020. [CrossRef] [PubMed]
14. Ueno, T.; Ikeda, T.; Tsukimura, N.; Ishijima, M.; Minamikawa, H.; Sugita, Y.; Yamada, M.; Wakabayashi, N.; Ogawa, T. Novel antioxidant capability of titanium induced by UV light treatment. *Biomaterials* **2016**, *108*, 177–186. [CrossRef]
15. Minamikawa, H.; Att, W.; Ikeda, T.; Hirota, M.; Ogawa, T. Long-Term Progressive Degradation of the Biological Capability of Titanium. *Materials* **2016**, *9*, 102. [CrossRef]
16. Hori, N.; Iwasa, F.; Tsukimura, N.; Sugita, Y.; Ueno, T.; Kojima, N.; Ogawa, T. Effects of UV photofunctionalization on the nanotopography enhanced initial bioactivity of titanium. *Acta Biomater.* **2011**, *7*, 3679–3691. [CrossRef] [PubMed]
17. Simoes, I.G.; Dos Reis, A.C.; Valente, M. Influence of surface treatment by laser irradiation on bacterial adhesion on surfaces of titanium implants and their alloys: Systematic review. *Saudi Dent. J.* **2023**, *35*, 111–124. [CrossRef] [PubMed]
18. Liu, K.; Cao, M.; Fujishima, A.; Jiang, L. Bio-inspired titanium dioxide materials with special wettability and their applications. *Chem. Rev.* **2014**, *114*, 10044–10094. [CrossRef]
19. Donohoe, E.; Kahatab, R.; Barrak, F. A systematic review comparing the macrophage inflammatory response to hydrophobic and hydrophilic sandblasted large grit, acid-etched titanium or titanium-zirconium surfaces during in vitro studies. *Clin. Exp. Dent. Res.* **2023**, *9*, 437–448. [CrossRef]
20. Arya, G.; Kumar, V. Evaluation of osseintegration between traditional and modified hydrophilic titanium dental implants—Systematic analysis. *Natl. J. Maxillofac. Surg.* **2020**, *11*, 176–181.
21. Hashimoto, K.; Fujishima, A.; Watanabe, S. *Mechanisms of Photocatalytic Activity*; Nihonjitugyou Publishing: Tokyo, Japan, 2003; pp. 122–135.
22. Wang, R.; Hashimoto, K.; Fujishima, A. Light-induced amphiphilic surfaces. *Nature* **1997**, *388*, 431–432. [CrossRef]
23. Aita, H.; Hori, N.; Takeuchi, M.; Suzuki, T.; Yamada, M.; Anpo, M.; Ogawa, T. The effect of ultraviolet functionalization of titanium on integration with bone. *Biomaterials* **2009**, *30*, 1015–1025. [CrossRef]
24. Law, K.Y. Definitions for Hydrophilicity, Hydrophobicity, and Superhydrophobicity: Getting the Basics Right. *J. Phys. Chem. Lett.* **2014**, *5*, 686–688. [CrossRef] [PubMed]
25. Att, W.; Hori, N.; Iwasa, F.; Yamada, M.; Ueno, T.; Ogawa, T. The effect of UV-photofunctionalization on the time-related bioactivity of titanium and chromium-cobalt alloys. *Biomaterials* **2009**, *30*, 4268–4276. [CrossRef] [PubMed]
26. de Avila, E.D.; Lima, B.P.; Sekiya, T.; Torii, Y.; Ogawa, T.; Shi, W.; Lux, R. Effect of UV-photofunctionalization on oral bacterial attachment and biofilm formation to titanium implant material. *Biomaterials* **2015**, *67*, 84–92. [CrossRef]
27. Hirota, M.; Ozawa, T.; Iwai, T.; Mitsudo, K.; Ogawa, T. UV-Mediated Photofunctionalization of Dental Implant: A Seven-Year Results of a Prospective Study. *J. Clin. Med.* **2020**, *9*, 2733. [CrossRef]
28. Soltanzadeh, P.; Ghassemi, A.; Ishijima, M.; Tanaka, M.; Park, W.; Iwasaki, C.; Hirota, M.; Ogawa, T. Success rate and strength of osseointegration of immediately loaded UV-photofunctionalized implants in a rat model. *J. Prosthet. Dent.* **2017**, *118*, 357–362. [CrossRef]
29. Ueno, T.; Yamada, M.; Suzuki, T.; Minamikawa, H.; Sato, N.; Hori, N.; Takeuchi, K.; Hattori, M.; Ogawa, T. Enhancement of bone-titanium integration profile with UV-photofunctionalized titanium in a gap healing model. *Biomaterials* **2010**, *31*, 1546–1557. [CrossRef]
30. Funato, A.; Tonotsuka, R.; Murabe, H.; Hirota, M.; Ogawa, T. A novel strategy for bone integration and regeneration-Photofunctionalization of dental implants and Ti mesh. *J. Cosmet. Dent.* **2014**, *29*, 74–86.
31. Chang, L.C. Clinical Applications of Photofunctionalization on Dental Implant Surfaces: A Narrative Review. *J. Clin. Med.* **2022**, *11*, 5823. [CrossRef]
32. Dini, C.; Nagay, B.E.; Magno, M.B.; Maia, L.C.; Barao, V.A.R. Photofunctionalization as a suitable approach to improve the osseointegration of implants in animal models-A systematic review and meta-analysis. *Clin. Oral Implant. Res.* **2020**, *31*, 785–802. [CrossRef]

33. Yin, C.; Zhang, T.; Wei, Q.; Cai, H.; Cheng, Y.; Tian, Y.; Leng, H.; Wang, C.; Feng, S.; Liu, Z. Surface treatment of 3D printed porous Ti6Al4V implants by ultraviolet photofunctionalization for improved osseointegration. *Bioact. Mater.* **2022**, *7*, 26–38. [CrossRef]
34. Ganeles, J.; Zollner, A.; Jackowski, J.; ten Bruggenkate, C.; Beagle, J.; Guerra, F. Immediate and early loading of Straumann implants with a chemically modified surface (SLActive) in the posterior mandible and maxilla: 1-year results from a prospective multicenter study. *Clin. Oral Implant. Res.* **2008**, *19*, 1119–1128. [CrossRef]
35. Wennerberg, A.; Galli, S.; Albrektsson, T. Current knowledge about the hydrophilic and nanostructured SLActive surface. *Clin. Cosmet. Investig. Dent.* **2011**, *3*, 59–67. [CrossRef]
36. Khandelwal, N.; Oates, T.W.; Vargas, A.; Alexander, P.P.; Schoolfield, J.D.; Alex McMahan, C. Conventional SLA and chemically modified SLA implants in patients with poorly controlled type 2 Diabetes mellitus—A randomized controlled trial. *Clin. Oral Impl. Res.* **2013**, *24*, 13–19. [CrossRef]
37. Schwarz, F.; Wieland, M.; Schwartz, Z.; Zhao, G.; Rupp, F.; Geis-Gerstorfer, J.; Schedle, A.; Broggini, N.; Bornstein, M.M.; Review: Potential of chemically modified hydrophilic surface characteristics to support tissue integration of titanium dental implants. *J. Biomed. Mater. Res. B Appl. Biomater.* **2009**, *88*, 544–557. [CrossRef] [PubMed]
38. Roessler, S.; Zimmermann, R.; Scharnweber, D.; Werner, C.; Worch, H. Characterization of oxide layers on Ti6Al4V and titanium by streaming potential and streaming current measurements. *Colloids Surf. B Biointerfaces* **2002**, *26*, 387–395. [CrossRef]
39. Pacha-Olivenza, M.; Gallardo-Moreno, A.; Méndez-Vilas, A.; Bruque, J.; González-Carrasco, J.L.; González-Martín, M. Effect of UV irradiation on the surface Gibbs energy of Ti6Al4V and thermally oxidized Ti6Al4V. *J. Colloid Interface Sci.* **2008**, *320*, 117–124. [CrossRef]
40. Att, W.; Hori, N.; Takeuchi, M.; Ouyang, J.; Yang, Y.; Anpo, M.; Ogawa, T. Time-dependent degradation of titanium osteoconductivity: An implication of biological aging of implant materials. *Biomaterials* **2009**, *30*, 5352–5363. [CrossRef] [PubMed]
41. Hori, N.; Att, W.; Ueno, T.; Sato, N.; Yamada, M.; Saruwatari, L.; Suzuki, T.; Ogawa, T. Age-dependent degradation of the protein adsorption capacity of titanium. *J. Dent. Res.* **2009**, *88*, 663–667. [CrossRef] [PubMed]
42. Buser, D.; Broggini, N.; Wieland, M.; Schenk, R.K.; Denzer, A.J.; Cochran, D.L.; Hoffmann, B.; Lussi, A.; Steinemann, S.G. Enhanced bone apposition to a chemically modified SLA titanium surface. *J. Dent. Res.* **2004**, *83*, 529–533. [CrossRef] [PubMed]
43. Iwasa, F.; Tsukimura, N.; Sugita, Y.; Kanuru, R.K.; Kubo, K.; Hasnain, H.; Att, W.; Ogawa, T. TiO2 micro-nano-hybrid surface to alleviate biological aging of UV-photofunctionalized titanium. *Int. J. Nanomed.* **2011**, *6*, 1327–1341.
44. Liu, X.; Zhao, X.; Li, B.; Cao, C.; Dong, Y.; Ding, C.; Chu, P.K. UV-irradiation-induced bioactivity on TiO2 coatings with nanostructural surface. *Acta Biomater.* **2008**, *4*, 544–552. [CrossRef] [PubMed]
45. Funato, A.; Moroi, H.; Ogawa, T. Guided bone regeneration assisted by tooth roots with periodontal ligament: Case reports of immediate and staged approaches to implant therapy. *Int. J. Esthet. Dent.* **2022**, *17*, 276–291.
46. Funato, A.; Yamada, M.; Ogawa, T. Success rate, healing time, and implant stability of photofunctionalized dental implants. *Int. J. Oral Maxillofac. Implant.* **2013**, *28*, 1261–1271. [CrossRef]
47. Gu, H.; Wang, C.; Gong, S.; Mei, Y.; Li, H.; Ma, W. Investigation on contact angle measurement methods and wettability transition of porous surfaces. *Surf. Coat. Technol.* **2016**, *292*, 72–77. [CrossRef]
48. Hierro-Oliva, M.; Gallardo-Moreno, A.M.; Rodríguez-Cano, A.; Bruque, J.M.; González-Martín, M.L. Experimental Approach towards the Water Contact Angle Value on the Biomaterial Alloy Ti6Al4V. In *Annales Universitatis Mariae Curie-Sklodowska*; Maria Curie-Skłodowska University: Lublin, Poland, 2015; p. 1.
49. Maciejewski, M.; Wojcieszak, D.; Mazur, M.; Zielinski, M.; Kaczmarek, D.; Domaradzki, J.; Prociow, E. Influence of Droplet Size and Surface Preparation of TiO_2 on Contact Angle Determination. In Proceedings of the 2010 International Students and Young Scientists Workshop" Photonics and Microsystems", Szklarska Poreba, Poland, 25–27 June 2010; IEEE: New York, NY, USA, 2010; pp. 48–51.
50. Zhou, W.; Peng, X.; Zhou, X.; Li, M.; Ren, B.; Cheng, L. Influence of bio-aging on corrosion behavior of different implant materials. *Clin. Implant. Dent. Relat. Res.* **2019**, *21*, 1225–1234. [CrossRef] [PubMed]
51. Suzuki, T.; Hori, N.; Att, W.; Kubo, K.; Iwasa, F.; Ueno, T.; Maeda, H.; Ogawa, T. Ultraviolet treatment overcomes time-related degrading bioactivity of titanium. *Tissue Eng. Part A* **2009**, *15*, 3679–3688. [CrossRef]
52. Chen, C.-J.; Ding, S.-J.; Chen, C.-C. Effects of surface conditions of titanium dental implants on bacterial adhesion. *Photomed. Laser Surg.* **2016**, *34*, 379–388. [CrossRef]
53. Oliveira, D.; Palmieri, A.; Carinci, F.; Bolfarini, C. Osteoblasts behavior on chemically treated commercially pure titanium surfaces. *J. Biomed. Mater. Res. Part A* **2014**, *102*, 1816–1822. [CrossRef]
54. Li, S.; Ni, J.; Liu, X.; Zhang, X.; Yin, S.; Rong, M.; Guo, Z.; Zhou, L. Surface characteristics and biocompatibility of sandblasted and acid-etched titanium surface modified by ultraviolet irradiation: An in vitro study. *J. Biomed. Mater. Res. Part B Appl. Biomater.* **2012**, *100*, 1587–1598. [CrossRef]
55. Strnad, G.; Chirila, N.; Petrovan, C.; Russu, O. Contact angle measurement on medical implant titanium based biomaterials. *Procedia Technol.* **2016**, *22*, 946–953. [CrossRef]
56. Li, Y.; Tian, Y.; Yang, C.; Zhang, D.; Liu, X. Laser-Induced Hydrophobicity on Ti-6Al-4V Surface. In Proceedings of the 2015 International Conference on Manipulation, Manufacturing and Measurement on the Nanoscale (3M-NANO), Changchun, China, 5–9 October 2015; IEEE: New York, NY, USA, 2015; pp. 153–158.
57. Hayashi, R.; Ueno, T.; Migita, S.; Tsutsumi, Y.; Doi, H.; Ogawa, T.; Hanawa, T.; Wakabayashi, N. Hydrocarbon Deposition Attenuates Osteoblast Activity on Titanium. *J. Dent. Res.* **2014**, *93*, 698–703. [CrossRef]

58. Tugulu, S.; Lowe, K.; Scharnweber, D.; Schlottig, F. Preparation of superhydrophilic microrough titanium implant surfaces by alkali treatment. *J. Mater. Sci. Mater. Med.* **2010**, *21*, 2751–2763. [CrossRef]
59. Vishnu, J.; Manivasagam, V.K.; Gopal, V.; Bartomeu Garcia, C.; Hameed, P.; Manivasagam, G.; Webster, T.J. Hydrothermal treatment of etched titanium: A potential surface nano-modification technique for enhanced biocompatibility. *Nanomedicine* **2019**, *20*, 102016. [CrossRef]
60. Dohan Ehrenfest, D.M.; Vazquez, L.; Park, Y.J.; Sammartino, G.; Bernard, J.P. Identification card and codification of the chemical and morphological characteristics of 14 dental implant surfaces. *J. Oral Implantol.* **2011**, *37*, 525–542. [CrossRef]
61. Ghassemi, A.; Ishijima, M.; Hasegawa, M.; Mohammadzadeh Rezaei, N.; Nakhaei, K.; Sekiya, T.; Torii, Y.; Hirota, M.; Park, W.; Miley, D.D.; et al. Biological and Physicochemical Characteristics of 2 Different Hydrophilic Surfaces Created by Saline-Storage and Ultraviolet Treatment. *Implant. Dent.* **2018**, *27*, 405–414. [CrossRef]
62. Massaro, C.; Rotolo, P.; De Riccardis, F.; Milella, E.; Napoli, A.; Wieland, M.; Textor, M.; Spencer, N.D.; Brunette, D.M. Comparative investigation of the surface properties of commercial titanium dental implants. Part I: Chemical composition. *J. Mater. Sci. Mater. Med.* **2002**, *13*, 535–548. [CrossRef]
63. Morra, M.; Cassinelli, C.; Bruzzone, G.; Carpi, A.; Di Santi, G.; Giardino, R.; Fini, M. Surface chemistry effects of topographic modification of titanium dental implant surfaces: 1. Surface analysis. *Int. J. Oral Maxillofac. Implant.* **2003**, *18*, 40–45.
64. Wennerberg, A.; Albrektsson, T. On implant surfaces: A review of current knowledge and opinions. *Int. J. Oral Maxillofac. Implant.* **2010**, *25*, 63–74.
65. Att, W.; Takeuchi, M.; Suzuki, T.; Kubo, K.; Anpo, M.; Ogawa, T. Enhanced osteoblast function on ultraviolet light-treated zirconia. *Biomaterials* **2009**, *30*, 1273–1280. [CrossRef]
66. Kitajima, H.; Komatsu, K.; Matsuura, T.; Ozawa, R.; Saruta, J.; Taleghani, S.R.; Cheng, J.; Ogawa, T. Impact of nano-scale trabecula size on osteoblastic behavior and function in a meso-nano hybrid rough biomimetic zirconia model. *J. Prosthodont. Res.* **2023**, *67*, 288–299. [CrossRef]
67. He, X.; Reichl, F.X.; Milz, S.; Michalke, B.; Wu, X.; Sprecher, C.M.; Yang, Y.; Gahlert, M.; Rohling, S.; Kniha, H.; et al. Titanium and zirconium release from titanium- and zirconia implants in mini pig maxillae and their toxicity in vitro. *Dent. Mater.* **2020**, *36*, 402–412. [CrossRef]
68. Flatebo, R.S.; Hol, P.J.; Leknes, K.N.; Kosler, J.; Lie, S.A.; Gjerdet, N.R. Mapping of titanium particles in peri-implant oral mucosa by laser ablation inductively coupled plasma mass spectrometry and high-resolution optical darkfield microscopy. *J. Oral Pathol. Med.* **2011**, *40*, 412–420. [CrossRef]
69. He, X.; Reichl, F.X.; Wang, Y.; Michalke, B.; Milz, S.; Yang, Y.; Stolper, P.; Lindemaier, G.; Graw, M.; Hickel, R.; et al. Analysis of titanium and other metals in human jawbones with dental implants—A case series study. *Dent. Mater.* **2016**, *32*, 1042–1051. [CrossRef]
70. Giljean, S.; Bigerelle, M.; Anselme, K.; Haidara, H. New insights on contact angle/roughness dependence on high surface energy materials. *Appl. Surf. Sci.* **2011**, *257*, 9631–9638. [CrossRef]
71. Park, J.H.; Olivares-Navarrete, R.; Baier, R.E.; Meyer, A.E.; Tannenbaum, R.; Boyan, B.D.; Schwartz, Z. Effect of cleaning and sterilization on titanium implant surface properties and cellular response. *Acta Biomater.* **2012**, *8*, 1966–1975. [CrossRef]
72. Bertolotti, R.L.; Hurst, V. Inhibition of corrosion during autoclave sterilization of carbon steel dental instruments. *J. Am. Dent. Assoc.* **1978**, *97*, 628–632. [CrossRef]
73. Choi, S.H.; Jeong, W.S.; Cha, J.Y.; Lee, J.H.; Lee, K.J.; Yu, H.S.; Choi, E.H.; Kim, K.M.; Hwang, C.J. Effect of the ultraviolet light treatment and storage methods on the biological activity of a titanium implant surface. *Dent. Mater.* **2017**, *33*, 1426–1435. [CrossRef]
74. Sugita, Y.; Saruta, J.; Taniyama, T.; Kitajima, H.; Hirota, M.; Ikeda, T.; Ogawa, T. UV-Pre-Treated and Protein-Adsorbed Titanium Implants Exhibit Enhanced Osteoconductivity. *Int. J. Mol. Sci.* **2020**, *21*, 4194. [CrossRef]
75. Tabuchi, M.; Ikeda, T.; Hirota, M.; Nakagawa, K.; Park, W.; Miyazawa, K.; Goto, S.; Ogawa, T. Effect of UV Photofunctionalization on Biologic and Anchoring Capability of Orthodontic Miniscrews. *Int. J. Oral Maxillofac. Implant.* **2015**, *30*, 868–879. [CrossRef] [PubMed]
76. Tabuchi, M.; Ikeda, T.; Nakagawa, K.; Hirota, M.; Park, W.; Miyazawa, K.; Goto, S.; Ogawa, T. Ultraviolet photofunctionalization increases removal torque values and horizontal stability of orthodontic miniscrews. *Am. J. Orthod. Dentofac. Orthop.* **2015**, *148*, 274–282. [CrossRef] [PubMed]
77. Hirota, M.; Ozawa, T.; Iwai, T.; Ogawa, T.; Tohnai, I. Implant Stability Development of Photofunctionalized Implants Placed in Regular and Complex Cases: A Case-Control Study. *Int. J. Oral Maxillofac. Implant.* **2016**, *31*, 676–686. [CrossRef] [PubMed]
78. Suzumura, T.; Matsuura, T.; Komatsu, K.; Ogawa, T. Decomposing Organic Molecules on Titanium with Vacuum Ultraviolet Light for Effective and Rapid Photofunctionalization. *J. Funct. Biomater.* **2022**, *14*, 11. [CrossRef]
79. Suzumura, T.; Matsuura, T.; Komatsu, K.; Ogawa, T. A Novel High-Energy Vacuum Ultraviolet Light Photofunctionalization Approach for Decomposing Organic Molecules around Titanium. *Int. J. Mol. Sci.* **2023**, *24*, 1978. [CrossRef]

Disclaimer/Publisher's Note: The statements, opinions and data contained in all publications are solely those of the individual author(s) and contributor(s) and not of MDPI and/or the editor(s). MDPI and/or the editor(s) disclaim responsibility for any injury to people or property resulting from any ideas, methods, instructions or products referred to in the content.

International Journal of *Molecular Sciences*

Review

Dental Pulp Stem Cells for Salivary Gland Regeneration—Where Are We Today?

David Muallah [1,†], Jan Matschke [2,†], Matthias Kappler [3,*], Lysann Michaela Kroschwald [2,4], Günter Lauer [2] and Alexander W. Eckert [5]

1 Department of Oral and Maxillofacial Surgery, University Hospital Hamburg-Eppendorf, Martinistraße 52, 20251 Hamburg, Germany
2 Department of Oral and Maxillofacial Surgery, Faculty of Medicine "Carl Gustav Carus", Technische Universität Dresden, Fetscherstraße 74, 01307 Dresden, Germany
3 Department of Oral and Maxillofacial Plastic Surgery, Martin Luther University Halle-Wittenberg, 06120 Halle, Germany
4 Center for Translational Bone, Joint and Soft Tissue Research, University Hospital "Carl Gustav Carus", Technische Universität Dresden, Fetscherstraße 74, 01307 Dresden, Germany
5 Department of Cranio Maxillofacial Surgery, Paracelsus Medical University, Breslauer Straße 201, 90471 Nuremberg, Germany
* Correspondence: matthias.kappler@uk-halle.de
† These authors contributed equally to this work.

Abstract: Xerostomia is the phenomenon of dry mouth and is mostly caused by hypofunction of the salivary glands. This hypofunction can be caused by tumors, head and neck irradiation, hormonal changes, inflammation or autoimmune disease such as Sjögren's syndrome. It is associated with a tremendous decrease in health-related quality of life due to impairment of articulation, ingestion and oral immune defenses. Current treatment concepts mainly consist of saliva substitutes and parasympathomimetic drugs, but the outcome of these therapies is deficient. Regenerative medicine is a promising approach for the treatment of compromised tissue. For this purpose, stem cells can be utilized due to their ability to differentiate into various cell types. Dental pulp stem cells are adult stem cells that can be easily harvested from extracted teeth. They can form tissues of all three germ layers and are therefore becoming more and more popular for tissue engineering. Another potential benefit of these cells is their immunomodulatory effect. They suppress proinflammatory pathways of lymphocytes and could therefore probably be used for the treatment of chronic inflammation and autoimmune disease. These attributes make dental pulp stem cells an interesting tool for the regeneration of salivary glands and the treatment of xerostomia. Nevertheless, clinical studies are still missing. This review will highlight the current strategies for using dental pulp stem cells in the regeneration of salivary gland tissue.

Keywords: dental pulp stem cells; DPSC; salivary glands; xerostomia; tissue engineering; differentiation; regenerative medicine; Sjogren syndrome

check for updates

Citation: Muallah, D.; Matschke, J.; Kappler, M.; Kroschwald, L.M.; Lauer, G.; Eckert, A.W. Dental Pulp Stem Cells for Salivary Gland Regeneration—Where Are We Today? *Int. J. Mol. Sci.* **2023**, *24*, 8664. https://doi.org/10.3390/ijms24108664

Academic Editor: Takayoshi Yamaza

Received: 27 February 2023
Revised: 2 May 2023
Accepted: 9 May 2023
Published: 12 May 2023

Copyright: © 2023 by the authors. Licensee MDPI, Basel, Switzerland. This article is an open access article distributed under the terms and conditions of the Creative Commons Attribution (CC BY) license (https://creativecommons.org/licenses/by/4.0/).

1. Introduction

The salivary glands (SG) play an essential role in the integrity of the orofacial system as the production of saliva is crucial for digestion, articulation and oral immune defense. Different reasons can lead to xerostomia, the phenomenon of dry mouth. Most cases of xerostomia are caused by tumors, radiotherapy, hormonal changes or autoimmune diseases and lead to a tremendous decrease in quality of life [1–9]. Dryness of the mouth causes dysphagia, increased incidence of caries, impaired articulation and an imbalance of the oral microbiome [10,11]. Current therapy concepts are based on saliva substitutes (oral rinses, gels, powders and sprays) and systemic medication (e.g., pilocarpine, cevimeline). However, none of these approaches provides a satisfying outcome [12].

Recent studies show that tissue engineering is a promising approach for SG regeneration [13–16]. The principle of tissue engineering is to use cells in combination with different biomaterials and biochemical/physicochemical factors to build tissues in vitro [16–20]. Not only primary cells but also stem cells can be utilized for this purpose. By using embryonic stem cells (ESC), Tanaka et al. succeeded in engineering the first fully functional SG organoid in 2018 [13]. Nevertheless, harvesting ESCs is difficult as they must be extracted from the inner cell mass of the blastocyst. This can cause the destruction of the blastocyst and thereby jeopardizes the life of the embryo. Thus, ESCs raise ethical issues and cannot be implemented into clinical practice yet.

An alternative to ESCs is adult stem cells, which can be harvested from a tissue specimen of the patient. A popular source of adult stem cells is the adipose tissue [21]. Adipose tissue-derived stem cells (AdSC) were shown to have the ability of transdifferentiating to acinar cells in vitro [22]. Furthermore, they seem to have a protective effect on SG tissue undergoing irradiation, which may be caused by antioxidative features [23–25]. Nevertheless, AdSCs are harvested by liposuction, which is associated with several severe complications, such as bowel perforation, pneumothorax and sciatic nerve injury [26–28].

Besides AdSC, bone marrow-derived stem cells (BMdSC) were intensively investigated throughout recent years and showed similar effects on compromised SGs like AdSC [29–31]. However, harvesting of BMdSCs requires a biopsy of the bone marrow or the application of drugs like filgrastim, which can cause side effects such as bone pain or other musculoskeletal symptoms [32].

Another promising approach is the use of induced pluripotent stem cells (iPS), which are generated from reprogrammed somatic cells [33]. The conversion from somatic cells to iPS was first conducted by delivering ESC-specific genes via retroviruses [34]. Unfortunately, the tumorigenicity of iPS and the low efficacy of the conversion are only two of many challenges of this technique, which inhibit the implementation of iPS in the clinic.

To implement stem cell therapy in the clinic, easy and harmless accessibility to the cells is essential. For this reason, dental pulp stem cells (DPSC), which originate from the neural crest, have been the focus of regenerative medicine since their first description in 2000 [35–38]. DPSCs can be easily isolated from the pulp of extracted third molars in many different ways. One way is the incubation of cut teeth in a culture flask, as shown in Figure 1B. This method brings the risk of contamination with oral bacteria, which is why the extracted teeth should be preserved in the antibiotic medium for at least 12 h after extraction. The advantage of this method is the weighting of the pulp tissue, which ensures direct contact of the cells with the culture flask's ground. This contact is necessary for the cells to attach to the ground and start migrating into the flask. To reduce the risk of contamination, the pulp can be peeled off the cracked tooth and put into the culture flask on its own (Figure 1C). This technique requires a higher level of experience, since the risk of the tissue floating away in the culture medium is higher.

The extraction of third molars is a routine intervention in maxillofacial surgery and can be conducted in a minimally invasive way and without general anesthesia. Most complications, such as swelling, pain and mild bleeding, are transient and resolve spontaneously within a few days. Severe complications are rare and can be avoided by selecting the right time and technique for extraction [39]. DPSCs can be cryopreserved, which makes it possible to store them and use them later on for autologous therapy when needed [40]. In contrast to other mesenchymal stem cells (MSC), DPSCs express transcription factors such as Oct-4, Sox2 and c-Myc, which are associated with pluripotency [41]. Still, they do not show tumor formation after transplantation as ESCs or induced pluripotent stem cells do [42]. Compared to BMdSCs, DPSCs show higher proliferation rates and a broader array of lineages [35,43]. They can be differentiated into tissues of all three germ sheets [44–47]. This extensive array of lineages makes DPSCs a precious tool for tissue engineering and regeneration of compromised SG tissue. Furthermore, DPSCs seem to have immunosuppressive effects by interfering with activated T-cells [48]. Therefore, they could be beneficial for the treatment of chronic inflammatory diseases such as rheumatoid arthritis, degenerative diseases of the nerval system, periodontitis or inflammatory bowel disease [49–52]. Similar to the aforementioned pathologies, Sjögren's syndrome is also characterized by chronic inflammation. Hence, it could be hypothesized that patients suffering from Sjögrens' syndrome would benefit from treatment with DPSC as well.

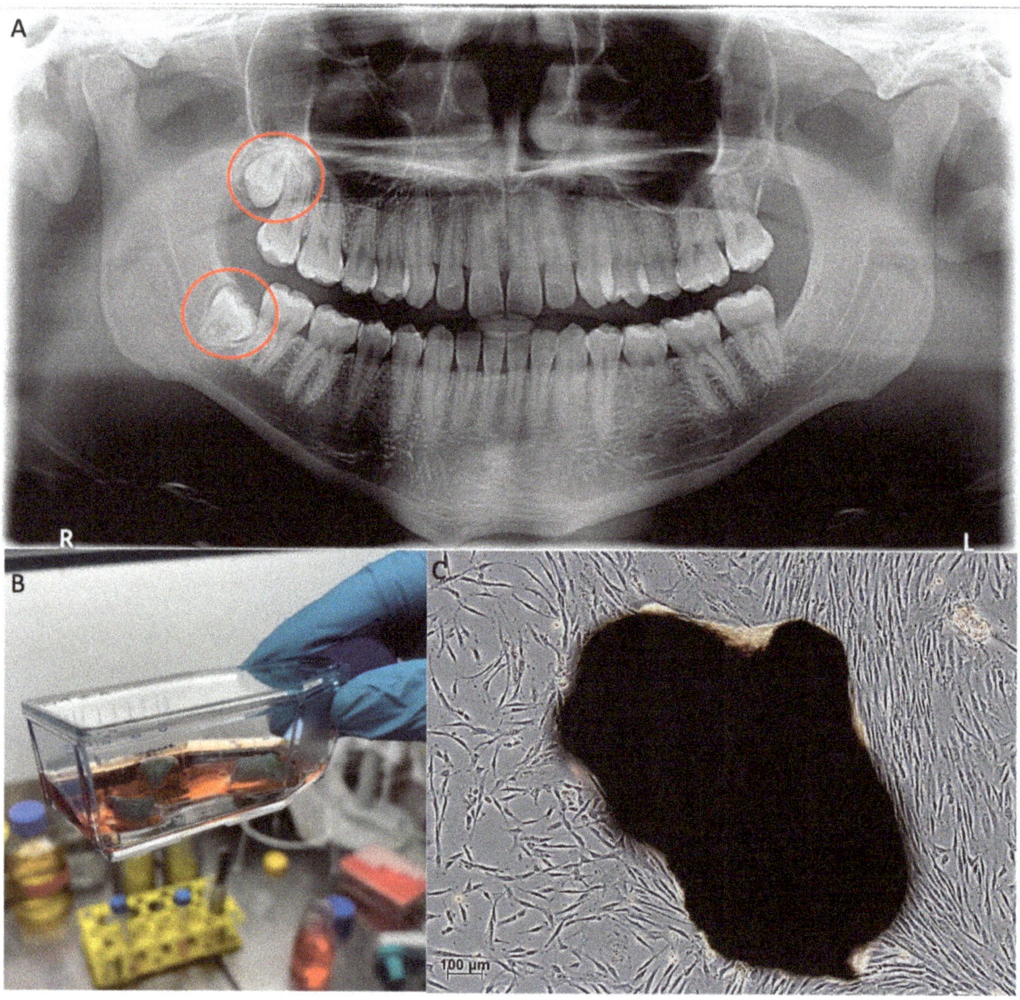

Figure 1. Harvesting of DPSCs; (**A**): Third molars in situ marked with red circles (panoramic X-ray); (**B**): Extracted third molars, split and digested in collagenase for explant culture; (**C**): DPSCs emigrating from dissolved pulp tissue (dark mass) and adhering to the culture flask investigated by light microscopy. This figure belongs to David Muallah, Department of Oral and Maxillofacial Surgery, University Hospital Hamburg-Eppendorf.

Taking these aspects into consideration, DPSCs may play an outstanding role in new approaches to regenerative medicine in the future [53]. Nevertheless, the use of DPSCs for the regeneration of SGs is still far away from clinical application. This review briefly describes the organogenesis of SGs to discuss frame conditions for regenerative approaches. An overview of the current literature and recent strategies for using DPSCs in regeneration of SGs will be provided. Subsequently, new aspects for further research will be discussed.

2. Organogenesis of Salivary Glands

The three major SGs are of different developmental origin. While the *Glandula submandibularis* and *Glandula sublingualis* derive from the endodermal germ sheet, the *Glandula parotis* originates from the ectoderm and therefore has the same origin as DPSCs [36–38]. As the first step of SG organogenesis, an epithelial placode comes to exist in the oral cavity during the seventh embryonic week. Subsequently, this placode infiltrates into the underlying mesodermal mesenchyme. Through dichotomous branching, a canalized system develops. The epithelial cells in the distal ends of the invaginating strands differentiate into acinar cells which produce primary saliva. During the morphogenesis the ectodermal cells continuously interact with the surrounding mesenchyme via several cytokines in both directions. Although this mechanism is not yet fully understood, FGF 10 was identified as one of these signaling molecules [44]. It is assumed that FGF 10 is expressed by cells of the mesenchyme and promotes the maturation of the epithelial gland tissue. But also vice versa, signals sent by the invaginating epithelium trigger the adjacent mesenchyme to differentiate into myoepithelial and stromal cells that surround the ducts and acini of the gland [54].

3. The Effect of DPSC on Primary Salivary Glands

The interaction between epithelial and mesenchymal cells led to different approaches to use DPSCs for SG regeneration. It could be shown that coculturing DPSCs with primary SG cells (SGC) on Matrigel™ increases the number and size of spontaneous acinus formation of the SGCs [55]. This in vitro observation by Reyes et al. was also confirmed by transplanting DPSCs and SGCs embedded in hyaluronic acid hydrogel subcutaneously into 2-month-old Rag1 null mice. Besides the typical acinar differentiation marker alpha amylase-1, other specific markers such as CD 44 and LAMP-1 were also increased compared to SGC-implantation alone [55]. It is hypothesized by many authors that DPSCs assume the role of embryonic mesenchyme that surrounds the invaginating epithelium during the organogenesis when they interact with primary acinar cells [55,56]. This assumption is even more strengthened by results observed when DPSCs were directly injected into compromised SGs of mice [57–59]. For instance, in a study by Yamamura et al., mice were exposed to irradiation to induce hyposalivation. Subsequently, DPSCs were injected into the submandibular glands. Eight weeks after irradiation, saliva flow was assessed. Mice treated with DPSCs showed a significantly higher saliva flow compared to the PBS control group [57]. Similar findings were reported for diabetic wistar rats by Narmada et al. and Suciadi et al. [58,59]. In summary, these approaches seem to prove the ability of DPSC to act as growth-supporting mesenchyme for acinar cells and thereby support the regeneration of compromised salivary glands.

4. The Immunomodulatory Effect of DPSC

BMdSCss, AdSC and umbilical cord-derived stem cells are known for having high immunomodulatory capacities as they are able to control inflammatory conditions [60–65]. It is believed that the immune response is regulated via cell–cell contact and/or paracrine production of soluble factors [66,67]. Some of these cells have even already reached phase I/II human trials [16,68,69].

According to DPSCs, only a few clinical trials have been conducted thus far and none of them aimed at the therapy of hyposalivation [70–75]. Nevertheless, DPSCs were reported to have immunomodulatory effects that even surpass those of other MSCs, which could be a precious tool in SG regeneration [76–78]. For instance, Ogata et al. found DPSCs to significantly surpass mesenchymal stem cells derived from bone marrow in a Sjögren's syndrome mouse model according to anti-inflammatory factors such as IL-10, the downregulation of T-helper 17 cells and the upregulation of regulatory T cells [76,78]. Similarly, Du et al. injected DPSCs into the tail vein of mice with induced Sjögren's syndrome [79]. The cells in this experiment were harvested from the pulp of exfoliated deciduous teeth, also known as stem cells from human exfoliated deciduous teeth (SHED). SHED were first isolated in 2003 and show similar characteristics as DPSCs [43]. It is noteworthy that they have an even higher proliferation rate when compared to DPSCs [80,81]. Du et al. reported that SHED have an anti-inflammatory and function-improving effect on damaged SGs of mice by migrating to the spleen and liver. The authors assume that pulp stem cells affect the SGs in an immunomodulatory way by influencing T-cell differentiation in these organs [79]. This assumption is substantiated by another study published in 2019, which provides evidence of the pulp stem cells' effect on T-cells [77]. Ji found a decreased differentiation of CD4+ T-cells into T-helper 17 cells and, subsequently, a decreased secretion of IL-17 and TNF-α after coculturing DPSCs with peripheral blood mononuclear cells. Furthermore, the DPSCs promoted the polarization of CD4+ T-cells into regulatory T-cells, which have immunosuppressive effects [77]. Rasha et al. observed increased salivary flow rates and a reduction of oxidative stress after injecting DPSCs into the tail veins of diabetic rats [82]. Nevertheless, these experimental settings do not reveal whether the cell–cell communication between DPSCs and immune cells happens in a paracrine or juxtacrine way as reported for other MSCs. While the juxtacrine communication would require direct cell contacts, paracrine communication could be carried out by proteins secreted by the DPSCs. This would raise the question of whether the immunomodulatory effect of DPSCs could also be provided by using only supernatants of DPSC cultures.

This approach was investigated by Takeuchi et al. [83]. Instead of using DPSCs, they injected conditioned supernatant of a DPSC culture intravenously. Mice with an induced defect of the *Glandula submandibularis* subsequently showed an increased regeneration of the SGs compared to the control group [83]. Nevertheless, conditioned supernatants contain several substances that are redundant and have no benefit for the aspired purpose such as antibiotics, fungicides, fetal bovine serum and HEPES. While irresponsible use of antibiotics can lead to bacterial resistance, other substances are discussed as being toxic or allergenic [84,85]. Therefore, exosomes became an object of interest for many researchers. Exosomes are small vesicles containing peptides and nucleic acids that are produced by cells for intercellular communication. These vesicles can be derived from conditioned cell supernatant via centrifugation. It has been found that MSC-derived exosomes could suppress T-cell activation and thereby stabilize an immune homeostasis [86]. BMdSC-derived exosomes were successfully used to save salivary glands from diabetic complications in rats [30]. Also, exosomes of DPSCs were shown to have the capacity of suppressing inflammation via facilitation of macrophages [87]. Compared to BMdSC-derived exosomes, their immunosuppressing effects are even higher [77]. Unfortunately, the reason for this remains unclear. It may be assumed that DPSC-derived exosomes contain different compositions of proteins, lipids, cytokines and RNAs that are responsible for their superior immunomodulatory effect. Thus, further investigations are necessary before they can be applied in the clinic.

However, these studies further aggravate the hypothesis of DPSCs supporting the regeneration of compromised salivary glands. In the future, one of many possible applications could be the injection of DPSC exosomes into the SGs of patients suffering from Sjögren's syndrome, which is associated with an infiltration of lymphocytes into the glands [78]. Moreover, other medical disciplines (e.g., rheumatology, plastic surgery, etc.) could benefit from immunomodulatory features of DPSC as well.

5. Differentiation of DPSCs into Acinar-like Cells

Nevertheless, the trials mentioned so far leave it unclear if DPSCs themselves could be differentiated into acinar cells to replace damaged SG tissue. A recent study by Yan et al. could clarify this question [15]. They induced the differentiation of DPSCs into acinar cells by performing coculture with primary cells of the submandibular gland. The cells were physically separated by a membrane that allowed the exchange of cytokines and other molecules but not the juxtacrine communication [15]. After 2 weeks, specific acinar markers such as amylase and cytokeratin 8 were observed in the DPSCs. They also shaped cobblestone-like islands, which are typical for acinar cells [15]. As a control group, fibroblasts were cocultured with acinar cells. The fibroblasts did not differentiate into acinar cells, which proves that the differentiation is an exclusive feature of the DPSCs. With this experimental setting, the authors could show that DPSCs not only support primary cells in growth and regeneration but that they can also be affected by primary cells and form SG-like tissue. Nevertheless, a detailed analysis of the cell signaling at the molecular level that induces the differentiation is still missing. Therefore, further experiments should be conducted to perform a protein analysis of the coculture's supernatants at different time points to retrace proteomic changes during the induction process.

While typical monolayers of cells as used by Yan et al. are two-dimensional arrangements, in situ cells are organized three-dimensionally [15]. Therefore, three-dimensional cell cultures, such as spheroids, more closely resemble in vivo conditions. These spheroids are usually produced by seeding cells on low-cell adhesion plates. A new way of producing DPSC spheroids was introduced by Adine et al. using a special 3D bioprinting technology [14]. To generate the spheroids, cells were incubated with a solution containing gold and iron oxide and subsequently printed using magnets beneath the well plate. The cells on the spheroids' surfaces could then be differentiated into SG-like cells using FGF 10. Furthermore, epithelial, ductal, myoepithelial and neural elements were detected in the spheroids after immunostaining. The organoids even produced α-amylase. Since the secretion of saliva is regulated by the autonomic nervous system, Adine et al. tried to stimulate the organoids with the neurotransmitter derivatives carbachol and isoproterenol. The cells reacted with intracellular calcium mobilization and a shift in transepithelial resistance, which suggests physiological integrity by an action potential [14]. In an ex vivo SG mouse model, the printed organoids even rescued epithelial growth after irradiation with a single dose of 7 Gy. Furthermore, the neural compartments of the ex vivo glands were integrated into the spheroids [88]. Nevertheless, an ex vivo model does not consider the potential immune response after the implantation of a graft. Still, this experiment shows the usefulness of 3D cultures for SG regeneration after irradiation.

Besides FGF 10, FGF 7 was also reported to induce acinar differentiation [89]. Akashi et al. reported an upregulated expression of acinar-specific markers such as aqua-porin 5 after treating DPSC with FGF 7 in vitro and in vivo [89]. During the organogenesis of SGs, an increased expression of FGF 7 and FGF 10 was observed in the embryonic mesenchyme surrounding the SG rudiments, which could explain the findings of these studies [13]. In conclusion, FGF 10 and FGF 7 seem to be potential additives for SG tissue engineering.

6. Discussion

DPSCs offer new aspects for SG tissue regeneration. They are easy to harvest and can be cryopreserved without losing their differentiation potential [90]. Compared to other stem cells, DPSCs have a higher proliferation rate, a broader array of lineages and a smaller risk of tumor formation [35,42,44–47]. Several studies have shown therapeutical effects of DPSCs, such as the differentiation to pancreatic tissue in rats with diabetes, anti-inflammatory effects in mice with rheumatoid arthritis, lupus erythematosus or COPD and even improved vascular function in patients with erectile dysfunction [49–51,72,91–94]. They either act by affecting the immune system or the primary cells or by differentiating into other cell types. Also, for SG tissue engineering, DPSCs hold application potential due to their ability to differentiate into acinar cells [14,15]. For this purpose, they could be embedded into a scaffold to replace damaged gland tissue in terms of a graft. Besides the transdifferentiation, DPSCs are able to affect acinar cells or immune cells after injection into compromised tissue or intravenously [58,59,79]. Even the supernatants of DPSC cultures offer useful effects on primary acinar cells as well as the immune system [78,83]. While the injection of DPSCs or their supernatants is technically easy, the transplantation of tissue-engineered grafts poses the challenge of nerval and vascular supply. While thin grafts could probably be supplied with nutrients and oxygen via diffusion from the adjacent tissues, for thicker grafts, which would accomplish a rather satisfying production of saliva, alternative steps such as microvascular surgery are necessary. Autologous submandibular gland transplantation as a therapy for *keratoconjunctivitis sicca* reveals some interesting aspects here. For this therapy, the whole submandibular gland is denervated and transplanted to the temporal region. While blood supply can directly be restored by microvascular surgery, a nerval connection cannot be implemented surgically. Still, the transplanted glands start to produce saliva after a few months and thereby improve the moistening of the eyes, which means that the glands become reinnervated after a hypofunctional period. The reinnervation was histologically proven in a rabbit model [95]. Zhang et al. hypothesized that the autonomic reinnervation originates from the auriculotemporal nerve, which runs through the temporal region close to the transplant, but also from the sympathetic plexus around the supplying arteries. According to these findings, tissue-engineered SG grafts should be transplanted close to a bigger nerve or ganglion (e.g., ganglion submandibulare or the lingual nerve) to increase the chance of innervation. Another approach could be to relocate a blood vessel (e.g., the facial artery) so that it runs directly through the graft. This may not only promote the reinnervation as shown by Zhang et al. but also the blood supply of the graft.

Besides cell differentiation, the composition of the scaffold/extracellular matrix is crucial for successful tissue-engineered grafts. None of the abovementioned studies investigated whether differentiated DPSCs stay acinar cells for long, which, indeed, depends on the surrounding extracellular matrix. Cells tend to dedifferentiate if they are not embedded in a supportive environment that matches their requirements. Thus, the perfect matrix for SG grafts needs to be functionalized with signaling molecules or growth factors such as FGF 10 or FGF 7, which were both shown to induce acinar differentiation in DPSCs and seem to play a key role in SG organogenesis [13,14,89]. Moreover, the optimal scaffold material for SG grafts needs to have mechanical properties that allow simple handling during surgery and that withstand the motions caused by speaking or eating after implantation of the graft. Furthermore, they should be degradable to be replaced by autologous tissue over time. Possible options are platelet-rich fibrin, hydrogels, collagen matrices or silk as they are already proved for tissue engineering in several studies and match the abovementioned criteria [96–101].

To sum up, although many in vitro assays and animal studies have already proven the value of DPSCs in SG regeneration, clinical studies are still missing (Table 1). For tissue engineering of SGs, useful materials for artificial extracellular matrices need to be studied. Nevertheless, DPSCs seem to be an outstanding tool for SG tissue regeneration since they are easy to harvest, suppress pathological immune reactions, regenerate compromised gland tissue and can be differentiated into the acinar cell lineage (Figure 2).

Table 1. Different mechanisms of DPSCs affecting salivary gland regeneration.

Mechanism	Author, Year
Effect on primary cells	
Increased development of acinar structures and expression of LAMP-1 and CD44 after coculture of human salivary gland cells with DPSCs	[55] Reyes et al., 2013
Increased saliva flow after DPSC injection into radiated salivary glands of mice	[57] Yamamura et al., 2013
Decreased acinar cell vacuolization and increased IL-10 serum levels after DPSC injection into diabetic rats' salivary glands	[58] Narmada et al., 2019
Increase of vascularization, TGF-β serum level and acinar cell number after DPSC injection into diabetic rats' salivary glands	[59] Suciadi et al., 2019
Immunomodulatory effects	
Decreased apoptotic cell number in salivary glands of diabetic rats after injection of DPSCs in tail veins; also reduced expression of ATG5 and Beclin-1 as well as suppression of Th1 and Tfh cells in spleen while increased number of Treg cells	[79] Du et al., 2019
Inhibition of CD4+T cells' differentiation into T helper 17 cells and reduction of IL-17 and TNF-α, promotion of Treg cells and increased release of IL-10 and TGF-β	[77] Ji et al., 2019
Downregulation of caspase-3 and upregulation of VEGF, decreased blood glucose, improved gland weight and salivary flow in diabetic rats after injection of DPSCs into the tail vein	[82] Al-Serwi et al., 2021
Treatment of mice in salivary gland duct ligation model with DPSC-conditioned medium leads to increased expression of CK5, AQP5	[83] Takeuchi et al., 2020
DPSC exosomes caused macrophages to transform from proinflammatory phenotype to anti-inflammatory phenotype	[87] Shen et al., 2020
Differentiation of DPSC to acinar cells	
Differentiation of DPSC via coculture with acinar cells in monolayer and expression of specific acinar morphology and markers such as CK8, amylase	[15] Yan et al., 2020
3D culture of DPSCs differentiated into acinar-like cells using FGF 10	[14] Adine et al., 2018
Differentiation of DPSC into acinar-like cells expressing AQP5 and αSMA after induction via FGF 7	[79] Akashi et al., 2021

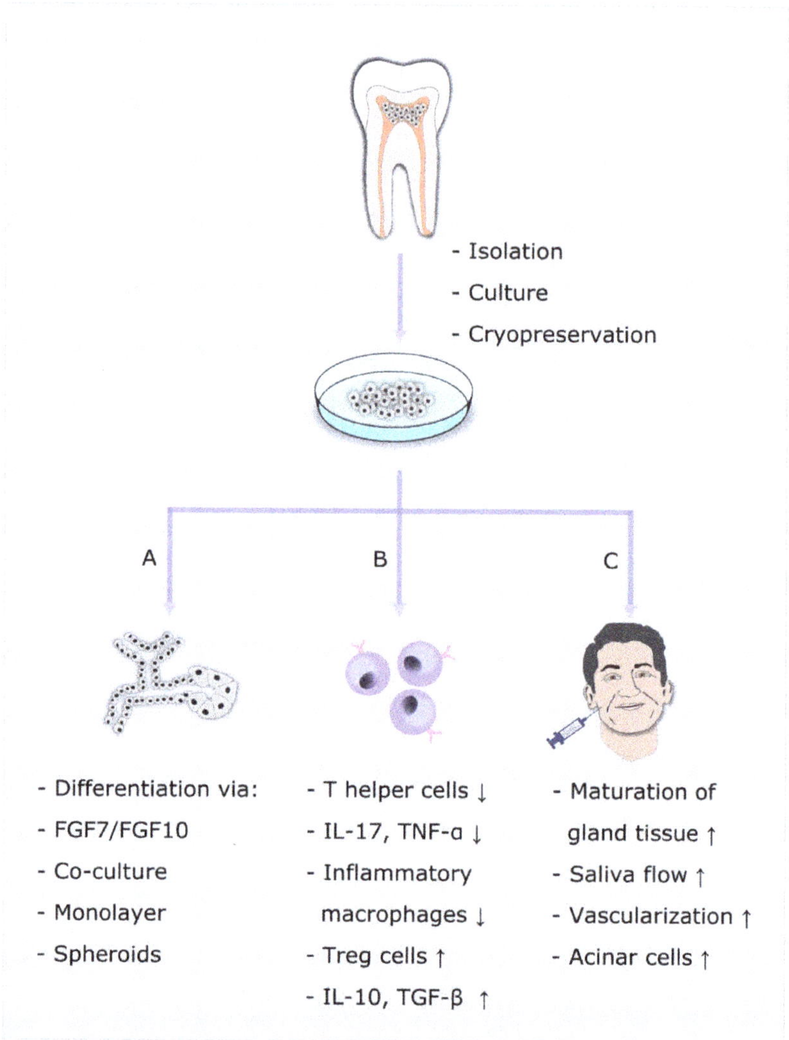

Figure 2. DPSCs for salivary gland regeneration. DPSCs can be isolated from the dental pulp, cultivated in vitro and cryopreserved. A: DPSCs can be differentiated toward acinar lineage by coculture or induction by FGF7/FGF10 in monolayers or as spheroids. B: DPSCs and DPSC-conditioned medium directly affect immune cells in various ways. C: DPSCs serve as supportive mesenchymal tissue after direct injection into salivary glands. This figure belongs to David Muallah, Department of Oral and Maxillofacial Surgery, University Hospital Hamburg-Eppendorf.

Author Contributions: D.M., J.M., M.K., L.M.K., G.L. and A.W.E. contributed to conception of the manuscript. D.M. and J.M. did the literature research. D.M. wrote the first draft of the manuscript. J.M., L.M.K. and A.W.E. wrote sections of the manuscript. All authors have read and agreed to the published version of the manuscript.

Funding: The authors received no funding.

Institutional Review Board Statement: Not applicable.

Informed Consent Statement: Not applicable.

Data Availability Statement: The data used for this review can be found on Pubmed data base and Google Scholar.

Acknowledgments: We acknowledge the financial support of the Open Access Publication Fund of the Martin Luther University Halle-Wittenberg.

Conflicts of Interest: The authors declare no conflict of interest.

References

1. Talha, B.; Swarnkar, S.A. Xerostomia. In *StatPearls [Internet]*; StatPearls Publishing: Treasure Island, FL, USA, 2020.
2. Hey, J.; Setz, J.; Gerlach, R.; Janich, M.; Hildebrandt, G.; Vordermark, D.; Gernhardt, C.R.; Kuhnt, T. Parotid gland-recovery after radiotherapy in the head and neck region–36 months follow-up of a prospective clinical study. *Radiat. Oncol.* **2011**, *6*, 125. [CrossRef]
3. Hedström, J.; Tuomi, L.; Finizia, C.; Olsson, C. Identifying organs at risk for radiation-induced late dysphagia in head and neck cancer patients. *Clin. Transl. Radiat. Oncol.* **2019**, *19*, 87–95. [CrossRef]
4. Murdoch-Kinch, C.-A.; Kim, H.M.; Vineberg, K.A.; Ship, J.A.; Eisbruch, A. Dose-effect relationships for the submandibular salivary glands and implications for their sparing by intensity modulated radiotherapy. *Int. J. Radiat. Oncol. Biol. Phys.* **2008**, *72*, 373–382. [CrossRef]
5. Wu, V.W.C.; Leung, K.Y. A Review on the Assessment of Radiation Induced Salivary Gland Damage after Radiotherapy. *Front. Oncol.* **2019**, *9*, 1090. [CrossRef] [PubMed]
6. Young, W.; Khan, F.; Brandt, R.; Savage, N.; Razek, A.A.; Huang, Q. Syndromes with salivary dysfunction predispose to tooth wear: Case reports of congenital dysfunction of major salivary glands, Prader-Willi, congenital rubella, and Sjögren's syndromes. *Oral Surg. Oral Med. Oral Pathol. Oral Radiol. Endod.* **2001**, *92*, 38–48. [CrossRef]
7. Dutt, P.; Chaudhary, S.; Kumar, P. Oral health and menopause: A comprehensive review on current knowledge and associated dental management. *Ann. Med. Health Sci. Res.* **2013**, *3*, 320–323. [CrossRef]
8. Kwon, H.-K.; Kim, J.M.; Shin, S.-C.; Sung, E.-S.; Kim, H.-S.; Park, G.C.; Cheon, Y.-I.; Lee, J.-C.; Lee, B.-J. The mechanism of submandibular gland dysfunction after menopause may be associated with the ferroptosis. *Aging* **2020**, *12*, 21376–21390. [CrossRef]
9. Agha-Hosseini, F.; Shirzad, N.; Moosavi, M.-S. Evaluation of Xerostomia and salivary flow rate in Hashimoto's Thyroiditis. *Med. Oral Patol. Oral Cir. Bucal* **2016**, *21*, e1–e5. [CrossRef]
10. Sroussi, H.Y.; Epstein, J.B.; Bensadoun, R.-J.; Saunders, D.P.; Lalla, R.V.; Migliorati, C.A.; Heaivilin, N.; Zumsteg, Z.S. Common oral complications of head and neck cancer radiation therapy: Mucositis, infections, saliva change, fibrosis, sensory dysfunctions, dental caries, periodontal disease, and osteoradionecrosis. *Cancer Med.* **2017**, *6*, 2918–2931. [CrossRef] [PubMed]
11. Gupta, N.; Pal, M.; Rawat, S.; Grewal, M.S.; Garg, H.; Chauhan, D.; Ahlawat, P.; Tandon, S.; Khurana, R.; Pahuja, A.K.; et al. Radiation-induced dental caries, prevention and treatment—A systematic review. *Natl. J. Maxillofac. Surg.* **2015**, *6*, 160–166. [CrossRef] [PubMed]
12. Villa, A.; Connell, C.L.; Abati, S. Diagnosis and management of xerostomia and hyposalivation. *Ther. Clin. Risk Manag.* **2014**, *11*, 45–51. [CrossRef]
13. Tanaka, J.; Ogawa, M.; Hojo, H.; Kawashima, Y.; Mabuchi, Y.; Hata, K.; Nakamura, S.; Yasuhara, R.; Takamatsu, K.; Irié, T.; et al. Generation of orthotopically functional salivary gland from embryonic stem cells. *Nat. Commun.* **2018**, *9*, 4216. [CrossRef]
14. Adine, C.; Ng, K.K.; Rungarunlert, S.; Souza, G.R.; Ferreira, J.N. Engineering innervated secretory epithelial organoids by magnetic three-dimensional bioprinting for stimulating epithelial growth in salivary glands. *Biomaterials* **2018**, *180*, 52–66. [CrossRef]
15. Yan, M.; Kluwe, L.; Friedrich, R.E.; Li, X.; Ren, G.; Smeets, R.; Gosau, M.; Liu, X. Inducing differentiation of human dental pulp cells toward acinar-lineage. *Am. J. Transl. Res.* **2020**, *12*, 5781–5788. [PubMed]
16. Chansaenroj, A.; Yodmuang, S.; Ferreira, J.N. Trends in Salivary Gland Tissue Engineering: From Stem Cells to Secretome and Organoid Bioprinting. *Tissue Eng. Part B Rev.* **2021**, *27*, 155–165. [CrossRef]
17. Zamproni, L.N.; Mundim, M.T.V.V.; Porcionatto, M.A. Neurorepair and Regeneration of the Brain: A Decade of Bioscaffolds and Engineered Microtissue. *Front. Cell Dev. Biol.* **2021**, *9*, 764. [CrossRef]
18. Jacob, G.; Shimomura, K.; Nakamura, N. Osteochondral Injury, Management and Tissue Engineering Approaches. *Front. Cell Dev. Biol.* **2020**, *8*, 1118. [CrossRef] [PubMed]
19. Leal, B.B.J.; Wakabayashi, N.; Oyama, K.; Kamiya, H.; Braghirolli, D.I.; Pranke, P. Vascular Tissue Engineering: Polymers and Methodologies for Small Caliber Vascular Grafts. *Front. Cardiovasc. Med.* **2021**, *7*, 376. [CrossRef]
20. Montero, P.; Flandes-Iparraguirre, M.; Musquiz, S.; Pérez Araluce, M.; Plano, D.; Sanmartín, C.; Orive, G.; Gavira, J.J.; Prosper, F.; Mazo, M.M. Cells, Materials, and Fabrication Processes for Cardiac Tissue Engineering. *Front. Bioeng. Biotechnol.* **2020**, *8*, 955. [CrossRef] [PubMed]
21. Mende, W.; Götzl, R.; Kubo, Y.; Pufe, T.; Ruhl, T.; Beier, J.P. The Role of Adipose Stem Cells in Bone Regeneration and Bone Tissue Engineering. *Cells* **2021**, *10*, 975. [CrossRef]

22. Dai, T.-Q.; Zhang, L.-L.; An, Y.; Xu, F.-F.; An, R.; Xu, H.-Y.; Liu, Y.-P.; Liu, B. In vitro transdifferentiation of adipose tissue-derived stem cells into salivary gland acinar-like cells. *Am. J. Transl. Res.* **2019**, *11*, 2908–2924. [PubMed]
23. Choi, J.-S.; An, H.-Y.; Shin, H.-S.; Kim, Y.-M.; Lim, J.-Y. Enhanced tissue remodelling efficacy of adipose-derived mesenchymal stem cells using injectable matrices in radiation-damaged salivary gland model. *J. Tissue Eng. Regen. Med.* **2018**, *12*, e695–e706. [CrossRef]
24. Saylam, G.; Bayır, Ö.; Gültekin, S.S.; Pınarlı, F.A.; Han, Ü.; Korkmaz, M.H.; Sancaktar, M.E.; Tatar, İ.; Sargon, M.F.; Tatar, E.Ç. Protective/restorative Role of the Adipose Tissue-derived Mesenchymal Stem Cells on the Radioiodine-induced Salivary Gland Damage in Rats. *Radiol. Oncol.* **2017**, *51*, 307–316. [CrossRef] [PubMed]
25. Lim, J.-Y.; Ra, J.C.; Shin, I.S.; Jang, Y.H.; An, H.-Y.; Choi, J.-S.; Kim, W.C.; Kim, Y.-M. Systemic transplantation of human adipose tissue-derived mesenchymal stem cells for the regeneration of irradiation-induced salivary gland damage. *PLoS ONE* **2013**, *8*, e71167. [CrossRef] [PubMed]
26. Bellini, E.; Grieco, M.P.; Raposio, E. A journey through liposuction and liposculture: Review. *Ann. Med. Surg.* **2017**, *24*, 53–60. [CrossRef]
27. Mentz, J.A.; Mentz, H.A.; Nemir, S. Pneumothorax as a Complication of Liposuction. *Aesthetic Surg. J.* **2020**, *40*, 753–758. [CrossRef] [PubMed]
28. Abdallah, I.E.; Ayoub, R.; Sawaya, R.; Saba, S.C. Iatrogenic sciatic nerve injury during liposuction and fat tissue grafting: A preventable surgical complication with devastating patient outcomes. *Patient Saf. Surg.* **2020**, *14*, 40. [CrossRef]
29. Elsaadany, B.; Zakaria, M.; Mousa, M.R. Transplantation of Bone Marrow-Derived Mesenchymal Stem Cells Preserve the Salivary Glands Structure after Head and Neck Radiation in Rats. *Open Access Maced. J. Med. Sci.* **2019**, *7*, 1588–1592. [CrossRef]
30. AbuBakr, N.; Haggag, T.; Sabry, D.; Salem, Z.A. Functional and histological evaluation of bone marrow stem cell-derived exosomes therapy on the submandibular salivary gland of diabetic Albino rats through TGFβ/Smad3 signaling pathway. *Heliyon* **2020**, *6*, e03789. [CrossRef]
31. Liang, L.; Wang, J.; Zhang, Y.; Shen, Z.; Zheng, J.; Li, J.; Su, Z.; Cai, J.; Jiang, W.; Sun, M. Transdifferentiation of bone marrow-derived mesenchymal stem cells into salivary gland-like cells using a novel culture method. *Biotechnol. Lett.* **2015**, *37*, 1505–1513. [CrossRef]
32. Aghedo, B.O.; Gupta, V. Filgrastim. In *StatPearls [Internet]*; StatPearls Publishing: Treasure Island, FL, USA, 2021.
33. Ono, H.; Obana, A.; Usami, Y.; Sakai, M.; Nohara, K.; Egusa, H.; Sakai, T. Regenerating Salivary Glands in the Microenvironment of Induced Pluripotent Stem Cells. *BioMed Res. Int.* **2015**, *2015*, 293570. [CrossRef] [PubMed]
34. Takahashi, K.; Yamanaka, S. Induction of pluripotent stem cells from mouse embryonic and adult fibroblast cultures by defined factors. *Cell* **2006**, *126*, 663–676. [CrossRef] [PubMed]
35. Gronthos, S.; Mankani, M.; Brahim, J.; Robey, P.G.; Shi, S. Postnatal human dental pulp stem cells (DPSCs) in vitro and in vivo. *Proc. Natl. Acad. Sci. USA* **2000**, *97*, 13625–13630. [CrossRef] [PubMed]
36. Janebodin, K.; Horst, O.V.; Ieronimakis, N.; Balasundaram, G.; Reesukumal, K.; Pratumvinit, B.; Reyes, M. Isolation and characterization of neural crest-derived stem cells from dental pulp of neonatal mice. *PLoS ONE* **2011**, *6*, e27526. [CrossRef]
37. Yamazaki, H.; Tsuneto, M.; Yoshino, M.; Yamamura, K.-I.; Hayashi, S.-I. Potential of dental mesenchymal cells in developing teeth. *Stem Cells* **2007**, *25*, 78–87. [CrossRef] [PubMed]
38. Chai, Y.; Jiang, X.; Ito, Y.; Bringas, P.J.; Han, J.; Rowitch, D.H.; Soriano, P.; McMahon, A.P.; Sucov, H.M. Fate of the mammalian cranial neural crest during tooth and mandibular morphogenesis. *Development* **2000**, *127*, 1671–1679. [CrossRef] [PubMed]
39. American Association of Oral and Maxillofacial Surgeons. Available online: https://www.aaoms.org/images/uploads/pdfs/management_third_molar_supporting_information.pdf (accessed on 10 August 2021).
40. Pilbauerová, N.; Suchánek, J. Cryopreservation of Dental Stem Cells. *Acta Med. Hradec Kral.* **2018**, *61*, 1–7. [CrossRef]
41. Liu, L.; Wei, X.; Ling, J.; Wu, L.; Xiao, Y. Expression pattern of Oct-4, Sox2, and c-Myc in the primary culture of human dental pulp derived cells. *J. Endod.* **2011**, *37*, 466–472. [CrossRef]
42. Gronthos, S.; Brahim, J.; Li, W.; Fisher, L.W.; Cherman, N.; Boyde, A.; DenBesten, P.; Robey, P.G.; Shi, S. Stem cell properties of human dental pulp stem cells. *J. Dent. Res.* **2002**, *81*, 531–535. [CrossRef]
43. Miura, M.; Gronthos, S.; Zhao, M.; Lu, B.; Fisher, L.W.; Robey, P.G.; Shi, S. SHED: Stem cells from human exfoliated deciduous teeth. *Proc. Natl. Acad. Sci. USA* **2003**, *100*, 5807–5812. [CrossRef]
44. Wells, K.L.; Gaete, M.; Matalova, E.; Deutsch, D.; Rice, D.; Tucker, A.S. Dynamic relationship of the epithelium and mesenchyme during salivary gland initiation: The role of Fgf10. *Biol. Open* **2013**, *2*, 981–989. [CrossRef] [PubMed]
45. Rosa, V.; Dubey, N.; Islam, I.; Min, K.-S.; Nör, J.E. Pluripotency of Stem Cells from Human Exfoliated Deciduous Teeth for Tissue Engineering. *Stem Cells Int.* **2016**, *2016*, 5957806. [CrossRef]
46. Cho, Y.-A.; Noh, K.; Jue, S.-S.; Lee, S.-Y.; Kim, E.-C. Melatonin promotes hepatic differentiation of human dental pulp stem cells: Clinical implications for the prevention of liver fibrosis. *J. Pineal Res.* **2015**, *58*, 127–135. [CrossRef] [PubMed]
47. Shi, X.; Mao, J.; Liu, Y. Pulp stem cells derived from human permanent and deciduous teeth: Biological characteristics and therapeutic applications. *Stem Cells Transl. Med.* **2020**, *9*, 445–464. [CrossRef] [PubMed]
48. Zhao, Y.; Wang, L.; Jin, Y.; Shi, S. Fas ligand regulates the immunomodulatory properties of dental pulp stem cells. *J. Dent. Res.* **2012**, *91*, 948–954. [CrossRef]

49. Földes, A.; Kádár, K.; Kerémi, B.; Zsembery, Á.; Gyires, K.; Zádori, Z.S.; Varga, G. Mesenchymal Stem Cells of Dental Origin-Their Potential for Antiinflammatory and Regenerative Actions in Brain and Gut Damage. *Curr. Neuropharmacol.* **2016**, *14*, 914–934. [CrossRef]
50. Lo Monaco, M.; Gervois, P.; Beaumont, J.; Clegg, P.; Bronckaers, A.; Vandeweerd, J.-M.; Lambrichts, I. Therapeutic Potential of Dental Pulp Stem Cells and Leukocyte- and Platelet-Rich Fibrin for Osteoarthritis. *Cells* **2020**, *9*, 980. [CrossRef]
51. Ishikawa, J.; Takahashi, N.; Matsumoto, T.; Yoshioka, Y.; Yamamoto, N.; Nishikawa, M.; Hibi, H.; Ishigro, N.; Ueda, M.; Furukawa, K.; et al. Factors secreted from dental pulp stem cells show multifaceted benefits for treating experimental rheumatoid arthritis. *Bone* **2016**, *83*, 210–219. [CrossRef]
52. Racz, G.Z.; Kadar, K.; Foldes, A.; Kallo, K.; Perczel-Kovach, K.; Keremi, B.; Nagy, A.; Varga, G. Immunomodulatory and potential therapeutic role of mesenchymal stem cells in periodontitis. *J. Physiol. Pharmacol. Off. J. Pol. Physiol. Soc.* **2014**, *65*, 327–339.
53. Al-Zer, H.; Apel, C.; Heiland, M.; Friedrich, R.E.; Jung, O.; Kroeger, N.; Eichhorn, W.; Smeets, R. Enrichment and Schwann Cell Differentiation of Neural Crest-derived Dental Pulp Stem Cells. *In Vivo* **2015**, *29*, 319–326.
54. Patel, V.N.; Hoffman, M.P. Salivary gland development: A template for regeneration. *Semin. Cell Dev. Biol.* **2014**, *25–26*, 52–60. [CrossRef]
55. Morayma Reyes, K.J. Neural Crest-Derived Dental Pulp Stem Cells Function as Ectomesenchyme to Support Salivary Gland Tissue Formation. *Dentistry* **2013**, *13*, 2161-1122. [CrossRef]
56. Gancheva, M.R.; Kremer, K.L.; Gronthos, S.; Koblar, S.A. Using Dental Pulp Stem Cells for Stroke Therapy. *Front. Neurol.* **2019**, *10*, 422. [CrossRef]
57. Yamamura, Y.; Yamada, H.; Sakurai, T.; Ide, F.; Inoue, H.; Muramatsu, T.; Mishima, K.; Hamada, Y.; Saito, I. Treatment of salivary gland hypofunction by transplantation with dental pulp cells. *Arch. Oral Biol.* **2013**, *58*, 935–942. [CrossRef]
58. Narmada, I.B.; Laksono, V.; Nugraha, A.P.; Ernawati, D.S.; Winias, S.; Prahasanti, C.; Dinaryanti, A.; Susilowati, H.; Hendrianto, E.; Ihsan, I.S.; et al. Regeneration of salivary gland defects of diabetic wistar rats post human dental pulp stem cells intraglandular transplantation on acinar cell vacuolization and interleukin-10 serum level. *Pesqui. Bras. Odontopediatria Clín. Integr.* **2019**, *19*, e5002. [CrossRef]
59. Suciadi, S.P.; Nugraha, A.P.; Ernawati, D.S.; Ayuningtyas, N.F.; Narmada, I.B.; Prahasanti, C.; Dinaryanti, A.; Ihsan, I.S.; Hendrinto, E.; Susilowati, H.; et al. The efficacy of human dental pulp stem cells in regenerating submandibular gland defects in diabetic wistar rats (*Rattus novergicus*). *Res. J. Pharm. Technol.* **2019**, *12*, 1573. [CrossRef]
60. Law, S.; Chaudhuri, S. Mesenchymal stem cell and regenerative medicine: Regeneration versus immunomodulatory challenges. *Am. J. Stem Cells* **2013**, *2*, 22–38.
61. Chen, S.; Cui, G.; Peng, C.; Lavin, M.F.; Sun, X.; Zhang, E.; Yang, Y.; Guan, Y.; Du, Z.; Shao, H. Transplantation of adipose-derived mesenchymal stem cells attenuates pulmonary fibrosis of silicosis via anti-inflammatory and anti-apoptosis effects in rats. *Stem Cell Res. Ther.* **2018**, *9*, 110. [CrossRef] [PubMed]
62. Stonesifer, C.; Corey, S.; Ghanekar, S.; Diamandis, Z.; Acosta, S.A.; Borlongan, C.V. Stem cell therapy for abrogating stroke-induced neuroinflammation and relevant secondary cell death mechanisms. *Prog. Neurobiol.* **2017**, *158*, 94–131. [CrossRef]
63. Al-Ghadban, S.; Bunnell, B.A. Adipose Tissue-Derived Stem Cells: Immunomodulatory Effects and Therapeutic Potential. *Physiology* **2020**, *35*, 125–133. [CrossRef]
64. Abdelmawgoud, H.; Saleh, A. Anti-inflammatory and antioxidant effects of mesenchymal and hematopoietic stem cells in a rheumatoid arthritis rat model. *Adv. Clin. Exp. Med.* **2018**, *27*, 873–880. [CrossRef] [PubMed]
65. Xu, Y.; Luo, H.; Chen, F.; Shi, Y.; Sun, M. Human umbilical cord mesenchymal stem cells polarize RAW264.7 macrophages to an anti-inflammatory subpopulation. *Int. J. Clin. Exp. Pathol.* **2018**, *11*, 1446–1452. [PubMed]
66. Prockop, D.J.; Oh, J.Y. Mesenchymal stem/stromal cells (MSCs): Role as guardians of inflammation. *Mol. Ther.* **2012**, *20*, 14–20. [CrossRef] [PubMed]
67. Ceccarelli, S.; Pontecorvi, P.; Anastasiadou, E.; Napoli, C.; Marchese, C. Immunomodulatory Effect of Adipose-Derived Stem Cells: The Cutting Edge of Clinical Application. *Front. Cell Dev. Biol.* **2020**, *8*, 236. [CrossRef] [PubMed]
68. Verstockt, B.; Ferrante, M.; Vermeire, S.; Van Assche, G. New treatment options for inflammatory bowel diseases. *J. Gastroenterol.* **2018**, *53*, 585–590. [CrossRef] [PubMed]
69. Riordan, N.H.; Hincapié, M.L.; Morales, I.; Fernández, G.; Allen, N.; Leu, C.; Madrigal, M.; Paz Rodríguez, J.; Novarro, N. Allogeneic Human Umbilical Cord Mesenchymal Stem Cells for the Treatment of Autism Spectrum Disorder in Children: Safety Profile and Effect on Cytokine Levels. *Stem Cells Transl. Med.* **2019**, *8*, 1008–1016. [CrossRef]
70. Nagpal, A.; Kremer, K.L.; Hamilton-Bruce, M.A.; Kaidonis, X.; Milton, A.G.; Levi, C.; Shi, S.; Carey, L.; Hillier, S.; Rose, M.; et al. TOOTH (The Open study of dental pulp stem cell Therapy in Humans): Study protocol for evaluating safety and feasibility of autologous human adult dental pulp stem cell therapy in patients with chronic disability after stroke. *Int. J. Stroke Off. J. Int. Stroke Soc.* **2016**, *11*, 575–585. [CrossRef] [PubMed]
71. Cubuk, S.; Oduncuoglu, B.F.; Alaaddinoglu, E.E. The effect of dental pulp stem cells and L-PRF when placed into the extraction sockets of impacted mandibular third molars on the periodontal status of adjacent second molars: A split-mouth, randomized, controlled clinical trial. *Oral Maxillofac. Surg.* **2023**, *27*, 59–68. [CrossRef]
72. Koga, S.; Horiguchi, Y. Efficacy of a cultured conditioned medium of exfoliated deciduous dental pulp stem cells in erectile dysfunction patients. *J. Cell. Mol. Med.* **2022**, *26*, 195–201. [CrossRef]

73. Ferrarotti, F.; Romano, F.; Gamba, M.N.; Quirico, A.; Giraudi, M.; Audagna, M.; Aimetti, M. Human intrabony defect regeneration with micrografts containing dental pulp stem cells: A randomized controlled clinical trial. *J. Clin. Periodontol.* **2018**, *45*, 841–850. [CrossRef]
74. Barbier, L.; Ramos, E.; Mendiola, J.; Rodriguez, O.; Santamaria, G.; Santamaria, J.; Arteagoitia, I. Autologous dental pulp mesenchymal stem cells for inferior third molar post-extraction socket healing: A split-mouth randomised clinical trial. *Med. Oral Patol. Oral Cir. Bucal* **2018**, *23*, e469–e477. [CrossRef]
75. Brizuela, C.; Meza, G.; Urrejola, D.; Quezada, M.A.; Concha, G.; Ramírez, V.; Angelopoulos, I.; Cadiz, M.I.; Tapia-Limonchi, R.; Khoury, M. Cell-Based Regenerative Endodontics for Treatment of Periapical Lesions: A Randomized, Controlled Phase I/II Clinical Trial. *J. Dent. Res.* **2020**, *99*, 523–529. [CrossRef] [PubMed]
76. Ogata, K.; Matsumura-Kawashima, M.; Moriyama, M.; Kawado, T.; Nakamura, S. Dental pulp-derived stem cell-conditioned media attenuates secondary Sjögren's syndrome via suppression of inflammatory cytokines in the submandibular glands. *Regen. Ther.* **2021**, *16*, 73–80. [CrossRef] [PubMed]
77. Ji, L.; Bao, L.; Gu, Z.; Zhou, Q.; Liang, Y.; Zheng, Y.; Xu, Y.; Zhang, X.; Feng, X. Comparison of immunomodulatory properties of exosomes derived from bone marrow mesenchymal stem cells and dental pulp stem cells. *Immunol. Res.* **2019**, *67*, 432–442. [CrossRef] [PubMed]
78. Matsumura-Kawashima, M.; Ogata, K.; Moriyama, M.; Murakami, Y.; Kawado, T.; Nakamura, S. Secreted factors from dental pulp stem cells improve Sjögren's syndrome via regulatory T cell-mediated immunosuppression. *Stem Cell Res. Ther.* **2021**, *12*, 182. [CrossRef] [PubMed]
79. Du, Z.H.; Ding, C.; Zhang, Q.; Zhang, Y.; Ge, X.Y.; Li, S.L.; Yu, G.Y. Stem cells from exfoliated deciduous teeth alleviate hyposalivation caused by Sjögren syndrome. *Oral Dis.* **2019**, *25*, 1530–1544. [CrossRef]
80. Kunimatsu, R.; Nakajima, K.; Awada, T.; Tsuka, Y.; Abe, T.; Ando, K.; Hiraki, T.; Kimura, A.; Tanimoto, K. Comparative characterization of stem cells from human exfoliated deciduous teeth, dental pulp, and bone marrow-derived mesenchymal stem cells. *Biochem. Biophys. Res. Commun.* **2018**, *501*, 193–198. [CrossRef]
81. Nakajima, K.; Kunimatsu, R.; Ando, K.; Hiraki, T.; Rikitake, K.; Tsuka, Y.; Abe, T.; Tanimoto, K. Success rates in isolating mesenchymal stem cells from permanent and deciduous teeth. *Sci. Rep.* **2019**, *9*, 16764. [CrossRef]
82. Al-Serwi, R.H.; El-Kersh, A.O.F.O.; El-Akabawy, G. Human dental pulp stem cells attenuate streptozotocin-induced parotid gland injury in rats. *Stem Cell Res. Ther.* **2021**, *12*, 577. [CrossRef]
83. Takeuchi, H.; Takahashi, H.; Tanaka, A. Effects of Human Dental Pulp Stem Cell-Derived Conditioned Medium on Atrophied Submandibular Gland after the Release from Ligation of the Main Excretory Duct in Mice. *J. Hard Tissue Biol.* **2020**, *29*, 183–192. [CrossRef]
84. le Roux, J.; Kleynhans, J.; Rubow, S. The use of HEPES-buffer in the production of gallium-68 radiopharmaceuticals–time to reconsider strict pharmacopoeial limits? *EJNMMI Radiopharm. Chem.* **2021**, *6*, 15. [CrossRef]
85. de Silva, R.; Dasanayake, W.M.D.K.; Wickramasinhe, G.D.; Karunatilake, C.; Weerasinghe, N.; Gunasekera, P.; Malavige, G.N. Sensitization to bovine serum albumin as a possible cause of allergic reactions to vaccines. *Vaccine* **2017**, *35*, 1494–1500. [CrossRef]
86. Baquir, B.; Hancock, R.E.W. Exosomes, your body's answer to immune health. *Ann. Transl. Med.* **2017**, *5*, 81. [CrossRef] [PubMed]
87. Shen, Z.; Kuang, S.; Zhang, Y.; Yang, M.; Qin, W.; Shi, X.; Lin, Z. Chitosan hydrogel incorporated with dental pulp stem cell-derived exosomes alleviates periodontitis in mice via a macrophage-dependent mechanism. *Bioact. Mater.* **2020**, *5*, 1113–1126. [CrossRef] [PubMed]
88. Dong, J.; Sakai, K.; Koma, Y.; Watanabe, J.; Liu, K.; Maruyama, H.; Sakaguchi, K.; Hibi, H. Dental pulp stem cell-derived small extracellular vesicle in irradiation-induced senescence. *Biochem. Biophys. Res. Commun.* **2021**, *575*, 28–35. [CrossRef]
89. Akashi, Y.; Nemoto, A.; Nakajima, K.; Kokubun, K.; Murakami, S.; Inoue, T.; Matsuzaka, K. The effect of fibroblast growth factor 7 on human dental pulp stem cells for differentiation to AQP5-positive and αSMA-positive cells in vitro and in vivo. *Clin. Exp. Dent. Res.* **2021**, *7*, 344–353. [CrossRef] [PubMed]
90. Morsczeck, C.; Reichert, T.E. Dental stem cells in tooth regeneration and repair in the future. *Expert Opin. Biol. Ther.* **2018**, *18*, 187–196. [CrossRef]
91. Croci, S.; Bonacini, M.; Dolci, G.; Massari, M.; Facciolongo, N.; Pignatti, E.; Pisciotta, A.; Carnevale, G.; Negro, A.; Cassone, G.; et al. Human Dental Pulp Stem Cells Modulate Cytokine Production in vitro by Peripheral Blood Mononuclear Cells from Coronavirus Disease 2019 Patients. *Front. Cell Dev. Biol.* **2020**, *8*, 609204. [CrossRef]
92. Gao, X.; Liu, Z.; Wang, Z. Dental Pulp Stem Cells Ameliorate Elastase-Induced Pulmonary Emphysema by Regulating Inflammation and Oxidative Stress. *J. Inflamm. Res.* **2023**, *16*, 1497–1508. [CrossRef]
93. Tang, X.; Li, W.; Wen, X.; Zhang, Z.; Chen, W.; Yao, G.; Chen, H.; Wang, D.; Shi, S.; Sun, L. Transplantation of dental tissue-derived mesenchymal stem cells ameliorates nephritis in lupus mice. *Ann. Transl. Med.* **2019**, *7*, 132. [CrossRef]
94. Inada, R.; Mendoza, H.Y.; Tanaka, T.; Horie, T.; Satomi, T. Preclinical study for the treatment of diabetes mellitus using β-like cells derived from human dental pulp stem cells. *Regen. Med.* **2022**, *17*, 905–913. [CrossRef] [PubMed]
95. Zhang, X.; Yang, N.; Liu, X.; Su, J.; Cong, X.; Wu, L.; Zhang, Y.; Yu, G. Autonomic reinnervation and functional regeneration in autologous transplanted submandibular glands in patients with severe keratoconjunctivitis sicca. *Int. J. Oral Sci.* **2018**, *10*, 14. [CrossRef]
96. Scheller, E.L.; Krebsbach, P.H.; Kohn, D.H. Tissue engineering: State of the art in oral rehabilitation. *J. Oral Rehabil.* **2009**, *36*, 368–389. [CrossRef] [PubMed]

97. Wang, Z.-S.; Feng, Z.-H.; Wu, G.-F.; Bai, S.-Z.; Dong, Y.; Chen, F.-M.; Zhao, Y.-M. The use of platelet-rich fibrin combined with periodontal ligament and jaw bone mesenchymal stem cell sheets for periodontal tissue engineering. *Sci. Rep.* **2016**, *6*, 28126. [CrossRef] [PubMed]
98. Duin, S.; Schütz, K.; Ahlfeld, T.; Lehmann, S.; Lode, A.; Ludwig, B.; Gelinsky, M. 3D Bioprinting of Functional Islets of Langerhans in an Alginate/Methylcellulose Hydrogel Blend. *Adv. Healthc. Mater.* **2019**, *8*, e1801631. [CrossRef]
99. Ahlfeld, T.; Köhler, T.; Czichy, C.; Lode, A.; Gelinsky, M. A Methylcellulose Hydrogel as Support for 3D Plotting of Complex Shaped Calcium Phosphate Scaffolds. *Gels* **2018**, *4*, 68. [CrossRef]
100. Kopp, A.; Smeets, R.; Gosau, M.; Friedrich, R.E.; Fuest, S.; Behbahani, M.; Barbeck, M.; Rutkowski, R.; Burg, S.; Kluwe, L.; et al. Production and Characterization of Porous Fibroin Scaffolds for Regenerative Medical Application. *In Vivo* **2019**, *33*, 757–762. [CrossRef]
101. Wu, D.T.; Munguia-Lopez, J.G.; Cho, Y.W.; Ma, X.; Song, V.; Zhu, Z.; Tran, S.D. Polymeric Scaffolds for Dental, Oral, and Craniofacial Regenerative Medicine. *Molecules* **2021**, *26*, 7043. [CrossRef] [PubMed]

Disclaimer/Publisher's Note: The statements, opinions and data contained in all publications are solely those of the individual author(s) and contributor(s) and not of MDPI and/or the editor(s). MDPI and/or the editor(s) disclaim responsibility for any injury to people or property resulting from any ideas, methods, instructions or products referred to in the content.

International Journal of *Molecular Sciences*

MDPI

Review

Inorganic Compounds as Remineralizing Fillers in Dental Restorative Materials: Narrative Review

Leena Ibraheem Bin-Jardan [1], Dalal Ibrahim Almadani [1], Leen Saleh Almutairi [1], Hadi A. Almoabid [1], Mohammed A. Alessa [1], Khalid S. Almulhim [2], Rasha N. AlSheikh [2], Yousif A. Al-Dulaijan [3], Maria S. Ibrahim [4], Afnan O. Al-Zain [5] and Abdulrahman A. Balhaddad [2,*]

[1] College of Dentistry, Imam Abdulrahman Bin Faisal University, P.O. Box 1982, Dammam 31441, Saudi Arabia; leena1421@hotmail.com (L.I.B.-J.); dalalmad@outlook.com (D.I.A.); leenalmutairiiau@gmail.com (L.S.A.); hadi44aziz@gmail.com (H.A.A.); x.mae@hotmail.com (M.A.A.)
[2] Department of Restorative Dental Sciences, College of Dentistry, Imam Abdulrahman Bin Faisal University, P.O. Box 1982, Dammam 31441, Saudi Arabia; ksalmulhim@iau.edu.sa (K.S.A.); ralsheikh@iau.edu.sa (R.N.A.)
[3] Department of Substitute Dental Sciences, College of Dentistry, Imam Abdulrahman Bin Faisal University, P.O. Box 1982, Dammam 31441, Saudi Arabia; yaaldulaijan@iau.edu.sa
[4] Department of Preventive Dental Sciences, College of Dentistry, Imam Abdulrahman Bin Faisal University, P.O. Box 1982, Dammam 31441, Saudi Arabia; msibrahim@iau.edu.sa
[5] Restorative Dentistry Department, Faculty of Dentistry, King Abdulaziz University Jeddah, P.O. Box 80209, Jeddah 21589, Saudi Arabia; alzain@kau.edu.sa
* Correspondence: abalhaddad@iau.edu.sa

Abstract: Secondary caries is one of the leading causes of resin-based dental restoration failure. It is initiated at the interface of an existing restoration and the restored tooth surface. It is mainly caused by an imbalance between two processes of mineral loss (demineralization) and mineral gain (remineralization). A plethora of evidence has explored incorporating several bioactive compounds into resin-based materials to prevent bacterial biofilm attachment and the onset of the disease. In this review, the most recent advances in the design of remineralizing compounds and their functionalization to different resin-based materials' formulations were overviewed. Inorganic compounds, such as nano-sized amorphous calcium phosphate (NACP), calcium fluoride (CaF_2), bioactive glass (BAG), hydroxyapatite (HA), fluorapatite (FA), and boron nitride (BN), displayed promising results concerning remineralization, and direct and indirect impact on biofilm growth. The effects of these compounds varied based on these compounds' structure, the incorporated amount or percentage, and the intended clinical application. The remineralizing effects were presented as direct effects, such as an increase in the mineral content of the dental tissue, or indirect effects, such as an increase in the pH around the material. In some of the reported investigations, inorganic remineralizing compounds were combined with other bioactive agents, such as quaternary ammonium compounds (QACs), to maximize the remineralization outcomes and the antibacterial action against the cariogenic biofilms. The reviewed literature was mainly based on laboratory studies, highlighting the need to shift more toward testing the performance of these remineralizing compounds in clinical settings.

Keywords: biofilm; bioactive; dental; secondary caries; resin composite

check for updates

Citation: Bin-Jardan, L.I.; Almadani, D.I.; Almutairi, L.S.; Almoabid, H.A.; Alessa, M.A.; Almulhim, K.S.; AlSheikh, R.N.; Al-Dulaijan, Y.A.; Ibrahim, M.S.; Al-Zain, A.O.; et al. Inorganic Compounds as Remineralizing Fillers in Dental Restorative Materials: Narrative Review. *Int. J. Mol. Sci.* **2023**, *24*, 8295. https://doi.org/10.3390/ijms24098295

Academic Editor: Marta Fernández-García

Received: 11 April 2023
Revised: 2 May 2023
Accepted: 3 May 2023
Published: 5 May 2023

Copyright: © 2023 by the authors. Licensee MDPI, Basel, Switzerland. This article is an open access article distributed under the terms and conditions of the Creative Commons Attribution (CC BY) license (https://creativecommons.org/licenses/by/4.0/).

1. Introduction

Tooth decay, often known as dental caries, is a societal and pervasive disease that affects people of every age and every group of the population worldwide. It results from mineral loss from susceptible tooth structures due to the acidic challenge induced by the cariogenic microorganisms that are capable of fermenting dietary carbohydrates [1,2]. The current concepts also identify it as an infectious, non-contagious, dental biofilm-mediated, and dental plaque-dependent disease [1,2]. Dental biofilms contain acid-producing bacteria that interact with the residual food or fermentable carbohydrates to demineralize the tooth structure, causing dysbiosis in minerals comprising the tooth structure [3]. The affected imbalance of minerals, namely phosphate and calcium, causes mineral loss in teeth. Thus, the continuous mineral loss may create cavitations associated with a higher rate of clinical intervention, starting from simple fillings and the probability of turning into an endodontic treatment due to neglecting this disease at its initial stages [4,5].

Dental caries could be treated by different means to restore the lost tooth structure, using non-invasive operative approaches to arrest active non-cavitated carious lesions and invasive approaches by removing and replacing the defective tooth structure with dental restorations [6]. Despite the increased use of resin-based composite restorations as the most used restorative material, several inherent physical properties could influence the integrity of the restoration margins, including polymerization shrinkage, modulus of elasticity, solubility, and water sorption [7]. Due to the resultant stresses caused by polymerization shrinkage, these polymeric materials are susceptible to marginal integrity failure, staining, microleakage, biofilm accumulation, and secondary caries [8]. As a result, dental restorations must be checked periodically to monitor and intervene in any failure at its early stage. Individuals with dental restorations placed within 6–36 months could be classified as medium to high caries-risk patients, and frequent dental visits every 6 to 12 months are needed [9–11]. During these visits, patients are examined for the onset of new carious lesions. Furthermore, old restorations must be monitored for biological and mechanical failures. Mechanical and physical failure reveals less concern, as usually the restoration can be repaired conservatively [9–11]. Opposingly, biological failure, such as secondary caries at the tooth-restoration interface, presents more challenges, as removing the secondary lesions is usually associated with more tooth destruction [9–11].

2. Statement of the Problem

In restorative dentistry, secondary caries is a complication following the placement of restorations leading to its failure, especially in polymeric restorative materials [12]. It is a developing lesion at the margin of an existing restoration, a common area for plaque accumulation and biofilm development [12]. The process of developing secondary caries has the same notion of primary carious lesions on any sound tooth, starting with demineralization and imbalance of net minerals, and followed by enzymatic breakdown of the tooth structures. However, a restoration or sealant margin might modify this process and assist in further tooth destruction [13]. In previous studies, the prevalence of secondary caries after polymeric restorative materials placement was as high as 60%. It was identified as the leading cause of resin-based composite degradation and replacement [14]. Secondary caries' existence around the margins is characteristically represented in two regions: the surface lesion; which grows perpendicularly to the tooth's surface adjacent to a restoration; and wall lesion, which extends perpendicularly along the interface of tooth/restoration [15].

It is reported that the caries-related bacteria in human saliva that induce both primary and secondary caries are comparable, most frequently *Streptococcus mutans*, *Lactobacilli*, and *Actinomyces naeslundii* [16]. The complexity and cariogenicity of oral biofilms contribute to the difficulty in developing effective restorative materials to render secondary caries [17,18]. Nonetheless, changes surrounding the resin-based composite restoration allow for the passage of salivary proteins and liquids saturated with foreign particles, rendering restorations to degrade biologically [19]. Esterase in human saliva can aggressively degrade ester groups in resin-based composites, accumulating monomer byproducts that may encourage biofilm growth [20]. When materials deteriorate and release their integrated agents, there are tremendous impacts on their long-term mechanical and physical performance [14]. Furthermore, surface features, such as surface roughness, significantly affect the adhesion phase of dental biofilms [14,21].

Throughout the restoration lifespan, under certain factors at the tooth-restoration interface, cycles of two contrary mineral loss and gain processes occur in demineralization and remineralization [22–26]. Understanding the mechanism of these two processes is essential when developing advanced approaches to control the onset of primary and secondary caries. The main participants in these cycles are calcium (Ca^{2+}), phosphate

(PO_4^{3-}), and fluoride (F^-) ions. The imbalance between pathological and protective factors may lead to more mineral loss and progression of dental caries [22–26], considering that the tooth-restoration interface is a plaque-stagnation area. Demineralization is the process of mineral loss from the tooth surface that starts and usually grows in the presence of intrinsic and extrinsic modifiable risk factors [26]. Acidic attack is the leading cause of the chemical demineralization of teeth, which happens in two ways—dietary acids and microbial attack [26–29].

Similar to primary carious lesions, demineralization begins when the bacteria metabolize fermentable carbohydrates, producing organic acids that spread around nano-sized aqueous spaces between the hexagonal crystal of the tooth [25,30,31]. This process is followed by dissolving the Ca^{2+} and PO_4^{3-} ions into those spaces, which leads to a much more acid-soluble structure than pure hydroxyapatite, resulting in the substitution of phosphate ions for carbonate ions in the crystal lattice [25,30,31]. Such substitution can produce defects and calcium-deficient regions, which present clinically as white spots [25,32–34]. One of the most critical factors in that dynamic process is the oral pH. The pH required to initiate mineral loss in root dentin ranges between 6–6.8, compared to 5.4 in the enamel [3,35], indicating that dentin and cementum are more acid soluble than enamel due to their higher magnesium and carbonate contents [3,35].

3. Remineralizing Dental Materials as a Strategy to Prevent Secondary Caries

The remineralization or the repairing process can be achieved to overcome the challenge of demineralization, either by host-related factors or clinical intervention. Remineralization has been studied for many decades, in an attempt to understand the mechanism and develop technologies that could help reverse incipient caries and prevent tooth demineralization [36]. Fluoride (F) has long been known to be effective in preventing caries by reducing tooth dissolution and enhancing tooth remineralization processes [36]. Fluoride has been seen in studies to directly deposit fluorapatite (FA) or fluoridated hydroxyapatite (FHA) over the affected tooth surface or to promote the transformation of other calcium phosphate phases to FA or FHA. As a result, FA and FHA formation can reduce the solubility of enamel and dentin [36–38]. However, studies had shown that the most significant effect was obtained only when 50% of the hydroxyl groups were replaced with fluoride, corresponding to the greatest lattice stability and low lattice-free energy [36–38].

Ca^{2+} and PO_4^{3-} ions, bioactive glass (BAG), and boron nitride (BN) can also be utilized in the remineralization process by depositing a hydroxyapatite layer over the affected tooth surface [36,39,40]. In comparing Ca with F ions, in vitro, mechanistic studies have shown that calcium is approximately twenty times more potent than phosphate in inhibiting enamel dissolution [36,39,40]. Nowadays, nanotechnology is one of the most inventive concepts, which has shown significant success. Nanomaterials exhibit superior antimicrobial activity and comparable physical properties compared to conventional materials [41–45]. This is most likely due to the nanoparticles' small size and high surface area, which can release high levels of ions at a low filler level [44–46]. Therefore, remineralizing nano-fillers in dental restorative materials can be an effective strategy to prevent secondary caries at the tooth-restoration interface. The release of ions from the restorations can prevent oral biofilm attachment and favor the process of remineralization at a specific micro-level site [41,45].

Until now, most restorative polymeric materials have had no bioactivity, exaggerating the risk of secondary caries development around resin-based composite restorations, which are considered a major limitation to the current treatment approach. To address these concerns, researchers have concentrated their efforts on designing antibacterial features that can reduce bacterial attachment to prevent further accumulation of biofilms and hinder demineralization [2,14,47]. The potential of a dental material to influence its biological environment favorably offers a way to extend longevity and clinical performance inside the oral cavity [47]. The advantages of integrating bioactive chemicals in dental polymeric formulations, which are thought to be essential for effectively managing caries around restorations, are currently a hot-spot area in dentistry. This review discusses the applications and functionalization of different remineralization approaches and their uses in different restorative materials. This narrative review focused on including only remineralizing compounds in different restorative materials. Other bioactive compounds, such as quaternary ammonium, organic agents, metallic particles, and nanotubes, with antibacterial properties, were not included as they were reviewed in some of the authors' previous papers [14,48]. All articles in English with no specific time frame were extracted from PubMed and Scopus and included in the review.

4. Remineralizing Fillers in Restorative Dental Materials

Several studies investigated incorporating different bioactive and antibacterial compounds into the resin matrix system to limit the onset of secondary caries. Two main approaches have been heavily explored; (1) the incorporation of antibacterial compounds or particles, and (2) the use of a remineralization approach that can neutralize the acidity induced by the oral biofilms and interfere with biofilm growth [47].

In the second approach, which is the focus of this review, several compounds have been implemented in different restorative materials (Figure 1), which are Nano-sized Amorphous Calcium Phosphate (NACP), Calcium Fluoride (CaF_2), Bioactive Glass (BAG), Hydroxyapatite (HA), Fluorapatite (FA), and Boron Nitride (BN). While the primary effect of these compounds relies on their ability to neutralize the acidity induced by the attached cariogenic biofilms and favor the remineralization process, their indirect biofilm inhibition has been observed in several studies, showing dual benefits in modulating the oral biofilms.

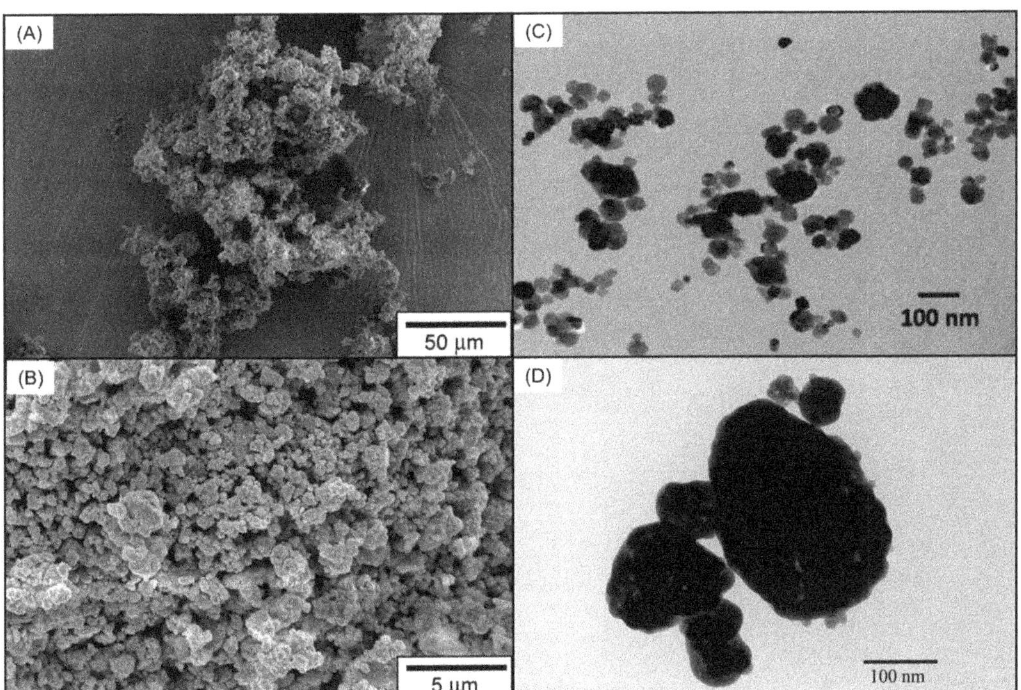

Figure 1. Different compounds that were incorporated in different resin-based materials to impart bioactivity and remineralize the surrounding dental tissues subjected to demineralization. (**A**,**B**) Scanning electron micrograph showing bioactive glass particles. Reprinted/adapted with permission from Ref. [49]. 2016, © Elsevier. Transmission electron microscope illustrating the size of (**C**) calcium fluoride (CaF_2) nanoparticles. Reprinted/adapted with permission from Ref. [50]. 2020, Mitwalli et al.

Another transmission electron microscope illustrating the size of (**D**) nano-sized amorphous calcium phosphate (NACP) fillers. Reprinted/adapted with permission from Ref. [51]. 2022, © Elsevier.

4.1. Nano-Sized Amorphous Calcium Phosphate (NACP) Fillers

The use of materials that release Ca^{2+} and $PO_4{}^{3-}$ ions is suggested to support the remineralization process. It is well known that Ca^{2+} and $PO_4{}^{3-}$ ions have multiple applications in dentistry since these two components form the inorganic portion of human teeth and bones [2]. The bioactivity of calcium phosphate (CaP) phases in dental materials has been examined. Different CaP compounds have been considered with a variation on the Ca/P molar ratio and the salt phase stability, including dicalcium phosphate dihydrate (DCPD), dicalcium phosphate anhydrous (DCPA), tetracalcium phosphate (TTCP), tricalcium phosphate (TCP) and amorphous calcium phosphate (ACP) [2,47,52–54]. Amorphous calcium phosphate (ACP) is the first phase that formed before reaching hydroxyapatite (HA), which is known to be the final thermodynamically stable product. ACP's lack of crystallinity structure, and it has high solubility, indicating that it contains a distinguished structure among CaP [2]. Nowadays, NACPs have been introduced with about a 100 nm particle size. These particles can neutralize the acidic environment caused by cariogenic bacteria and adjust the oral pH, subsequently promoting the remineralization process [55,56].

Historically, several resin-based materials containing calcium orthophosphate phases (DCPA, DCPD, ACP, and TCP) were investigated [57]. In enamel lesions, results showed that a resin-based composite with ACP-releasing ions at 40 wt.% released 0.74 mmol\L of calcium and 0.54 mmol/L of phosphate ions [57]. These released ions participated in remineralizing 14% of enamel lesions in a period of 30 days in comparison with another material containing commercial fluoride cement, which achieved 4% of remineralization in the same period [57]. Another study showed that in a period of 5 weeks, resin cement comprising DCPA and TTCP at 73–78 wt.% released 0.05–0.1 mmol/L of phosphate as well as 0.3–0.5 mmol/L of calcium, which were capable of remineralizing the demineralized dentin lesions by 38–47% [57]. The main drawback observed in these compounds was the low amount of ion release and the inferior mechanical properties [2]. Thus, nanotechnology via NACP was advanced enough to overcome these drawbacks and tailor different restorative materials with a high amount of ion release and excellent mechanical properties.

4.1.1. NACP in Resin-Based Composite Restorations

NACP synthesized via the spray-drying technique was heavily evaluated in multiple studies. Early investigations compared different weight percentages of NACP in resin-based composite formulations [56,58]. Resin-based composite, composed of bisphenol glycidyl dimethacrylate (BisGMA), triethylene glycol dimethacrylate (TEGDMA), and glass fillers, was modified to contain 10 to 40 wt.% of NACP [56]. The flexural strength value was reduced as the NACP concentration increased. Still, the values of the formulations were higher than 80 MPa, the minimum suggested by the International Organization for Standardization (ISO). It was found that the 10 wt.% NACP resin-based composite demonstrated reduced Ca^{2+} and $PO_4{}^{3-}$ ion release compared to the other formulations [56]. The same was observed in another study, where 10 and 15 wt.% NACP resin-based composite did not reveal a high ion release [58]. As a result, recent investigations designing NACP resin-based composites have focused on incorporating 20 wt.% or more of NACP into resin-based composite formulations.

In one study, using BisGMA and TEGDMA as a resin matrix system, 30 wt.% of NACP and 35 wt.% of glass fillers were mixed with and without different small fractions of silver nanoparticles [59]. It was found that combining NACP and silver nanoparticles resulted in a significant inhibition against multi-species biofilm formation, bacterial metabolic activities, and lactic acid production [59]. An in situ experiment was held in 2013 to test the NACP resin-based composite inside the oral cavity [60]. Resin-based composites containing 20 wt.% of NACP were used to fill cavities prepared in extracted bovine teeth and mounted in removable appliances. Participants placed the appliances inside their oral cavities for 14 weeks. Then, the biofilm formation and mineral loss of the enamel surface at the tooth restoration interface via transverse microradiography were assessed. Less biofilm formation, but not statistically significant, was observed over the NACP resin-based composite with a higher amount of Ca^{2+} and PO_4^{3-} ions in the biofilms. Regarding the mineral loss, the lesion depth around the NACP resin-based composite was significantly less than the control group [60]. Such findings may reveal the capabilities of NACP resin-based composites to hinder enamel demineralization at the tooth-restoration interface.

The ability of NACP resin-based composites to recharge the resin matrix with Ca^{2+} and PO_4^{3-} ions to allow frequent ion release is a primary concern. While it is worth saying that the ion release process will be initiated only with low pH at the risk of demineralization, having rechargeable NACP resin-based composites will be beneficial to assure long-term bioactivity [47], especially among high caries risk patients. The rechargeable resin matrix was composed of pyromellitic glycerol dimethacrylate (PMGDM) and ethoxylated bisphenol A dimethacrylate (EBPADMA) at a 1:1 ratio [61,62]. The amount of Ca^{2+} and PO_4^{3-} ion re-release was significantly higher among different cycles of recharges compared to the NACP resin-based composite that contained BisGMA and TEGDMA as a resin matrix [61]. When the NACP rechargeable resin-based composite was combined with DMAHDM, a significant biofilm reduction of 2- to 3-log was observed [62]. The synergetic antibacterial and remineralization action was also observed when NACP was mixed with a protein-repelling agent, named methacryloyloxyethyl phosphorylcholine (MPC) [52]. However, some of the main drawbacks of this rechargeable resin matrix are the reduced mechanical properties compared to other formulations [61,62], suggesting the need for further characterization to enforce the strength of this formulation.

In most of the earliest investigations, the synergistic effect of 20% NACP and DMAHDM was not reflected. However, increasing the DMAHDM concentration from 3 to 5 wt.% allowed the synergistic effect of these two bioactive compounds to be clearly observed. Increasing the DMAHDM concentration was associated with an increased surface charge density by around 2-fold [63], allowing the resin-based composite to interact more aggressively with the bacterial membrane. Adding the NACP to a resin-based composite containing 3 and 5% of DMAHDM resulted in a 1-log additional reduction compared to the formulations with no NACP. The overall biofilm decrease in total microorganisms, total *streptococci*, total *lactobacilli*, and *mutans streptococci* compared to the control was a reduction of around 2- to 5-log. Significant inhibition was also observed concerning these formulations' metabolic activities and lactic acid production [63]. This strong synergetic effect of 20% NACP and 3–5% DMAHDM was also potent after one year of water aging. The mechanical and antibacterial properties were slightly reduced but still sustained after aging [64].

This formulation with an increased DMAHDM concentration was investigated against highly cariogenic multi-species plaque-derived biofilms transferred from root carious lesions [65]. Resin-based composites containing 3 and 5% of DMAHDM without NACP were able to inhibit total microorganisms, total *streptococci*, total *lactobacilli*, and *mutans streptococci* by only 2-log compared to the control. When the NACP was added (Figure 2), significant inhibition of 4- to 6-log reduction was achieved, emphasizing the potency of this combination. The same trend was observed when the lactic acid production was quantified [65]. Incorporating DMAHDM alone inhibited lactic acid production by 36 to 58%. When NACP was combined with DMAHDM, a reduction in lactic acid production of more than a 90% was observed (Figure 3). In this study, it was found that the incorporation of neither DMAHDM nor NACP influenced the degree of conversion of the synthesized formulations [65].

Similar findings were observed when the NACP-DMAHDM resin-based composites were challenged using anaerobic biofilms isolated from deep periodontal pockets [66]. In this study, the DMAHDM-NACP resin-based composites substantially restrained the growth of anaerobic microorganisms by 3- to 5-log. Additionally, significant inhibition was seen when the metabolic activities and polysaccharide production were measured [66].

Such findings may suggest that these formulations can prevent the growth of periodontal pathogens around the margins of subgingival extending restorations.

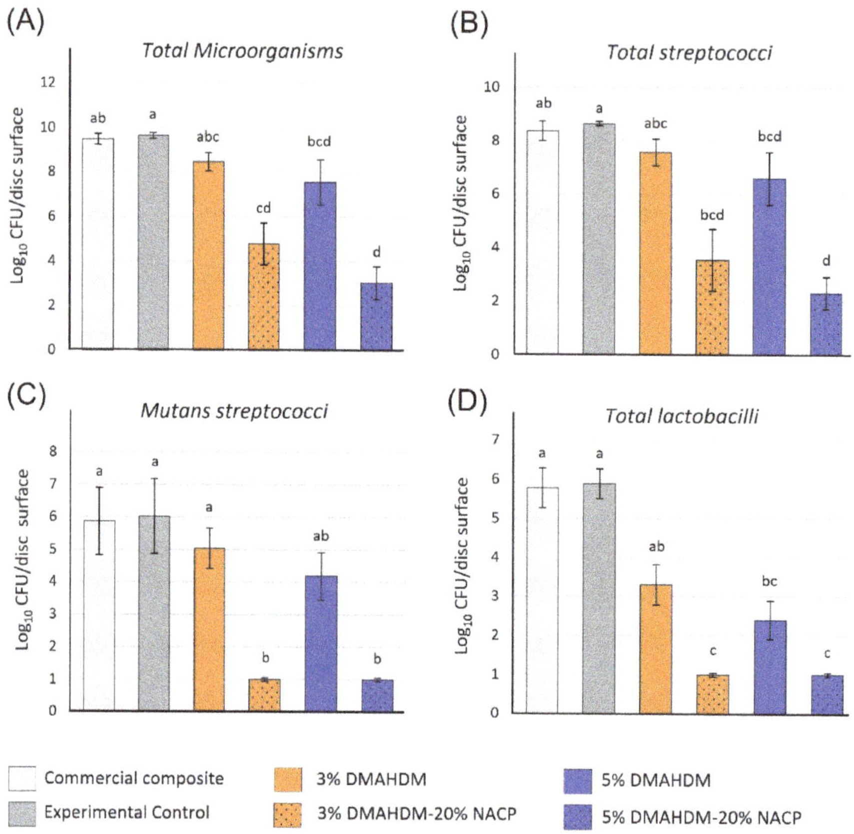

Figure 2. The biofilm inhibition of the NACP-DMAHDM against (**A**) Total microorganisms, (**B**) Total streptococci, (**C**) mutans streptococci, and (**D**) Total lactobacilli. More biofilm inhibition of 4- to 6-log reduction was observed when the DMAHDM was combined with the nano-sized amorphous calcium phosphate (NACP) fillers. Values indicated by different letters are statistically different from each other ($p < 0.05$). Reprinted/adapted with permission from Ref. [65]. 2020, Balhaddad et al.

Recent investigations evaluated the incorporation of NACP into a low-shrinkage-stress resin matrix consisting of urethane dimethacrylate (UDMA) and triethylene glycol divinylbenzyl ether (TEG-DVBE) [67,68]. This formulation was designed to minimize the stress induced by the resin shrinkage at the tooth-restoration interface. This formulation effectively inhibited the *S. mutans* biofilms without affecting the restoration's mechanical properties and the polymerization kinetics [67]. The high amount of Ca^{2+} and PO_4^{3-} ion release could significantly preserve the enamel microhardness after microbial demineralization compared to the control samples with no NACP [67]. This approach with a high concentration of DMAHDM up to 5% was capable of inducing potent antibacterial action against multi-species biofilm, resulting in a 2- to 5-log biofilm reduction [68]. The antibacterial activity was sustained after 20,000 cycles of thermocycling aging, equivalent to two years of clinical service [68]. These findings suggest that this combinatory approach of DMAHDM and NACP can preserve its bioactivity after aging.

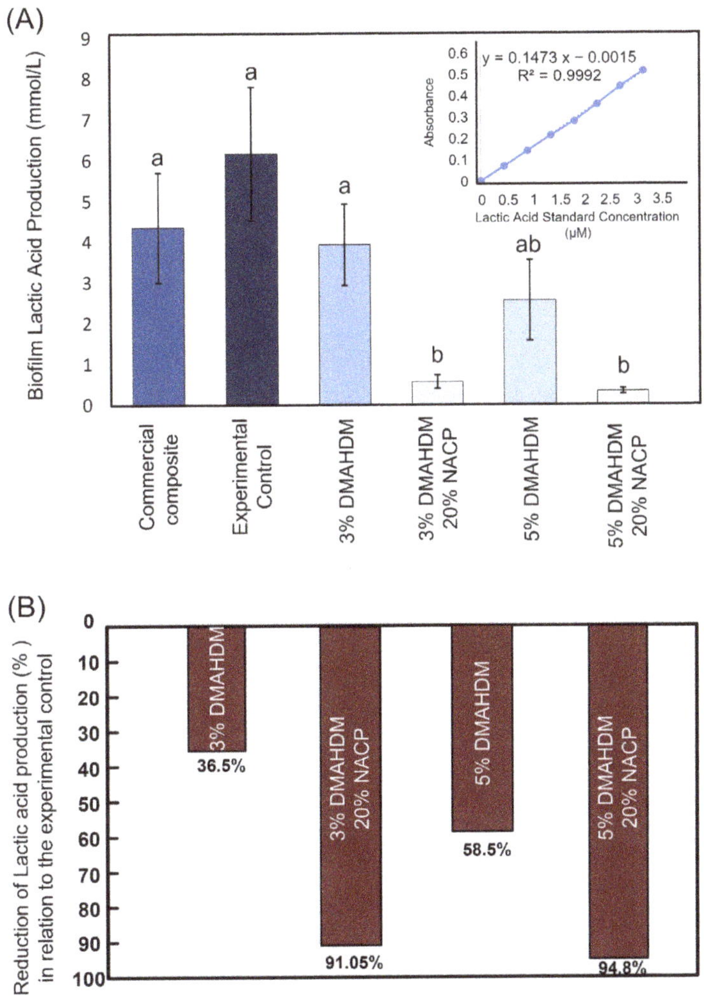

Figure 3. (**A**) The amount of lactic acid produced by multi-species cariogenic biofilms. Higher values indicate more lactic acid production. Values indicated by different letters are statistically different from each other ($p < 0.05$). (**B**) The percentage of lactic acid production inhibition shows the capabilities of nano-sized amorphous calcium phosphate (NACP) fillers to prevent demineralization and promote the remineralization process. Reprinted/adapted with permission from ref. [65]. 2020, Balhaddad et al.

4.1.2. NACP in Resin-Based Pit and Fissure Sealants

Dental sealants serve as a protective physical barrier against plaque accumulation in deep pits and fissures in the occlusal surface of teeth [17]. Pit and fissure sealants can therefore successfully prevent cavities and reduce the need for future restorations. The therapeutic bioactivity of resin-based pit and fissure sealants could be greatly enhanced and lead to caries prevention by using remineralizing agents [17]. Ibrahim et al. conducted a series of investigations to evaluate the mechanical and bioactive properties of dental sealants containing different mass fractions of NACP and 5 wt.% DMAHDM. It was found that sealants containing 20 wt.% of NACP and 5 wt.% of DMAHDM demonstrated massive Ca^{2+} and PO_4^{3-} ion release without compromising the mechanical properties of the sealant [69]. The unique resin matrix compositions, containing PMGDM, EBPADMA, BisGMA, and 2-hydroxyethyl methacrylate (HEMA), permitted frequent cycles of ion recharge and release, allowing long-term bioactivity of the formulation [69].

The exact formulation of 20 wt.% NACP and 5% DMAHDM was challenged with *S. mutans* biofilms [70]. Only when combined with DMAHDM did NACP-containing sealants reduced the *S. mutans* biofilm by around 4-log. The cariogenic biofilm's metabolic activities, lactic acid, and polysaccharide production were also significantly reduced [70]. The same formulation was also found effective in eradicating *Candida albicans* growth and activities [71], one of the possible contributing microorganisms in the pathogenesis of early childhood caries. Qualitative analyses via scanning electron microscopy/energy-dispersive X-ray spectroscopy (SEM-EDX) revealed that enamel surfaces restored with NACP-containing sealant revealed higher microhardness and presented more elevated Ca^{2+} and PO_4^{3-} ions following chemical demineralization (Figure 4). Furthermore, polarized light microscopy (PLM) images showed less demineralized surface area around the enamel restored with NACP-containing sealant than the control (Figure 5) [72]. In a more challenging condition, saliva-derived biofilms secluded from high-caries risk pediatric patients were grown over the sealants [73]. Using sealants containing NACP and DMAHDM, biofilm development, lactic acid production, and metabolic activities were all significantly reduced [73].

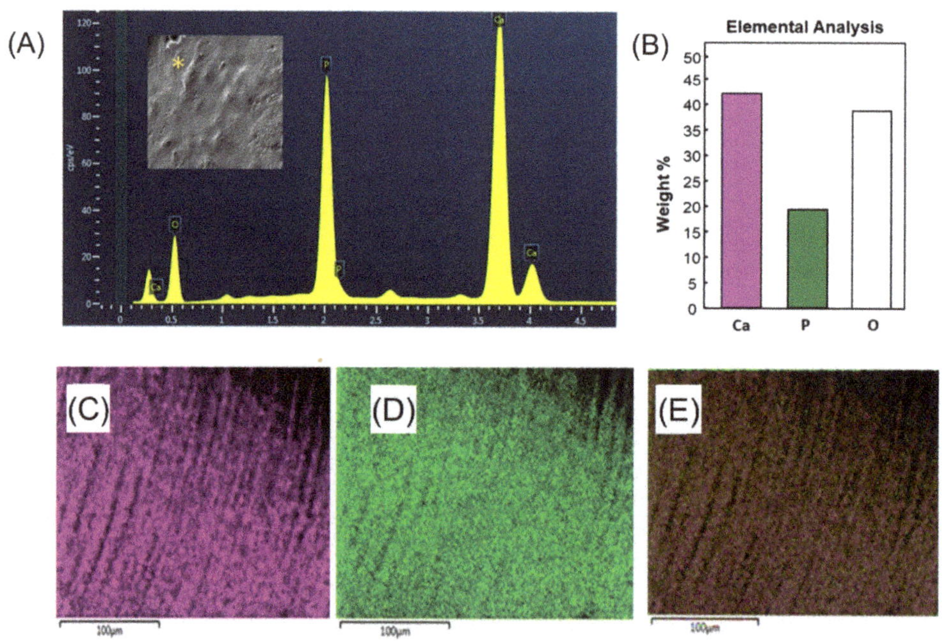

Figure 4. Scanning electron microscopy/energy-dispersive X-ray spectroscopy (SEM-EDX) illustrating the mineral contents of enamel restored with pit and fissure sealant containing nano-sized amorphous calcium phosphate (NACP) fillers. (**A**) SEM-EDX spectrum highlighting the mineral contents of the enamel surface restored with NACP-containing sealant. (**B**) Percentage of elemental concentration in weight of calcium, phosphate, and oxygen within the enamel restored with NACP-containing sealant. EDX mapping of elemental (**C**) calcium and (**D**) phosphate. (**E**) EDX mapping shows the overlay of C and D images, where the calcium is indicated in the pink color, and the phosphate is displayed in the green color. Reprinted/adapted with permission from Ref. [72], 2020, © Elsevier.

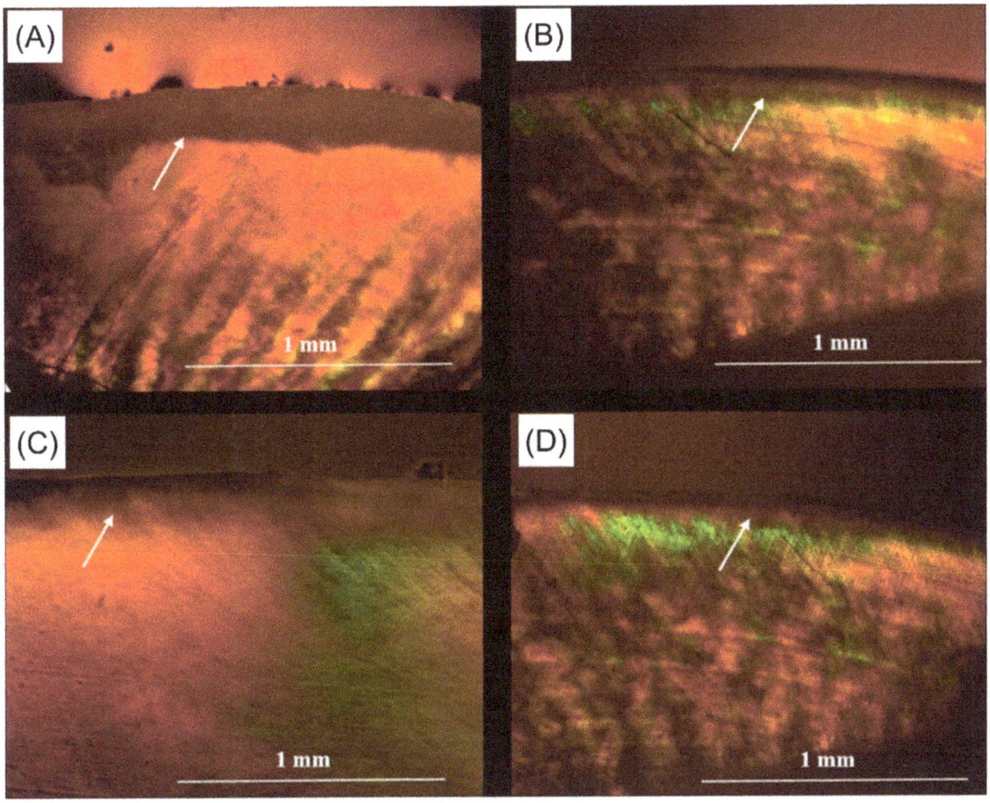

Figure 5. Polarized light photomicrograph showing dark bands (white arrows) at the enamel surface restored with parental resin-based sealant (**A**,**B**) and NAC-containing resin-based sealants (**C**,**D**), which represent the demineralized areas. It can be observed that enamel restored with the NAC-containing resin-based sealants is associated with a narrow dark band, compared to the wide one in the control group, suggesting the capabilities of these bioactive resin-based sealants to resist demineralization and mineral loss. Reprinted/adapted with permission from Ref. [72], 2020, © Elsevier.

4.1.3. NACP in Dental Adhesives

Resin-based adhesives with remineralizing properties decrease the risk of white spots and secondary caries around orthodontic brackets and dental restorations. NACP as adhesive fillers is a promising approach to limit such consequences by releasing significant amounts of Ca^{2+} and $PO_4{}^{3-}$ ions [74]. A study found that a 30% NACP resin-based composite prevented an acid attack and increased the pH of the solution from a pH of 4 which is cariogenic, to a safe pH of 6.5, most likely due to NACP alkaline nature [74]. The bonding agent is predicted to benefit from the 30% NACP inclusion in the form of antibacterial, acid neutralization, and remineralizing properties [74]. When conjugated with silver nanoparticles, NACP-containing adhesive reduced the biofilm growth, metabolic activities, and lactic acid production of multi-species cariogenic biofilms by more than 50% [74]. In another investigation, NACP-containing adhesives with and without DMAHDM preserved the bonding strength after 12 months of aging, greater than what was observed in the control groups [75]. The DMAHDM was inserted in the dental adhesive to convey antibacterial properties and the remineralization capabilities induced by the NACP fillers [75].

Using a rechargeable resin matrix in dental adhesives was also attempted [76,77]. When combined with MPC or DMAHDM, rechargeable NACP dental adhesives were found to release high amounts of Ca^{2+} and PO_4^{3-} ions in low pH and inhibit the growth and activities of highly cariogenic biofilms [76,77]. It is worth saying that reachability is less critical among dental adhesives than resin-based restorations or sealants, as the adhesive layer is usually sealed and protected from the external environment. In a recent interesting investigation, the bonding strength of an NACP-containing adhesive was significantly improved by imparting magnetic particles and using a magnetic field during the bonding procedure [78]. The bonding strength was increased by around 30–40% compared to the control. When DMAHDM as an antibacterial monomer was added, the designed adhesive demonstrated high bonding strength with excellent antibacterial and remineralization properties [78]. These findings may encourage using a combinatory approach to creating highly bioactive resin-based materials with several desirable properties.

Moreover, a study compared the effect of an NACP-containing adhesive with a control group in a challenging condition with an *S. mutans* biofilm [79]. Results showed that the control group achieved minimal remineralization, while the NACP-containing adhesive achieved a high rate of remineralization via the massive release of Ca^{2+} and PO_4^{3-} ions. Furthermore, the NACP-containing adhesive reduced the lactic acid production and minimized the biofilm growth of the *S. mutans* biofilm [79]. The performance of the NACP dental adhesive was also investigated using dentin as a bonding substrate [80].

The growth of white spot lesions surrounding orthodontic brackets is one of the biggest obstacles during orthodontic treatment [81]. The onset of these lesions compromises the mechanical properties and the esthetic appearance of teeth [81]. The incorporation of 40 wt.% of NACP into an orthodontic adhesive was attempted [82]. When this adhesive was applied to attach orthodontic brackets to premolars, the bonding strength was comparable to the control. Furthermore, when an antibacterial monomer, named 2-methacryloxylethyl dodecyl methyl ammonium bromide (MAE-DB), was added to the NACP-adhesive, the orthodontic adhesive reduced the growth of *S. mutans* biofilms and preserved the enamel microhardness [82]. Such an approach could be beneficial to minimize the onset of white spot lesions around orthodontic brackets.

4.1.4. NACP in Resin-Based Dental Cements

Resin-based dental cements are used to bond or lute indirect or fixed restorations, such as crowns and bridges, to the tooth structure. Secondary caries around fixed restorations is a clinical concern, especially when these restorations are close to the gingival margins [83]. The onset of secondary caries affects the clinical longevity of the placed restorations, leading to the restorations' replacement and sometimes tooth extraction [84]. Therefore, imparting bioactive restorative materials into dental cements may minimize the biological failure of fixed restorations. In one study, a resin-based dental cement was designed to contain 25 wt.% of NACP and different mass fractions of DMAHDM, ranging between 3–5 wt.% [85]. The bonding strength, flexural strength, elastic modulus, and film thickness of the designed formulations were comparable to the control. Additionally, a high amount of Ca^{2+} and PO_4^{3-} ion release was observed. When the formulations were challenged with *S. mutans* biofilms, a reduction in 3-log was noted when the NACP was combined with 5% DMAHDM [85].

The rechargeability of bioactive resin-based cement was achieved in another investigation, where the resin matrix was composed of PMGDM and EBPADMA [86]. Several cycles of recharge and re-release were achieved with an excellent amount of Ca^{2+} and PO_4^{3-} ion release. This rechargeable formulation also effectively inhibited the growth and activities of *S. mutans* biofilms [86]. These findings may embark on new avenues to minimize the failure of indirect restorations due to secondary caries.

4.2. Calcium Fluoride (CaF_2) Fillers

Fluoride-releasing dental materials have been frequently studied and used in dentistry due to their working mechanisms, which significantly affect the progression of dental caries [87–89]. Caries prevention is achieved through the adsorption of fluoride ions, which occurs on HA crystals' surfaces, which in turn, prevent crystals dissolution in an acidic cariogenic medium [87,90]. Fluoride ions fight dental caries through acid resistance, fluorapatite formation, inhibition of bacterial growth in the oral cavity, remineralization process promotion, and demineralization process inhibition [87,91–93]. The addition of fluoride to resin-based materials can promote the prevention of secondary caries formation; it presents in different addition forms, such as organic fluoride, inorganic salts, and leachable glasses. Sodium fluoride (NaF) and Tin(II) fluoride (SnF_2) have been used as water-soluble salts, and more recently, CaF_2 particles have been used [87,91–94]. Due to their functions as labile reservoirs for the calcium (Ca^{2+}) and fluoride (F^-) ions and their ability to enhance the remineralization effects of the F regimen without increasing the F level, CaF_2 particles are of great interest in the prevention of dental caries.

4.2.1. CaF_2 in Resin-Based Composite Restorations

Incorporating calcium fluoride (CaF_2) particles into a resin-based composite was attempted in several investigations. CaF_2 particles at the load of 30% were associated with a high amount of Ca^{2+} and F^- ion release without affecting the material's mechanical properties [95]. In another investigation, 15 wt.% of CaF_2 nanoparticles were incorporated into a resin-based composite system containing BisGMA and TEGDMA as a resin matrix and glass as co-fillers [50]. This formulation was achieved with and without the addition of DMAHDM and MPC. When CaF_2 nanoparticles were incorporated into the parental formulation, good mechanical properties were achieved. However, adding either DMAHDM or MPC to the formulation significantly reduced the mechanical properties, but the values were comparable to the commercial control. All the designed formulations were associated with a high amount of F^- and Ca^{2+} ion release, showing the remineralization potential of these formulations. When the antibacterial properties were assessed, CaF_2 nanoparticles were associated with minor antibacterial properties, but when the DMAHDM was added, significant antibacterial performance was observed (Figure 6). The bioactive formulation containing CaF_2 nanoparticles and DMAHDM reduced the multi-species biofilm by around 4-log, illustrating this formulation's potent antibacterial and remineralization capabilities [50].

One study functionalized CaF_2 nanoparticles into a rechargeable resin-based composite matrix to allow the long-term release of ions [96]. The resin matrix was composed of PMGDM and EBPADMA. This formulation allowed multiple cycles of ion release and recharge, demonstrating that long-term ion release is possible. The only concerns related to this formulation are the reduced mechanical properties compared to the other resin matrix formulations, indicating the need for further characterization to engineer a mechanically sustained rechargeable formulation [96]. The synergetic effect of CaF_2 nanoparticles and chlorhexidine was investigated in one study, where a 3-log inhibition against *S. mutans* biofilms was observed [97] and the metabolic activities and lactic acid production were considerably diminished [97]. Current findings suggest that CaF_2 nanoparticles may contribute to remineralizing the tooth structure and reduce the amount of lactic acid production induced by the cariogenic species. The significant inhibition related to the biofilm colony-forming units (CFUs) or metabolic activities was only observed when the CaF_2 nanoparticles were combined with another bioactive agent. Such findings suggest using several bioactive materials to maximize the protection of restored teeth.

4.2.2. CaF_2 in Resin-Based Pit and Fissure Sealants

For many years in preventive dentistry, dental practitioners have used glass ionomer as a fluoride-releasing sealant to occlude food stagnation areas and release F^- ions to the sealed tooth structure [98,99]. The main limitations of glass ionomer sealants were related to the low mechanical performance and the low retention rate. Therefore, highly flowable resin-based sealants were also used as an alternative material [98,99]. One of the main limitations of resin-based sealants is the lack of bioactivity and the development of carious lesions around the sealants. As a result, imparting bioactive compounds to resin-based sealants is a promising strategy to overcome this limitation.

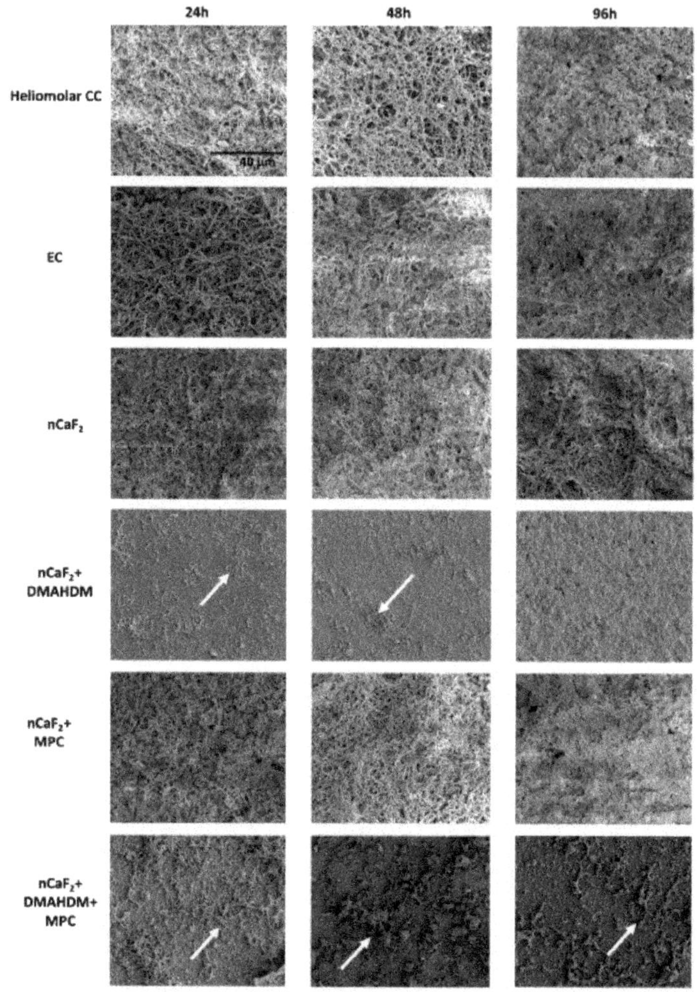

Figure 6. Scanning electron microscope images showing the growth of cariogenic multi-species saliva-derived biofilms over different bioactive resin-based composites. White arrows are showing the biofilm colonies over the resin-based composites. Combining calcium fluoride nanoparticles and DMAHDM was associated with the least biofilm formation. Reprinted/adapted with permission from Ref. [50]. 2020, Mitwalli et al.

One study attempted to incorporate CaF_2 nanoparticles into a resin-based sealant at a load of 20 wt.% [100]. The resin matrix was composed of BisGMA and TEGDMA at the ratio of 1:1. The fillers were composed of barium boro-aluminosilicate glass particles, which were silanized with 4% 3-methacryloxypropyltrimethoxysilane, 2% n-propylamine, and CaF_2 nanoparticles. DMAHDM was added to the formulation at 5 wt.%. The overall matrix-to-filler ratio was 1:1. Adding CaF_2 nanoparticles and DMAHDM did not compromise the design formulations' flowability and enamel shear bond strength. A significant F^- ion release was observed, and the amount of ion release was slightly reduced when DMAHDM was added to the formulation. Combining the CaF_2 nanoparticles and DMAHDM significantly reduced the CFUs, metabolic activities, and lactic acid production of the *S. mutans* biofilms [100]. Such formulation may minimize the onset of caries around sealants and assure good mechanical and bonding properties during the clinical service inside the oral cavity.

4.2.3. CaF_2 in Dental Adhesives

A few investigations discussed the design of CaF_2-dental adhesives. The addition of CaF_2 fillers allowed dental adhesives to release a high amount of F^- ions without compromising the bonding strength of the material [101]. In another investigation, a dental adhesive containing zinc, calcium, fluoride, and bioglass compounds was engineered [102]. This bioactive adhesive demonstrated good bonding properties and polymerization kinetics with a high amount of ion release, revealing the high potential to remineralize tooth structure subjected to demineralization attacks. The zinc-calcium-fluoride-bioglass adhesive inhibited the *S. mutans* biofilms [102].

Incorporating CaF_2 nanoparticles in an orthodontic adhesive as a remineralizing agent to prevent the onset of white spot lesions was attempted [103]. This incorporation was achieved in two different resin matrix systems. The first was composed of HEMA and BisGMA. In contrast, the other was composed of PMGDM and EBPADMA, both at the ratio of 1:1. The two dental adhesives released a high amount of F^- ions with a high amount of re-release upon recharge compared to a resin-modified glass ionomer (RMGI). All the formulations did not induce high cytotoxic effects against human gingival fibroblasts. The only concern is the reduced enamel bonding strength compared to the control, which may necessitate further characterization in future investigations [103].

4.2.4. CaF_2 in Resin-Based Dental Cements

The incorporation of CaF_2 nanoparticles into fixed prostheses cement was achieved in one study [104]. The resin matrix was composed of PMGDM and EBPADMA, both at a ratio of 1:1 to allow rechargeability features. CaF_2 nanoparticles were incorporated either alone at the load of 25 wt.%, or combined with NACP fillers, 12.5 wt.% each. The dentin shear bond strength in both formulations was higher than the control. The film thickness, flexural strength, and elastic modulus values were within the normal range. The amount of calcium and fluoride release and phosphate in the formulation containing NACP was very high and was sustained for up to 70 days. Both formulations could re-release the same amount of ions following three recharge cycles (Figure 7) [104]. Such cements may provide an advanced approach to control demineralization around fixed prostheses.

4.3. *Bioactive Glass (BAG) Fillers*

Bioactive glass (BAG) has been proven to have antibacterial properties against oral bacteria and the potential to remineralize oral hard tissues [49]. Depending on the ratio of calcium oxide to phosphorus pentoxide, the bioactivity of BAG can be controlled [105,106]. The antibacterial activity of BAG is ascribed to the part that releases ions (e.g., calcium and phosphate), which are poisonous to bacteria and induce the neutralization of the surrounding acidic environment [49]. Although the first BAG was produced over 40 years ago, research into its possible use in resin-based composites has recently begun. Thus, the precise mechanism of the BAG antimicrobial effect could have been illustrated better [49]. However, the high precipitation of Ca^{2+} and PO_4^{3-} ions may induce a neutralization effect and indirectly interfere with the growth of cariogenic biofilms [107].

4.3.1. BAG in Resin-Based Composite Restorations

In several studies, BAG was added to resin-based composite restorations to induce acid neutralization [17]. It has been shown that a resin-based composite containing BAG can fulfill the mechanical property demands needed for dental restorations [105,106]. In addition, BAG can reduce fluid movements in the tubules, thereby reducing dentinal hypersensitivity [105,106]. In one study, resin-based composites containing BAG with and without fluoride have been advocated to reduce the degradation of dentin [108]. After 30 days of storage in artificial saliva, the two resin-based composite systems reduced the solubility of C-terminal cross-linked telopeptide (ICTP) and C-terminal telopeptide (CTX), preventing dentin degradation and demineralization. Such observation may reveal BAG's capabilities to inhibit the activation of matrix metalloproteinases (MMPs). The remineralization effect was seen following the 30 days of storage. High participation of calcium and phosphate minerals was observed over the teeth restored with the BAG-containing resin-based composite. The amount of remineralization was greater in the group containing both BAG and fluoride [108].

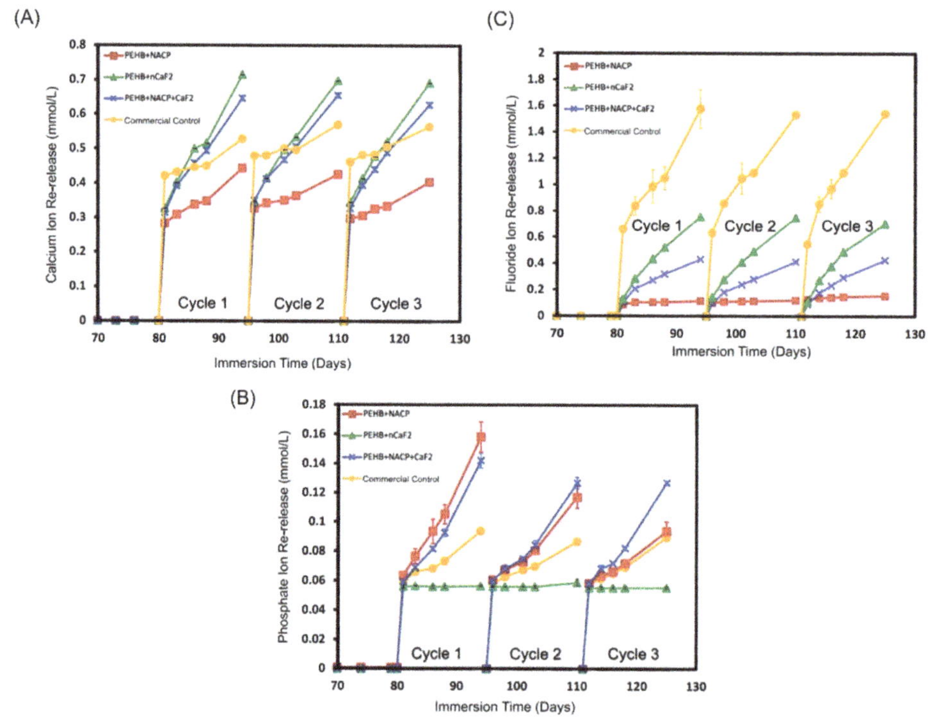

Figure 7. Resin-based dental cement containing 25 wt.% of calcium fluoride (CaF_2) or 12.5 wt.% of CaF_2 and 12.5 wt.% of nano-sized amorphous calcium phosphate (NACP) fillers, releasing high amounts of (**A**) calcium, (**B**) phosphate, and (**C**) fluoride ions. After 80 days of continuous release, the remineralizing resin-based cements were recharged for three consecutive cycles, indicating their capabilities of long-term clinical service inside the oral cavity. Reprinted/adapted with permission from Ref. [104]. 2022, © Elsevier.

BAG as a filler can contribute to the inhibition of microbial activity. In one investigation, two resin-based composites, BAG-free resin-based composite as a control and resin-based composite containing 15 wt.% of BAG, were tested to see their effectiveness on gap formation and depth of the bacterial penetration relying on optical micrograph. For the control, the bacterial penetration was very deep, reaching the bottom of the cavity prepared with a high number of microorganisms [49], while the 15 wt.% BAG resin-based composites reduced the bacterial penetration by an average of 40%. (Figure 8) This implies that the release of ions from BAG can inhibit the biofilm's growth and spread by regulating the gap environment [49].

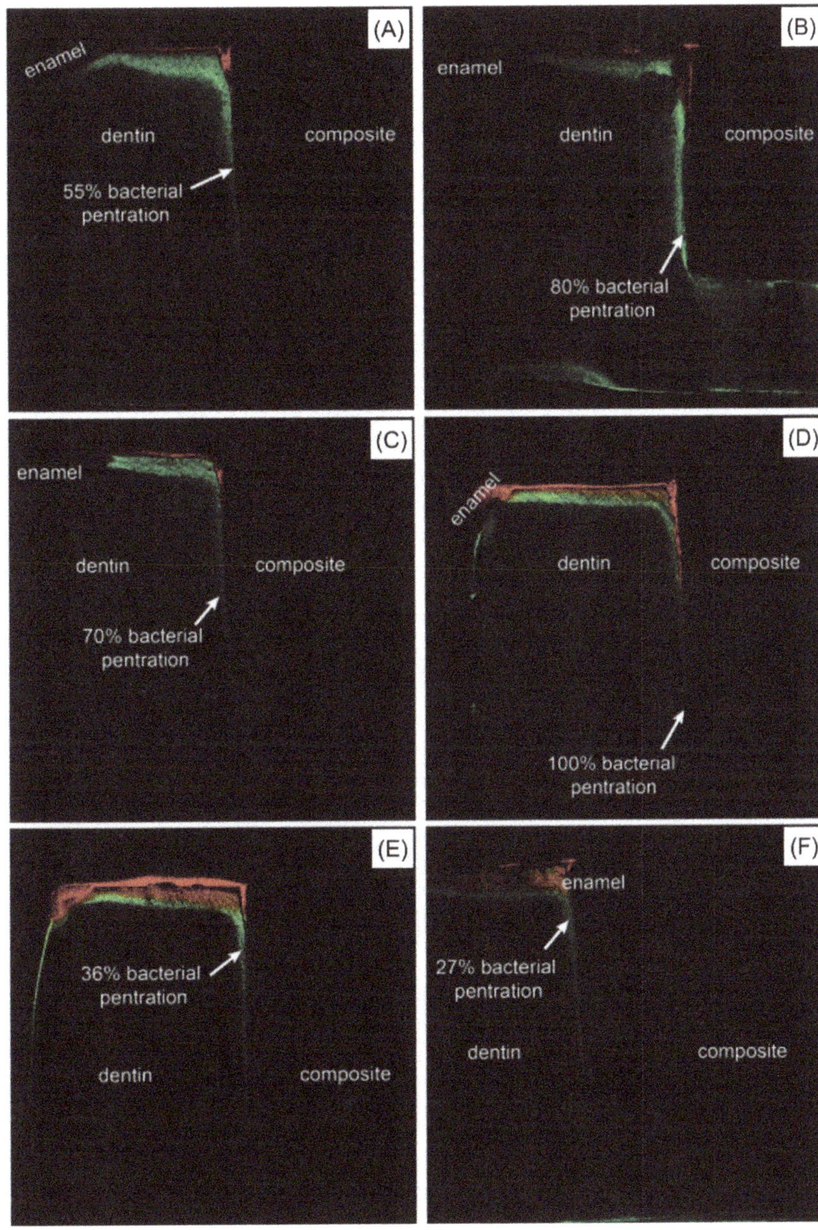

Figure 8. Different fluorescence images of six different samples (**A–F**) showing the capabilities of a resin-based composite containing 15 wt.% of bioactive glass (BAG) to prevent bacterial penetration (red color) and dentin demineralization (green color) by an average of 40%, compared to the control that allowed 100% bacterial penetration. Reprinted/adapted with permission from Ref. [49]. 2016, © Elsevier.

Resin-based composite containing BAG inhibited the growth of *S. mutans* biofilms. BAG was incorporated into resin-based composites at 5, 10, and 30 wt.%., where the resin matrix was composed of BisGMA and TEGDMA at the ratio of 70:30 [109]. It was found that resin-based composites containing 10 and 30 wt.% of BAD reduced the *S. mutans* by 1- and 2-log, respectively. However, loading at 30 wt.% significantly reduced compressive and flexural strength [109]. In another investigation, a synergetic effect was observed when BAG and magnesium oxide nanoparticles were combined in a resin-based composite formulation [110]. This combination could reduce the growth of *S. mutans* biofilms. However, as in previous studies, increasing the BAG concentration negatively affected the material's mechanical properties [110].

4.3.2. BAG in Resin-Based Pit and Fissure Sealants and Dental Adhesives

Pits and fissure sealants containing BAG had high marginal adaptability and retention and a high preventive effect, which decreases bacterial infiltration in plaque stagnation areas. In one study, BAG was incorporated into resin-based sealants in different mass fractions ranging between 12.5 and 50 wt.% [111]. While increasing BAG concentrations were associated with an increased neutralization effect, the mechanical properties were significantly reduced, indicating that BAG, as most bioactive fillers, can only be incorporated in small mass fractions [111]. It was found that enamel surfaces restored with sealants containing BAG were more resistant to demineralization challenges [112].

Orthodontic adhesives might also be reinforced by BAG, which is proven to prevent the demineralization process from taking place in the oral cavity [113]. In one study, a commercial orthodontic adhesive was modified to contain 1, 3, and 5 wt.% of BAG. As the concentration of BAG was increased in the orthodontic adhesives, a greater anti-demineralization action was seen without affecting the adhesive's biocompatibility and bonding strength. This is because the ion-buffering impact of BAGs-released ions prevents a drop in intraoral pH, which is responsible for the chemical anti-demineralization effect [113]. Another investigation illustrated that antibacterial and remineralization effects were observed when 10 to 15 wt.% of BAG was incorporated into an orthodontic adhesive [114]. It was found that around 200–300 μm of the tooth structure around BAG-containing orthodontic adhesives did not undergo demineralization, as the mineral deposition from the BAG particles preserved the tooth structure. Higher microhardness of teeth was observed when BAG-containing adhesives were used, and BAG's incorporation did not affect the formulated adhesives' biocompatibility and bonding strength [114].

Several reports found that orthodontic adhesives containing BAG may prevent tooth demineralization around orthodontic brackets [115,116]. In one study, BAG nanoparticles were doped with gallium and incorporated into an orthodontic adhesive of 1, 3, and 5 wt.% [117]. In a dose-dependent manner, higher Ca^{2+} and $PO_4{}^{3-}$ ion release and greater antibacterial properties were observed as the concentration of BAG-gallium increased [117]. Another investigation illustrated that 4-methacryloxyethyl trimellitic anhydride/methyl methacrylate-tri-n-butyl borane (4-META/MMA-TBB)-based resin containing various amounts (0–50%) of BAG as an orthodontic adhesive was capable of releasing high amounts of Ca^{2+} and $PO_4{}^{3-}$ ions, and preserving enamel hardness following chemical demineralization [118]. This was accomplished without losing the intended adhesive's shear bond strength [118].

4.4. *Hydroxyapatite (HA) and Fluorapatite (FA)*

Hydroxyapatite ($Ca_{10}(PO_4)_6OH)_2$ has been utilized extensively in biomedical and dental applications because of its similarities to the primary mineral components of hard tissues in the human body, such as bone, dental enamel, and dentin, as well as its biocompatibility, bioactivity, and low solubility in moist conditions [119]. The substation of the hydroxyl group in enamel apatite could also happen with F^- ions, forming fluorapatite (FA) or Fluorohydroxyapatite (FHA). Fluoride is known to have superior acidic substance resistance, lower solubility, and greater hardness. Meanwhile, biocompatibility between FA and HP remained similar [120].

Early investigations incorporating HA and FA into restorative materials aimed to improve the mechanical properties of the designed materials. It was found that glass ionomer cements containing HA and FA particles were associated with higher fracture toughness and flexural and tensile strength [121,122]. When incorporated into resin-based composite formulation, HA could improve the flexural and compressive strength, with the highest strength observed at 20 wt.% loading [123]. Recently, the incorporation of HA and FA into resin-based composites was attempted as a strategy to impart bioactivity in these polymeric materials. In one study, HA nanowires were synthesized via a hydrothermal technique and soaked in dopamine [124]. Then, dopamine-coated HA was combined with silver nanoparticles and incorporated into a resin matrix composed of BisGMA and TEGDMA. Deposition of calcium and silver elements was observed over the material's surface, which contributed to the inhibition of *S. mutans* biofilms. Increasing the dopamine-coated HA-silver concentration was associated with more biofilm reduction and greater strength [124]. Doping HA with zinc-strontium [125] or titanium dioxide [126] was also attempted as a strategy to improve the bioactivity of resin-based composite restorations.

HA was incorporated into pit and fissure sealants as a remineralization strategy [127]. Incorporating 10 or 30 wt.% of HA did not dramatically affect the bond strength, depth of cure, and degree of conversion of the formulated sealants. A high release of Ca^{2+} and PO_4^{3-} ions was observed, and a scanning electron microscope visualized more remineralized areas over enamel surfaces adjacent to HA sealants. Orthodontic adhesives can also be treated with 5% HA nanoparticles to limit bacterial development and reduce the growth of cariogenic bacteria [128]. Another study indicates that when incorporating nano-FA or nano-FHA in a resin-based orthodontic adhesive, the fluoride release properties were increased, quadrupling the amount of fluoride after 70 days when compared to the control group [129].

4.5. Boron Nitride (BN)

Boron nitride (BN) has several medical applications. Due to its high chemical stability, BN has been used as an alternative to graphene and its derivatives [130]. BN also has excellent biocompatibility and functionality, giving a wide variety in pharmaceutical drug design [131]. Initial investigations utilizing BN in dentistry aimed to improve the mechanical properties of the designed material [132,133]. Its capabilities to deposit minerals have led many researchers to incorporate it for bioactivity purposes. In one study, BN nanotubes were incorporated into a dental adhesive at different mass fractions ranging between 0.05 to 0.15 wt.% [134]. The resin matrix was composed of BisGMA and HEMA in mass ratios of 66.6 and 33.3 wt.%, respectively. As BN nanotube concentration increased, greater microhardness and ultimate strength values were observed without affecting the degree of conversion. Micro-Raman spectroscopy and scanning electron microscopy images showed mineral deposition over the adhesive surfaces, indicating the capabilities of these nanotubes to remineralize the surrounding hard dental tissues [134].

BN nanosheets were modified with zinc oxide nanoparticles and incorporated into a resin-based composite formulation in another study [135]. At the load of 0.5 wt.%, BN nanosheets-zinc oxide NPs significantly reduced the *S. mutans* biofilms without affecting the material's mechanical properties [135]. Similar results were observed when BN nanotubes were incorporated into a resin-based sealant [136]. The resin matrix comprised 90 wt.% of TEGDMA and 10 wt.% of BisGMA. Then, 0.1 and 0.2 wt.% of BN nanotubes were added and subjected to different mechanical and physical assessments. The addition of BN nanotubes did not affect the biocompatibility and polymerization kinetics of the sealants and the ultimate tensile strength. When immersed in artificial saliva, sealants containing BN nanotubes demonstrated mineral deposition in a dose-dependent manner (Figure 9) [136]. BN as a potential remineralizing agent is a new avenue in restorative dentistry. Therefore, more research papers may conduct further investigations to evaluate the different applications of this material.

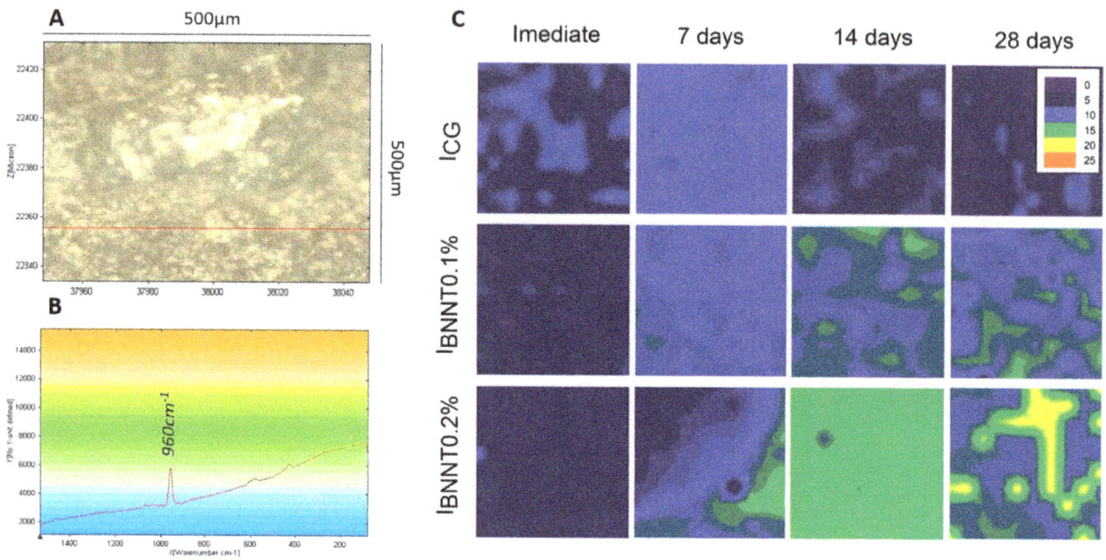

Figure 9. Evaluation of the mineral deposition using Raman analysis after immersion in artificial saliva for different time points. (**A**) Resin-based sealant showing the scanned surface (500 µm × 500 µm). (**B**) Phosphate ion (PO_4^{3-}) peak at 960 cm^{-1}. (**C**) As in the legends, color changes from blue to orange indicate more phosphate deposition. More phosphate deposition was observed as more boron nitride nanotubes were incorporated. Reprinted/adapted with permission from Ref. [136]. 2019, Bohns et al.

5. Future Perspectives

Several studies discussed ion-releasing materials as a strategy to prevent the onset of secondary caries around dental restorations. While the available data is promising, further investigation will be beneficial to overcome the drawbacks of the weak material's mechanical and physical properties and tailor different restorative materials according to the intended applications. The focus of the previous reports was to investigate the bioactivity of such formulations. This led these reports to conduct a basic mechanical evaluation of the designed formulations. It is important to realize that materials with high bioactivity would fail mechanically due to stress-induced fractures if the mechanical properties were poor [14,47,48]. Properties such as bonding strength, microhardness, compressive strength, water sorption and solubility, and color characteristics were ignored in some investigations. Structuring a comprehensive mechanical and physical evaluation in future studies will be essential to obtain more valuable information concerning the performance of such formulations [14,47,48].

It is very important to subject materials' mechanical and physical properties for long-term evaluation. Dental restorative materials are subjected to cyclic load and fatigue inside the oral cavity due to the force of mastication and frequent exposure to oral fluids and consumable beverages [14,47,48]. As a result, restorative materials may show surface and body degradation over time, affecting the designed formulation's integrity. The same can be applied concerning the materials' bioactivity, as this feature might decay following aging [64]. Therefore, newly designed remineralizing formulations must undergo comprehensive evaluation at immediate testing and after actual or artificial aging.

One of the main drawbacks in the reported investigations is that the listed materials' polymerization properties are not yet fully understood, such as the material cross-link density, degree of conversion, and depth of cure. Optimum polymerization and cross-linking are essential to ensure suitable the materials' mechanical and physical properties [137].

Several reports illustrated that resin-based materials with under-achieved polymerization are more susceptible to clinical failure due to the high risk of resin matrix degradation that can weaken the material and facilitate biofilm adhesion [138–140]. Therefore, evaluating the polymerization kinetics of such materials can allow further improvement of the materials' properties, characterization, and ion release capabilities.

Most of the reported studies tested the designed materials in vitro. Having these materials tested in a more representable environment, in situ or in vivo, is highly needed, as the complexity of the oral biofilm and the influence of host-related factors can be experimented with, and different results are expected compared to the in vitro settings [141]. Therefore, future studies may consider adopting a translational clinical setting to conduct further investigations concerning the clinical performance of ion-releasing polymeric materials inside the oral cavity.

6. Conclusions

A plethora of evidence suggests that implementing ion-releasing restorative materials in restorative dentistry may minimize the biological failure of these materials due to the onset of secondary caries. The most common remineralizing fillers used in resin-based material formulations were Nano-sized Amorphous Calcium Phosphate (NACP), Calcium Fluoride (CaF_2), Bioactive Glass (BAG), Hydroxyapatite (HA), Fluorapatite (FA), and Boron Nitride (BN). Most of the reported studies focused on formulating resin-based composite formulations, with fewer reports concerning the design of resin-based sealants, dental adhesives, and crown cement. The released ions from the designed bioactive formulations may neutralize the acidity around the placed materials, restore the lost minerals from the tooth structure, and indirectly modulate the oral biofilms. In addition, synergetic antibiofilm inhibition was observed when the remineralizing compounds were combined with other bioactive compounds, such as quaternary ammonium, suggesting a dual action against the onset of secondary caries. Future investigations may consider further evaluation and characterizations of the designed materials to understand the mechanical and antibacterial performance of these materials comprehensively. Furthermore, clinical translational models are needed to test these bioactive formulations inside the oral cavity.

Author Contributions: Conceptualization, K.S.A., R.N.A., Y.A.A.-D., M.S.I. and A.O.A.-Z.; methodology, L.I.B.-J., D.I.A., L.S.A., H.A.A., M.A.A. and A.A.B.; writing—original draft preparation, L.I.B.-J., D.I.A., L.S.A., H.A.A., M.A.A. and A.A.B.; writing—review and editing, K.S.A., R.N.A., Y.A.A.-D., M.S.I. and A.O.A.-Z. All authors have read and agreed to the published version of the manuscript.

Funding: This research received no external funding.

Institutional Review Board Statement: Not applicable.

Informed Consent Statement: Not applicable.

Data Availability Statement: Not applicable.

Conflicts of Interest: The authors declare no conflict of interest.

References

1. Machiulskiene, V.; Campus, G.; Carvalho, J.C.; Dige, I.; Ekstrand, K.R.; Jablonski-Momeni, A.; Maltz, M.; Manton, D.J.; Martignon, S.; Martinez-Mier, E.A.; et al. Terminology of Dental Caries and Dental Caries Management: Consensus Report of a Workshop Organized by ORCA and Cariology Research Group of IADR. *Caries Res.* **2020**, *54*, 7–14. [CrossRef] [PubMed]
2. Balhaddad, A.A.; Kansara, A.A.; Hidan, D.; Weir, M.D.; Xu, H.H.K.; Melo, M.A.S. Toward Dental Caries: Exploring Nanoparticle-Based Platforms and Calcium Phosphate Compounds for Dental Restorative Materials. *Bioact. Mater.* **2019**, *4*, 43–55. [CrossRef] [PubMed]
3. AlQranei, M.S.; Balhaddad, A.A.; Melo, M.A.S. The Burden of Root Caries: Updated Perspectives and Advances on Management Strategies. *Gerodontology* **2021**, *38*, 136–153. [CrossRef] [PubMed]
4. Garcia, I.M.; Balhaddad, A.A.; Ibrahim, M.S.; Weir, M.D.; Xu, H.H.K.; Collares, F.M.; Melo, M.A.S. Antibacterial Response of Oral Microcosm Biofilm to Nano-Zinc Oxide in Adhesive Resin. *Dent. Mater.* **2021**, *37*, e182–e193. [CrossRef]

5. Mokeem, L.S.; Balhaddad, A.A.; Garcia, I.M.; Collares, F.M.; Melo, M.A.S. Benzyldimethyldodecyl Ammonium Chloride Doped Dental Adhesive: Impact on Core's Properties, Biosafety, and Antibacterial/Bonding Performance after Aging. *J. Funct. Biomater.* **2022**, *13*, 190. [CrossRef]
6. Frencken, J.E.; Peters, M.C.; Manton, D.J.; Leal, S.C.; Gordan, V.V.; Eden, E. Minimal Intervention Dentistry for Managing Dental Caries—A Review: Report of a FDI Task Group. *Int. Dent. J.* **2012**, *62*, 223–243. [CrossRef]
7. Dennison, J.B.; Sarrett, D.C. Prediction and Diagnosis of Clinical Outcomes Affecting Restoration Margins. *J. Oral Rehabil.* **2012**, *39*, 301–318. [CrossRef]
8. Albeshir, E.G.; Alsahafi, R.; Albluwi, R.; Balhaddad, A.A.; Mitwalli, H.; Oates, T.W.; Hack, G.D.; Sun, J.; Weir, M.D.; Xu, H.H.K. Low-Shrinkage Resin Matrices in Restorative Dentistry-Narrative Review. *Materials* **2022**, *15*, 2951. [CrossRef]
9. Mills, C.; Patel, P. Adopting Caries Risk Assessment in All Practice Environments. *Gen. Dent.* **2016**, *64*, 66–72.
10. Featherstone, J.D.B.; Chaffee, B.W. The Evidence for Caries Management by Risk Assessment (CAMBRA®). *Adv. Dent. Res.* **2018**, *29*, 9–14. [CrossRef]
11. Yeung, C.A. Dental Caries: Caries Risk Assessment Tools. *Br. Dent. J.* **2018**, *224*, 2. [CrossRef]
12. Askar, H.; Krois, J.; Göstemeyer, G.; Bottenberg, P.; Zero, D.; Banerjee, A.; Schwendicke, F. Secondary Caries: What Is It, and How It Can Be Controlled, Detected, and Managed? *Clin. Oral Investig.* **2020**, *24*, 1869–1876. [CrossRef]
13. Ionescu, A.C.; Hahnel, S.; Delvecchio, P.; Ilie, N.; Moldovan, M.; Zambelli, V.; Bellani, G.; Brambilla, E. Microbiological Models for Accelerated Development of Secondary Caries in Vitro. *J. Dent.* **2022**, *127*, 104333. [CrossRef]
14. Mitwalli, H.; Alsahafi, R.; Balhaddad, A.A.; Weir, M.D.; Xu, H.H.K.; Melo, M.A.S. Emerging Contact-Killing Antibacterial Strategies for Developing Anti-Biofilm Dental Polymeric Restorative Materials. *Bioengineering* **2020**, *7*, 83. [CrossRef]
15. Hollanders, A.C.C.; Kuper, N.K.; Maske, T.T.; Huysmans, M.-C.D.N.J.M. Secondary Caries in Situ Models: A Systematic Review. *Caries Res.* **2018**, *52*, 454–462. [CrossRef]
16. Jokstad, A. Secondary Caries and Microleakage. *Dent. Mater.* **2016**, *32*, 11–25. [CrossRef]
17. AlGhannam, M.I.; AlAbbas, M.S.; AlJishi, J.A.; AlRuwaili, M.A.; AlHumaid, J.; Ibrahim, M.S. Remineralizing Effects of Resin-Based Dental Sealants: A Systematic Review of In Vitro Studies. *Polymers* **2022**, *14*, 779. [CrossRef]
18. Al-Dulaijan, Y.A.; Balhaddad, A.A. Prospects on Tuning Bioactive and Antimicrobial Denture Base Resin Materials: A Narrative Review. *Polymer* **2022**, *15*, 54. [CrossRef]
19. Pinna, R.; Usai, P.; Filigheddu, E.; Garcia-Godoy, F.; Milia, E. The Role of Adhesive Materials and Oral Biofilm in the Failure of Adhesive Resin Restorations. *Am. J. Dent.* **2017**, *30*, 285–292.
20. Finer, Y.; Santerre, J.P. Salivary Esterase Activity and Its Association with the Biodegradation of Dental Composites. *J. Dent. Res.* **2004**, *83*, 22–26. [CrossRef]
21. Feng, X. Cause of secondary caries and prevention. *Hua Xi Kou Qiang Yi Xue Za Zhi* **2014**, *32*, 107–110. [CrossRef] [PubMed]
22. Featherstone, J.D. Prevention and Reversal of Dental Caries: Role of Low Level Fluoride. *Community Dent. Oral Epidemiol.* **1999**, *27*, 31–40. [CrossRef] [PubMed]
23. Featherstone, J.D. The Science and Practice of Caries Prevention. *J. Am. Dent. Assoc.* **2000**, *131*, 887–899. [CrossRef] [PubMed]
24. Featherstone, J.D.B. The Caries Balance: The Basis for Caries Management by Risk Assessment. *Oral Health Prev. Dent.* **2004**, *2* (Suppl. 1), 259–264.
25. Featherstone, J.D.B. Dental Caries: A Dynamic Disease Process. *Aust. Dent. J.* **2008**, *53*, 286–291. [CrossRef]
26. Abou Neel, E.A.; Aljabo, A.; Strange, A.; Ibrahim, S.; Coathup, M.; Young, A.M.; Bozec, L.; Mudera, V. Demineralization-Remineralization Dynamics in Teeth and Bone. *Int. J. Nanomed.* **2016**, *11*, 4743–4763. [CrossRef]
27. Hara, A.T.; Carvalho, J.C.; Zero, D.T. Causes of Dental Erosion: Extrinsic Factors. In *Dental Erosion and Its Clinical Management*; Amaechi, B.T., Ed.; Springer International Publishing: Cham, Switzerland, 2015; pp. 69–96, ISBN 978-3-319-13993-7.
28. Kwang, S.; Abbott, P. The Presence and Distribution of Bacteria in Dentinal Tubules of Root Filled Teeth. *Int. Endod. J.* **2014**, *47*, 600–610. [CrossRef]
29. Scaramucci, T.; Carvalho, J.C.; Hara, A.T.; Zero, D.T. Causes of Dental Erosion: Intrinsic Factors. In *Dental Erosion and Its Clinical Management*; Amaechi, B.T., Ed.; Springer International Publishing: Cham, Switzerland, 2015; pp. 35–67, ISBN 978-3-319-13993-7.
30. Arifa, M.K.; Ephraim, R.; Rajamani, T. Recent Advances in Dental Hard Tissue Remineralization: A Review of Literature. *Int. J. Clin. Pediatr. Dent.* **2019**, *12*, 139–144. [CrossRef]
31. González-Cabezas, C. The Chemistry of Caries: Remineralization and Demineralization Events with Direct Clinical Relevance. *Dent. Clin. N. Am.* **2010**, *54*, 469–478. [CrossRef]
32. Zhang, C.; Campbell, S.D.; Dickens, S.H.; Yang, B. Remineralization of Natural Human Carious Dentin Lesions with an Experimental Whisker-Reinforced Atraumatic Restorative Treatment Composite. *J. Prosthodont.* **2019**, *28*, 920–926. [CrossRef]
33. Paula, A.B.P.; Fernandes, A.R.; Coelho, A.S.; Marto, C.M.; Ferreira, M.M.; Caramelo, F.; do Vale, F.; Carrilho, E. Therapies for White Spot Lesions-A Systematic Review. *J. Evid. Based. Dent. Pract.* **2017**, *17*, 23–38. [CrossRef]
34. Lei, C.; Li, J.; Xu, H.H.K.; Zhou, X. Demineralization and Remineralization. In *Dental Caries: Principles and Management*; Zhou, X., Ed.; Springer: Berlin/Heidelberg, Germany, 2016; pp. 71–83, ISBN 978-3-662-47450-1.
35. Moradian-Oldak, J.; George, A. Biomineralization of Enamel and Dentin Mediated by Matrix Proteins. *J. Dent. Res.* **2021**, *100*, 1020–1029. [CrossRef]
36. Li, X.; Wang, J.; Joiner, A.; Chang, J. The Remineralisation of Enamel: A Review of the Literature. *J. Dent.* **2014**, *42* (Suppl. 1), S12–S20. [CrossRef]

37. Kim, M.-J.; Lee, M.-J.; Kim, K.-M.; Yang, S.-Y.; Seo, J.-Y.; Choi, S.-H.; Kwon, J.-S. Enamel Demineralization Resistance and Remineralization by Various Fluoride-Releasing Dental Restorative Materials. *Materials* **2021**, *14*, 4554. [CrossRef]
38. Cury, J.A.; Tenuta, L.M.A. Enamel Remineralization: Controlling the Caries Disease or Treating Early Caries Lesions? *Braz. Oral Res.* **2009**, *23* (Suppl. 1), 23–30. [CrossRef]
39. Tanaka, M.; Kadoma, Y. Comparative Reduction of Enamel Demineralization by Calcium and Phosphate in Vitro. *Caries Res.* **2000**, *34*, 241–245. [CrossRef]
40. Lynch, R.J.M.; ten Cate, J.M. The Anti-Caries Efficacy of Calcium Carbonate-Based Fluoride Toothpastes. *Int. Dent. J.* **2005**, *55*, 175–178. [CrossRef]
41. Kasraei, S.; Sami, L.; Hendi, S.; Alikhani, M.-Y.; Rezaei-Soufi, L.; Khamverdi, Z. Antibacterial Properties of Composite Resins Incorporating Silver and Zinc Oxide Nanoparticles on Streptococcus Mutans and Lactobacillus. *Restor. Dent. Endod.* **2014**, *39*, 109–114. [CrossRef]
42. Cao, W.; Zhang, Y.; Wang, X.; Chen, Y.; Li, Q.; Xing, X.; Xiao, Y.; Peng, X.; Ye, Z. Development of a Novel Resin-Based Dental Material with Dual Biocidal Modes and Sustained Release of Ag+ Ions Based on Photocurable Core-Shell AgBr/Cationic Polymer Nanocomposites. *J. Mater. Sci. Mater. Med.* **2017**, *28*, 103. [CrossRef]
43. Fernandes, G.L.; Delbem, A.C.B.; do Amaral, J.G.; Gorup, L.F.; Fernandes, R.A.; de Souza Neto, F.N.; Souza, J.A.S.; Monteiro, D.R.; Hunt, A.M.A.; Camargo, E.R.; et al. Nanosynthesis of Silver-Calcium Glycerophosphate: Promising Association against Oral Pathogens. *Antibiotics* **2018**, *7*, 52. [CrossRef]
44. Melo, M.A.S.; Guedes, S.F.F.; Xu, H.H.K.; Rodrigues, L.K.A. Nanotechnology-Based Restorative Materials for Dental Caries Management. *Trends Biotechnol.* **2013**, *31*, 459–467. [CrossRef] [PubMed]
45. Song, W.; Ge, S. Application of Antimicrobial Nanoparticles in Dentistry. *Molecules* **2019**, *24*, 1033. [CrossRef] [PubMed]
46. Saafan, A.; Zaazou, M.H.; Sallam, M.K.; Mosallam, O.; El Danaf, H.A. Assessment of Photodynamic Therapy and Nanoparticles Effects on Caries Models. *Open Access Maced. J. Med. Sci.* **2018**, *6*, 1289–1295. [CrossRef]
47. Mokeem, L.; Balhaddad, A.A.; Garcia, I.M.; Collares, F.M.; Melo, M.A.S. Chapter 18—Nanoparticle-Based Antimicrobial for Dental Restorative Materials. In *Emerging Nanomaterials and Nano-Based Drug Delivery Approaches to Combat Antimicrobial Resistance*; Saravanan, M., Barabadi, H., Mostafavi, E., Webster, T., Eds.; Micro and Nano Technologies; Elsevier: Amsterdam, The Netherlands, 2022; pp. 661–700, ISBN 978-0-323-90792-7.
48. Balhaddad, A.A.; Garcia, I.M.; Mokeem, L.; Alsahafi, R.; Collares, F.M.; Sampaio de Melo, M.A. Metal Oxide Nanoparticles and Nanotubes: Ultrasmall Nanostructures to Engineer Antibacterial and Improved Dental Adhesives and Composites. *Bioengineering* **2021**, *8*, 146. [CrossRef]
49. Khvostenko, D.; Hilton, T.J.; Ferracane, J.L.; Mitchell, J.C.; Kruzic, J.J. Bioactive Glass Fillers Reduce Bacterial Penetration into Marginal Gaps for Composite Restorations. *Dent. Mater.* **2016**, *32*, 73–81. [CrossRef]
50. Mitwalli, H.; Balhaddad, A.A.; AlSahafi, R.; Oates, T.W.; Melo, M.A.S.; Xu, H.H.K.; Weir, M.D. Novel CaF2 Nanocomposites with Antibacterial Function and Fluoride and Calcium Ion Release to Inhibit Oral Biofilm and Protect Teeth. *J. Funct. Biomater.* **2020**, *11*, 56. [CrossRef]
51. Albeshir, E.G.; Balhaddad, A.A.; Mitwalli, H.; Wang, X.; Sun, J.; Melo, M.A.S.; Weir, M.D.; Xu, H.H.K. Minimally-Invasive Dentistry via Dual-Function Novel Bioactive Low-Shrinkage-Stress Flowable Nanocomposites. *Dent. Mater.* **2022**, *38*, 409–420. [CrossRef]
52. Al-Dulaijan, Y.A.; Weir, M.D.; Melo, M.A.S.; Sun, J.; Oates, T.W.; Zhang, K.; Xu, H.H.K. Protein-Repellent Nanocomposite with Rechargeable Calcium and Phosphate for Long-Term Ion Release. *Dent. Mater.* **2018**, *34*, 1735–1747. [CrossRef]
53. Xu, H.H.K.; Weir, M.D.; Sun, L. Calcium and Phosphate Ion Releasing Composite: Effect of PH on Release and Mechanical Properties. *Dent. Mater.* **2009**, *25*, 535–542. [CrossRef]
54. Cheng, L.; Weir, M.D.; Xu, H.H.K.; Antonucci, J.M.; Kraigsley, A.M.; Lin, N.J.; Lin-Gibson, S.; Zhou, X. Antibacterial Amorphous Calcium Phosphate Nanocomposites with a Quaternary Ammonium Dimethacrylate and Silver Nanoparticles. *Dent. Mater.* **2012**, *28*, 561–572. [CrossRef]
55. Liang, K.; Wang, S.; Tao, S.; Xiao, S.; Zhou, H.; Wang, P.; Cheng, L.; Zhou, X.; Weir, M.D.; Oates, T.W.; et al. Dental Remineralization via Poly(Amido Amine) and Restorative Materials Containing Calcium Phosphate Nanoparticles. *Int. J. Oral Sci.* **2019**, *11*, 15. [CrossRef]
56. Moreau, J.L.; Sun, L.; Chow, L.C.; Xu, H.H.K. Mechanical and Acid Neutralizing Properties and Bacteria Inhibition of Amorphous Calcium Phosphate Dental Nanocomposite. *J. Biomed. Mater. Res.* **2011**, *98B*, 80–88. [CrossRef]
57. Rodrigues, M.C.; Natale, L.C.; Arana-Chaves, V.E.; Braga, R.R. Calcium and Phosphate Release from Resin-Based Materials Containing Different Calcium Orthophosphate Nanoparticles. *J. Biomed. Mater. Res. B Appl. Biomater.* **2015**, *103*, 1670–1678. [CrossRef]
58. Xu, H.H.K.; Moreau, J.L.; Sun, L.; Chow, L.C. Nanocomposite Containing Amorphous Calcium Phosphate Nanoparticles for Caries Inhibition. *Dent. Mater.* **2011**, *27*, 762–769. [CrossRef]
59. Cheng, L.; Weir, M.D.; Xu, H.H.K.; Antonucci, J.M.; Lin, N.J.; Lin-Gibson, S.; Xu, S.M.; Zhou, X. Effect of Amorphous Calcium Phosphate and Silver Nanocomposites on Dental Plaque Microcosm Biofilms. *J. Biomed. Mater. Res. B Appl. Biomater.* **2012**, *100*, 1378–1386. [CrossRef]
60. Ma, M.; Md, W.; Lk, R.; Hh, X. Novel Calcium Phosphate Nanocomposite with Caries-Inhibition in a Human in Situ Model. *Dent. Mater. Off. Publ. Acad. Dent. Mater.* **2013**, *29*, 231–240. [CrossRef]

61. Zhang, L.; Weir, M.D.; Chow, L.C.; Antonucci, J.M.; Chen, J.; Xu, H.H.K. Novel Rechargeable Calcium Phosphate Dental Nanocomposite. *Dent. Mater.* **2016**, *32*, 285–293. [CrossRef]
62. Al-Dulaijan, Y.A.; Cheng, L.; Weir, M.D.; Melo, M.A.S.; Liu, H.; Oates, T.W.; Wang, L.; Xu, H.H.K. Novel Rechargeable Calcium Phosphate Nanocomposite with Antibacterial Activity to Suppress Biofilm Acids and Dental Caries. *J. Dent.* **2018**, *72*, 44–52. [CrossRef]
63. Balhaddad, A.A.; Ibrahim, M.S.; Weir, M.D.; Xu, H.H.K.; Melo, M.A.S. Concentration Dependence of Quaternary Ammonium Monomer on the Design of High-Performance Bioactive Composite for Root Caries Restorations. *Dent. Mater.* **2020**, *36*, e266–e278. [CrossRef]
64. Balhaddad, A.A.; Mokeem, L.S.; Weir, M.D.; Xu, H.; Melo, M.A.S. Sustained Antibacterial Effect and Wear Behavior of Quaternary Ammonium Contact-Killing Dental Polymers after One-Year of Hydrolytic Degradation. *Appl. Sci.* **2021**, *11*, 3718. [CrossRef]
65. Balhaddad, A.A.; Ibrahim, M.S.; Garcia, I.M.; Collares, F.M.; Weir, M.D.; Xu, H.H.; Melo, M.A.S. Pronounced Effect of Antibacterial Bioactive Dental Composite on Microcosm Biofilms Derived From Patients With Root Carious Lesions. *Front. Mater.* **2020**, *7*, 583861. [CrossRef]
66. Balhaddad, A.A.; Garcia, I.M.; Mokeem, L.; Ibrahim, M.S.; Collares, F.M.; Weir, M.D.; Xu, H.H.K.; Melo, M.a.S. Bifunctional Composites for Biofilms Modulation on Cervical Restorations. *J. Dent. Res.* **2021**, *100*, 1063–1071. [CrossRef] [PubMed]
67. Bhadila, G.; Wang, X.; Zhou, W.; Menon, D.; Melo, M.A.S.; Montaner, S.; Oates, T.W.; Weir, M.D.; Sun, J.; Xu, H.H.K. Novel Low-Shrinkage-Stress Nanocomposite with Remineralization and Antibacterial Abilities to Protect Marginal Enamel under Biofilm. *J. Dent.* **2020**, *99*, 103406. [CrossRef] [PubMed]
68. Filemban, H.; Bhadila, G.; Wang, X.; Melo, M.A.S.; Oates, T.W.; Hack, G.D.; Lynch, C.D.; Weir, M.D.; Sun, J.; Xu, H.H.K. Effects of Thermal Cycling on Mechanical and Antibacterial Durability of Bioactive Low-Shrinkage-Stress Nanocomposite. *J. Dent.* **2022**, *124*, 104218. [CrossRef]
69. Ms, I.; Fd, A.; Ya, A.-D.; Md, W.; Tw, O.; Hhk, X.; Mas, M. Tuning Nano-Amorphous Calcium Phosphate Content in Novel Rechargeable Antibacterial Dental Sealant. *Materials* **2018**, *11*, 1544. [CrossRef]
70. Ibrahim, M.S.; Ibrahim, A.S.; Balhaddad, A.A.; Weir, M.D.; Lin, N.J.; Tay, F.R.; Oates, T.W.; Xu, H.H.K.; Melo, M.A.S. A Novel Dental Sealant Containing Dimethylaminohexadecyl Methacrylate Suppresses the Cariogenic Pathogenicity of Streptococcus Mutans Biofilms. *Int. J. Mol. Sci.* **2019**, *20*, 3491. [CrossRef]
71. Ibrahim, M.S.; Balhaddad, A.A.; Garcia, I.M.; Hefni, E.; Collares, F.M.; Martinho, F.C.; Weir, M.D.; Xu, H.H.K.; Melo, M.A.S. Tooth Sealing Formulation with Bacteria-Killing Surface and on-Demand Ion Release/Recharge Inhibits Early Childhood Caries Key Pathogens. *J. Biomed. Mater. Res. B Appl. Biomater.* **2020**, *108*, 3217–3227. [CrossRef]
72. Ibrahim, M.S.; Balhaddad, A.A.; Garcia, I.M.; Collares, F.M.; Weir, M.D.; Xu, H.H.K.; Melo, M.A.S. PH-Responsive Calcium and Phosphate-Ion Releasing Antibacterial Sealants on Carious Enamel Lesions in Vitro. *J. Dent.* **2020**, *97*, 103323. [CrossRef]
73. Ibrahim, M.S.; Garcia, I.M.; Vila, T.; Balhaddad, A.A.; Collares, F.M.; Weir, M.D.; Xu, H.H.K.; Melo, M.A.S. Multifunctional Antibacterial Dental Sealants Suppress Biofilms Derived from Children at High Risk of Caries. *Biomater. Sci.* **2020**, *8*, 3472–3484. [CrossRef]
74. Melo, M.A.S.; Cheng, L.; Zhang, K.; Weir, M.D.; Rodrigues, L.K.A.; Xu, H.H.K. Novel Dental Adhesives Containing Nanoparticles of Silver and Amorphous Calcium Phosphate. *Dent. Mater.* **2013**, *29*, 199–210. [CrossRef]
75. Wu, J.; Zhou, C.; Ruan, J.; Weir, M.D.; Tay, F.; Sun, J.; Melo, M.A.S.; Oates, T.W.; Chang, X.; Xu, H.H.K. Self-Healing Adhesive with Antibacterial Activity in Water-Aging for 12 Months. *Dent. Mater.* **2019**, *35*, 1104–1116. [CrossRef]
76. al-Qarni, F.; Weir, M.; Melo, M.A.; Al-Dulaijan, Y.; Almulhim, K.S.; Xu, H.H.K. Novel Calcium Phosphate Ion-Rechargeable and Antibacterial Adhesive to Inhibit Dental Caries. *Clin. Oral Investig.* **2022**, *26*, 313–323. [CrossRef]
77. Al-Qarni, F.D.; Tay, F.; Weir, M.D.; Melo, M.A.S.; Sun, J.; Oates, T.W.; Xie, X.; Xu, H.H.K. Protein-Repelling Adhesive Resin Containing Calcium Phosphate Nanoparticles with Repeated Ion-Recharge and Re-Releases. *J. Dent.* **2018**, *78*, 91–99. [CrossRef]
78. Li, Y.; Hu, X.; Xia, Y.; Ji, Y.; Ruan, J.; Weir, M.D.; Lin, X.; Nie, Z.; Gu, N.; Masri, R.; et al. Novel Magnetic Nanoparticle-Containing Adhesive with Greater Dentin Bond Strength and Antibacterial and Remineralizing Capabilities. *Dent. Mater.* **2018**, *34*, 1310–1322. [CrossRef]
79. Fan, M.; Yang, J.; Xu, H.H.K.; Weir, M.D.; Tao, S.; Yu, Z.; Liu, Y.; Li, M.; Zhou, X.; Liang, K.; et al. Remineralization Effectiveness of Adhesive Containing Amorphous Calcium Phosphate Nanoparticles on Artificial Initial Enamel Caries in a Biofilm-Challenged Environment. *Clin. Oral. Investig.* **2021**, *25*, 5375–5390. [CrossRef]
80. Tao, S.; Su, Z.; Xiang, Z.; Xu, H.H.K.; Weir, M.D.; Fan, M.; Yu, Z.; Zhou, X.; Liang, K.; Li, J. Nano-Calcium Phosphate and Dimethylaminohexadecyl Methacrylate Adhesive for Dentin Remineralization in a Biofilm-Challenged Environment. *Dent. Mater.* **2020**, *36*, e316–e328. [CrossRef]
81. Cao, L.; Wu, J.; Zhang, Q.; Baras, B.; Bhadila, G.; Li, Y.; Melo, M.A.S.; Weir, M.D.; Bai, Y.; Zhang, N.; et al. Novel Protein-Repellent and Antibacterial Resins and Cements to Inhibit Lesions and Protect Teeth. *Int. J. Polym. Sci.* **2019**, *2019*, 5602904. [CrossRef]
82. Liu, Y.; Zhang, L.; Niu, L.-N.; Yu, T.; Xu, H.H.K.; Weir, M.D.; Oates, T.W.; Tay, F.R.; Chen, J.-H. Antibacterial and Remineralizing Orthodontic Adhesive Containing Quaternary Ammonium Resin Monomer and Amorphous Calcium Phosphate Nanoparticles. *J. Dent.* **2018**, *72*, 53–63. [CrossRef]
83. Behr, M.; Zeman, F.; Baitinger, T.; Galler, J.; Koller, M.; Handel, G.; Rosentritt, M. The Clinical Performance of Porcelain-Fused-to-Metal Precious Alloy Single Crowns: Chipping, Recurrent Caries, Periodontitis, and Loss of Retention. *Int. J. Prosthodont.* **2014**, *27*, 153–160. [CrossRef]

84. Srimaneepong, V.; Heboyan, A.; Zafar, M.S.; Khurshid, Z.; Marya, A.; Fernandes, G.V.O.; Rokaya, D. Fixed Prosthetic Restorations and Periodontal Health: A Narrative Review. *J. Funct. Biomater.* **2022**, *13*, 15. [CrossRef]
85. AlSahafi, R.; Balhaddad, A.A.; Mitwalli, H.; Ibrahim, M.S.; Melo, M.A.S.; Oates, T.W.; Xu, H.H.K.; Weir, M.D. Novel Crown Cement Containing Antibacterial Monomer and Calcium Phosphate Nanoparticles. *Nanomaterials* **2020**, *10*, 2001. [CrossRef] [PubMed]
86. AlSahafi, R.; Mitwalli, H.; Alhussein, A.; Melo, M.A.S.; Martinho, F.; Lynch, C.D.; Oates, T.W.; Xu, H.H.K.; Weir, M.D. Novel Rechargeable Nanostructured Calcium Phosphate Crown Cement with Long-Term Ion Release and Antibacterial Activity to Suppress Saliva Microcosm Biofilms. *J. Dent.* **2022**, *122*, 104140. [CrossRef] [PubMed]
87. Dai, Q.; Weir, M.D.; Ruan, J.; Liu, J.; Gao, J.; Lynch, C.D.; Oates, T.W.; Li, Y.; Chang, X.; Xu, H.H.K. Effect of Co-Precipitation plus Spray-Drying of Nano-CaF2 on Mechanical and Fluoride Properties of Nanocomposite. *Dent. Mater.* **2021**, *37*, 1009–1019. [CrossRef] [PubMed]
88. Tyas, M.J. Clinical Evaluation of Glass-Ionomer Cement Restorations. *J. Appl. Oral Sci.* **2006**, *14*, 10–13. [CrossRef]
89. ten Cate, J.M.; Buijs, M.J.; Miller, C.C.; Exterkate, R.A.M. Elevated Fluoride Products Enhance Remineralization of Advanced Enamel Lesions. *J. Dent. Res.* **2008**, *87*, 943–947. [CrossRef]
90. Ten Cate, M.J. Contemporary Perspective on the Use of Fluoride Products in Caries Prevention. *Br. Dent. J.* **2013**, *214*, 161–167. [CrossRef]
91. Wiegand, A.; Buchalla, W.; Attin, T. Review on Fluoride-Releasing Restorative Materials–Fluoride Release and Uptake Characteristics, Antibacterial Activity and Influence on Caries Formation. *Dent. Mater.* **2007**, *23*, 343–362. [CrossRef]
92. Xu, H.H.K.; Moreau, J.L.; Sun, L.; Chow, L.C. Novel CaF(2) Nanocomposite with High Strength and Fluoride Ion Release. *J. Dent. Res.* **2010**, *89*, 739–745. [CrossRef]
93. Weir, M.D.; Moreau, J.L.; Levine, E.D.; Strassler, H.E.; Chow, L.C.; Xu, H.H.K. Nanocomposite Containing CaF2 Nanoparticles: Thermal Cycling, Wear and Long-Term Water-Aging. *Dent. Mater.* **2012**, *28*, 642–652. [CrossRef]
94. Francois, P.; Fouquet, V.; Attal, J.; Dursun, E. Commercially Available Fluoride-Releasing Restorative Materials: A Review and a Proposal for Classification. *Materials* **2020**, *10*, 2313. [CrossRef]
95. Xu, H.H.K.; Moreau, J.L.; Sun, L.; Chow, L.C. Strength and Fluoride Release Characteristics of a Calcium Fluoride Based Dental Nanocomposite. *Biomaterials* **2008**, *29*, 4261–4267. [CrossRef]
96. Mitwalli, H.; AlSahafi, R.; Alhussein, A.; Oates, T.W.; Melo, M.A.S.; Xu, H.H.K.; Weir, M.D. Novel Rechargeable Calcium Fluoride Dental Nanocomposites. *Dent. Mater.* **2022**, *38*, 397–408. [CrossRef]
97. Cheng, L.; Weir, M.D.; Xu, H.H.K.; Kraigsley, A.M.; Lin, N.J.; Lin-Gibson, S.; Zhou, X. Antibacterial and Physical Properties of Calcium-Phosphate and Calcium-Fluoride Nanocomposites with Chlorhexidine. *Dent. Mater.* **2012**, *28*, 573–583. [CrossRef]
98. Salar, D.V.; García-Godoy, F.; Flaitz, C.M.; Hicks, M.J. Potential Inhibition of Demineralization in Vitro by Fluoride-Releasing Sealants. *J. Am. Dent. Assoc.* **2007**, *138*, 502–506. [CrossRef]
99. Kashbour, W.; Gupta, P.; Worthington, H.V.; Boyers, D. Pit and fissure sealants versus fluoride varnishes for preventing dental decay in the permanent teeth of children and adolescents. *Cochrane Database Syst Rev.* **2020**, *11*, CD003067. [CrossRef]
100. Fei, X.; Li, Y.; Weir, M.D.; Baras, B.H.; Wang, H.; Wang, S.; Sun, J.; Melo, M.A.S.; Ruan, J.; Xu, H.H.K. Novel Pit and Fissure Sealant Containing Nano-CaF_2 and Dimethylaminohexadecyl Methacrylate with Double Benefits of Fluoride Release and Antibacterial Function. *Dent. Mater.* **2020**, *36*, 1241–1253. [CrossRef]
101. Gholam, M.; Hameed, M. Fluoride Release from Dental Adhesive Reinforced with Fluorapatite or Calcium Fluoride. *Int. J. Med. Res. Health Sci.* **2018**, *7*, 137–143.
102. Yao, C.; Ahmed, M.H.; Li, X.; Nedeljkovic, I.; Vandooren, J.; Mercelis, B.; Zhang, F.; Van Landuyt, K.L.; Huang, C.; Van Meerbeek, B. Zinc-Calcium-Fluoride Bioglass-Based Innovative Multifunctional Dental Adhesive with Thick Adhesive Resin Film Thickness. *ACS Appl. Mater. Interfaces* **2020**, *12*, 30120–30135. [CrossRef]
103. Yi, J.; Dai, Q.; Weir, M.D.; Melo, M.A.S.; Lynch, C.D.; Oates, T.W.; Zhang, K.; Zhao, Z.; Xu, H.H.K. A Nano-CaF2-Containing Orthodontic Cement with Antibacterial and Remineralization Capabilities to Combat Enamel White Spot Lesions. *J. Dent.* **2019**, *89*, 103172. [CrossRef]
104. AlSahafi, R.; Mitwalli, H.; Alhussein, A.; Balhaddad, A.A.; Alquria, T.A.; Melo, M.A.S.; Lynch, C.D.; Oates, T.W.; Zhang, K.; Xu, H.H.K.; et al. Novel Rechargeable Nano-Calcium Phosphate and Nano-Calcium Fluoride Resin Cements. *J. Dent.* **2022**, *126*, 104312. [CrossRef]
105. Sawant, K.; Pawar, A. Bioactive Glass in Dentistry: A Systematic Review. *Saudi J. Oral Sci.* **2020**, *7*, 3. [CrossRef]
106. Khalid, M.D.; Khurshid, Z.; Zafar, M.S.; Farooq, I.; Khan, R.S.; Najmi, A. Bioactive Glasses and Their Applications in Dentistry. *J. Pak. Dent. Assoc.* **2017**, *26*, 32–38. [CrossRef]
107. Par, M.; Gubler, A.; Attin, T.; Tarle, Z.; Tarle, A.; Tauböck, T.T. Ion Release and Hydroxyapatite Precipitation of Resin Composites Functionalized with Two Types of Bioactive Glass. *J. Dent.* **2022**, *118*, 103950. [CrossRef] [PubMed]
108. Tezvergil-Mutluay, A.; Seseogullari-Dirihan, R.; Feitosa, V.P.; Cama, G.; Brauer, D.S.; Sauro, S. Effects of Composites Containing Bioactive Glasses on Demineralized Dentin. *J. Dent. Res.* **2017**, *96*, 999–1005. [CrossRef] [PubMed]
109. Korkut, E.; Torlak, E.; Altunsoy, M. Antimicrobial and Mechanical Properties of Dental Resin Composite Containing Bioactive Glass. *J. Appl. Biomater. Funct. Mater.* **2016**, *14*, e296–e301. [CrossRef] [PubMed]

110. Tian, J.; Wu, Z.; Wang, Y.; Han, C.; Zhou, Z.; Guo, D.; Lin, Y.; Ye, Z.; Fu, J. Multifunctional Dental Resin Composite with Antibacterial and Remineralization Properties Containing NMgO-BAG. *J. Mech. Behav. Biomed. Mater.* **2023**, *141*, 105783. [CrossRef]
111. Yang, S.-Y.; Piao, Y.-Z.; Kim, S.-M.; Lee, Y.-K.; Kim, K.-N.; Kim, K.-M. Acid Neutralizing, Mechanical and Physical Properties of Pit and Fissure Sealants Containing Melt-Derived 45S5 Bioactive Glass. *Dent. Mater.* **2013**, *29*, 1228–1235. [CrossRef]
112. Yang, S.-Y.; Kwon, J.-S.; Kim, K.-N.; Kim, K.-M. Enamel Surface with Pit and Fissure Sealant Containing 45S5 Bioactive Glass. *J. Dent. Res.* **2016**, *95*, 550–557. [CrossRef]
113. Lee, S.-M.; Yoo, K.-H.; Yoon, S.-Y.; Kim, I.-R.; Park, B.-S.; Son, W.-S.; Ko, C.-C.; Son, S.-A.; Kim, Y.-I. Enamel Anti-Demineralization Effect of Orthodontic Adhesive Containing Bioactive Glass and Graphene Oxide: An In-Vitro Study. *Materials* **2018**, *11*, 1728. [CrossRef]
114. Kim, Y.-M.; Kim, D.-H.; Song, C.W.; Yoon, S.-Y.; Kim, S.-Y.; Na, H.S.; Chung, J.; Kim, Y.-I.; Kwon, Y.H. Antibacterial and Remineralization Effects of Orthodontic Bonding Agents Containing Bioactive Glass. *Korean J. Orthod.* **2018**, *48*, 163–171. [CrossRef]
115. Manfred, L.; Covell, D.A.; Crowe, J.J.; Tufekci, E.; Mitchell, J.C. A Novel Biomimetic Orthodontic Bonding Agent Helps Prevent White Spot Lesions Adjacent to Brackets. *Angle Orthod.* **2013**, *83*, 97–103. [CrossRef]
116. Firzok, H.; Zahid, S.; Asad, S.; Manzoor, F.; Khan, A.S.; Shah, A.T. Sol-Gel Derived Fluoridated and Non-Fluoridated Bioactive Glass Ceramics-Based Dental Adhesives: Compositional Effect on Re-Mineralization around Orthodontic Brackets. *J. Non-Cryst. Solids* **2019**, *521*, 119469. [CrossRef]
117. Song, H.-K.; Yoo, K.-H.; Yoon, S.-Y.; Na, H.S.; Chung, J.; Son, W.-S.; Lee, S.-M.; Kim, Y.-I. In Vitro Effect of Gallium-Doped Bioactive Glass on Enamel Anti-Demineralization and Bond Strength of Orthodontic Resins. *Appl. Sci.* **2019**, *9*, 4918. [CrossRef]
118. Kohda, N.; Iijima, M.; Kawaguchi, K.; Toshima, H.; Muguruma, T.; Endo, K.; Mizoguchi, I. Inhibition of Enamel Demineralization and Bond-Strength Properties of Bioactive Glass Containing 4-META/MMA-TBB-Based Resin Adhesive. *Eur. J. Oral Sci.* **2015**, *123*, 202–207. [CrossRef]
119. Pepla, E.; Besherat, L.K.; Palaia, G.; Tenore, G.; Migliau, G. Nano-Hydroxyapatite and Its Applications in Preventive, Restorative and Regenerative Dentistry: A Review of Literature. *Ann. Di Stomatol.* **2014**, *5*, 108–114. [CrossRef]
120. Pajor, K.; Pajchel, L. Link to external site, this link will open in a new window Hydroxyapatite and Fluorapatite in Conservative Dentistry and Oral Implantology—A Review. *Materials* **2019**, *12*, 2683. [CrossRef]
121. Moshaverinia, A.; Ansari, S.; Movasaghi, Z.; Billington, R.W.; Darr, J.A.; Rehman, I.U. Modification of Conventional Glass-Ionomer Cements with N-Vinylpyrrolidone Containing Polyacids, Nano-Hydroxy and Fluoroapatite to Improve Mechanical Properties. *Dent. Mater.* **2008**, *24*, 1381–1390. [CrossRef]
122. Gu, Y.W.; Yap, A.U.J.; Cheang, P.; Khor, K.A. Effects of Incorporation of HA/ZrO(2) into Glass Ionomer Cement (GIC). *Biomaterials* **2005**, *26*, 713–720. [CrossRef]
123. Liu, F.; Sun, B.; Jiang, X.; Aldeyab, S.S.; Zhang, Q.; Zhu, M. Mechanical Properties of Dental Resin/Composite Containing Urchin-like Hydroxyapatite. *Dent. Mater.* **2014**, *30*, 1358–1368. [CrossRef]
124. Ai, M.; Du, Z.; Zhu, S.; Geng, H.; Zhang, X.; Cai, Q.; Yang, X. Composite Resin Reinforced with Silver Nanoparticles-Laden Hydroxyapatite Nanowires for Dental Application. *Dent. Mater.* **2017**, *33*, 12–22. [CrossRef]
125. Li, Y.; Zhang, D.; Wan, Z.; Yang, X.; Cai, Q. Dental Resin Composites with Improved Antibacterial and Mineralization Properties via Incorporating Zinc/Strontium-Doped Hydroxyapatite as Functional Fillers. *Biomed. Mater.* **2022**, *17*, 045002. [CrossRef] [PubMed]
126. Zhao, Y.; Zhang, H.; Hong, L.; Zou, X.; Song, J.; Han, R.; Chen, J.; Yu, Y.; Liu, X.; Zhao, H.; et al. A Multifunctional Dental Resin Composite with Sr-N-Doped TiO2 and n-HA Fillers for Antibacterial and Mineralization Effects. *Int. J. Mol. Sci.* **2023**, *24*, 1274. [CrossRef] [PubMed]
127. Utneja, S.; Talwar, S.; Nawal, R.R.; Sapra, S.; Mittal, M.; Rajain, A.; Verma, M. Evaluation of Remineralization Potential and Mechanical Properties of Pit and Fissure Sealants Fortified with Nano-Hydroxyapatite and Nano-Amorphous Calcium Phosphate Fillers: An in Vitro Study. *J. Conserv. Dent.* **2018**, *21*, 681–690. [CrossRef] [PubMed]
128. Sodagar, A.; Akhavan, A.; Hashemi, E.; Arab, S.; Pourhajibagher, M.; Sodagar, K.; Kharrazifard, M.J.; Bahador, A. Evaluation of the Antibacterial Activity of a Conventional Orthodontic Composite Containing Silver/Hydroxyapatite Nanoparticles. *Prog. Orthod.* **2016**, *17*, 40. [CrossRef]
129. Lin, J.; Zhu, J.; Gu, X.; Wen, W.; Li, Q.; Fischer-Brandies, H.; Wang, H.; Mehl, C. Effects of Incorporation of Nano-Fluorapatite or Nano-Fluorohydroxyapatite on a Resin-Modified Glass Ionomer Cement. *Acta Biomater.* **2011**, *7*, 1346–1353. [CrossRef]
130. Merlo, A.; Mokkapati, V.R.S.S.; Pandit, S.; Mijakovic, I. Boron Nitride Nanomaterials: Biocompatibility and Bio-Applications. *Biomater. Sci.* **2018**, *6*, 2298–2311. [CrossRef]
131. Genchi, G.G.; Ciofani, G. Bioapplications of Boron Nitride Nanotubes. *Nanomedicine* **2015**, *10*, 3315–3319. [CrossRef]
132. Lee, B.; Kwon, J.-S.; Khalid, M.W.; Kim, K.-M.; Kim, J.; Lim, K.M.; Hong, S.H. Boron Nitride Nanoplatelets as Reinforcement Material for Dental Ceramics. *Dent. Mater.* **2020**, *36*, 744–754. [CrossRef]
133. Alqahtani, M. Effect of Hexagonal Boron Nitride Nanopowder Reinforcement and Mixing Methods on Physical and Mechanical Properties of Self-Cured PMMA for Dental Applications. *Materials* **2020**, *13*, 2323. [CrossRef]
134. Degrazia, F.W.; Leitune, V.C.B.; Samuel, S.M.W.; Collares, F.M. Boron Nitride Nanotubes as Novel Fillers for Improving the Properties of Dental Adhesives. *J. Dent.* **2017**, *62*, 85–90. [CrossRef]

135. Alansy, A.S.; Saeed, T.A.; Al-Attab, R.; Guo, Y.; Yang, Y.; Liu, B.; Fan, Z. Boron Nitride Nanosheets Modified with Zinc Oxide Nanoparticles as Novel Fillers of Dental Resin Composite. *Dent. Mater.* **2022**, *38*, e266–e274. [CrossRef]
136. Bohns, F.R.; Degrazia, F.W.; de Souza Balbinot, G.; Leitune, V.C.B.; Samuel, S.M.W.; García-Esparza, M.A.; Sauro, S.; Collares, F.M. Boron Nitride Nanotubes as Filler for Resin-Based Dental Sealants. *Sci. Rep.* **2019**, *9*, 7710. [CrossRef]
137. Maktabi, H.; Balhaddad, A.A.; Alkhubaizi, Q.; Strassler, H.; Melo, M.A.S. Factors Influencing Success of Radiant Exposure in Light-Curing Posterior Dental Composite in the Clinical Setting. *Am. J. Dent.* **2018**, *31*, 320–328.
138. Balhaddad, A.A.; Garcia, I.M.; Maktabi, H.; Ibrahim, M.S.; Alkhubaizi, Q.; Strassler, H.; Collares, F.M.; Melo, M.A.S. Errors in Light-Emitting Diodes Positioning When Curing Bulk Fill and Incremental Composites: Impact on Properties after Aging. *Restor. Dent. Endod.* **2021**, *46*, e51. [CrossRef]
139. Maktabi, H.; Ibrahim, M.S.; Balhaddad, A.A.; Alkhubaizi, Q.; Garcia, I.M.; Collares, F.M.; Strassler, H.; Fugolin, A.P.P.; Pfeifer, C.S.; Melo, M.A.S. Improper Light Curing of Bulkfill Composite Drives Surface Changes and Increases *S. mutans* Biofilm Growth as a Pathway for Higher Risk of Recurrent Caries around Restorations. *Dent. J.* **2021**, *9*, 83. [CrossRef]
140. Maktabi, H.; Ibrahim, M.; Alkhubaizi, Q.; Weir, M.; Xu, H.; Strassler, H.; Fugolin, A.P.P.; Pfeifer, C.S.; Melo, M.A.S. Underperforming Light Curing Procedures Trigger Detrimental Irradiance-Dependent Biofilm Response on Incrementally Placed Dental Composites. *J. Dent.* **2019**, *88*, 103110. [CrossRef]
141. Ibrahim, M.S.; Garcia, I.M.; Kensara, A.; Balhaddad, A.A.; Collares, F.M.; Williams, M.A.; Ibrahim, A.S.; Lin, N.J.; Weir, M.D.; Xu, H.H.K.; et al. How We Are Assessing the Developing Antibacterial Resin-Based Dental Materials? A Scoping Review. *J. Dent.* **2020**, *99*, 103369. [CrossRef]

Disclaimer/Publisher's Note: The statements, opinions and data contained in all publications are solely those of the individual author(s) and contributor(s) and not of MDPI and/or the editor(s). MDPI and/or the editor(s) disclaim responsibility for any injury to people or property resulting from any ideas, methods, instructions or products referred to in the content.

International Journal of *Molecular Sciences*

MDPI

Article

In Vitro and In Vivo Biocompatibility of Natural and Synthetic *Pseudomonas aeruginosa* Pyomelanin for Potential Biomedical Applications

Mateusz M. Urbaniak [1,2], Małgorzata Gazińska [3], Karolina Rudnicka [1], Przemysław Płociński [1], Monika Nowak [1] and Magdalena Chmiela [1,*]

1 Department of Immunology and Infectious Biology, Faculty of Biology and Environmental Protection, University of Łódź, 90-237 Łódź, Poland
2 The Bio-Med-Chem Doctoral School, University of Lodz and Lodz Institutes of the Polish Academy of Sciences, 90-237 Łódź, Poland
3 Department of Engineering and Technology of Polymers, Faculty of Chemistry, Wrocław University of Science and Technology (WUST), 50-370 Wrocław, Poland
* Correspondence: magdalena.chmiela@biol.uni.lodz.pl

Abstract: Bacteria are the source of many bioactive compounds, including polymers with various physiological functions and the potential for medical applications. Pyomelanin from *Pseudomonas aeruginosa*, a nonfermenting Gram-negative bacterium, is a black–brown negatively charged extracellular polymer of homogentisic acid produced during L-tyrosine catabolism. Due to its chemical properties and the presence of active functional groups, pyomelanin is a candidate for the development of new antioxidant, antimicrobial and immunomodulatory formulations. This work aimed to obtain bacterial water-soluble (Pyo_{sol}), water-insoluble (Pyo_{insol}) and synthetic (sPyo) pyomelanin variants and characterize their chemical structure, thermosensitivity and biosafety in vitro and in vivo (*Galleria mallonella*). FTIR analysis showed that aromatic ring connections in the polymer chains were dominant in Pyo_{sol} and sPyo, whereas Pyo_{insol} had fewer C_{ar}-C_{ar} links between rings. The differences in chemical structure influence the solubility of various forms of pyomelanins, their thermal stability and biological activity. Pyo_{sol} and Pyo_{insol} showed higher biological safety than sPyo. The obtained results qualify Pyo_{sol} and Pyo_{insol} for evaluation of their antimicrobial, immunomodulatory and proregenerative activities.

Keywords: pyomelanin; *Pseudomonas aeruginosa*; chemical structure; biocompatibility

check for updates

Citation: Urbaniak, M.M.; Gazińska, M.; Rudnicka, K.; Płociński, P.; Nowak, M.; Chmiela, M. In Vitro and In Vivo Biocompatibility of Natural and Synthetic *Pseudomonas aeruginosa* Pyomelanin for Potential Biomedical Applications. *Int. J. Mol. Sci.* **2023**, *24*, 7846. https://doi.org/10.3390/ijms24097846

Academic Editor: Marco Tatullo

Received: 27 January 2023
Revised: 17 April 2023
Accepted: 24 April 2023
Published: 25 April 2023

Copyright: © 2023 by the authors. Licensee MDPI, Basel, Switzerland. This article is an open access article distributed under the terms and conditions of the Creative Commons Attribution (CC BY) license (https://creativecommons.org/licenses/by/4.0/).

1. Introduction

Pseudomonas aeruginosa, a nonfermenting Gram-negative bacterium that is widespread in the environment and is an opportunistic pathogen, produces a variety of pigments, including pyocyanin, pyorubin, pyoverdine and pyomelanin. Pyomelanin is a negatively charged extracellular polymer of homogentisic acid (HGA) that is produced during L-tyrosine catabolism [1]. L-tyrosine is converted to 4-hydroxyphenylopyruvic acid (4-HPPA) by tyrosine aminotransferase, and then 4-HPPA is biotransformed to HGA by 4-hydroxyphenylpyruvate dioxygenase. Pyomelanin then forms spontaneously when secreted HGA autoxidizes to benzoquinone acetic acid, which then undergoes polymerization to form pyomelanin chains. Similar to chemically synthesized polymers of HGA, pyomelanin is a dark brown to black pigment [2,3]. The chemical structure of melanins, including pyomelanin, determines their oxidizing (quinone groups) and reducing (hydroxyquinone groups) properties that allow them to accept or donate electrons [4,5].

The primary function of microbial melanins is to protect cells from UV radiation [6]. In addition, pyomelanin increases the effectiveness of bacterial adhesion to surfaces, thus supporting the formation of biofilms and the extracellular transfer of electrons [7]. The production of pyomelanin by bacteria in the marine bacterial genus *Pseudoalteromonas* is induced during biofilm formation and under heat stress, suggesting that this pigment is involved in the adaptation of these bacteria to grow in a hostile ecological niche [8]. Pyomelanin is produced by clinical *P. aeruginosa* isolates from patients with chronic lung infections. It remains unclear whether pyomelanin is relevant to the pathogenicity of these bacteria; its production is linked to a reduction in oxidative stress, which may be an important trait to improve bacterial survival within pulmonary macrophages [9,10]. Additionally, melanins of *Aspergillus fumigatus* protect the fungus from host reactive oxygen intermediates [11]. A study by Fonseca et al. revealed that metal binding and metal reduction by pyomelanin are necessary for iron acquisition [7]. Due to the antimicrobial and antioxidant activities of microbial melanins, including pyomelanin produced by *P. aeruginosa,* they can be used in the food industry to coat packaging to prolong food expiration dates [5]. Bacterial melanins can be exploited as dyes and colorants and used in cosmetics as sunscreen and oxidative stress-reducing substances, whereas in agriculture, they may help to retain metals and improve iron availability in the soil [12]. These compounds are promising candidates for the design of new drugs, including antitumor drugs, vaccine antigen carriers, adjuvants and optoacoustic imaging contrasts [13–15]. The treatment of infections with pyomelanin with an accompanying inflammatory reaction driven by oxidative stress has also been considered [16]. The increased sensitivity of several bacterial pathogens to antibiotics in the milieu of pyomelanin has also been demonstrated, as well as the antimicrobial activity of melanin itself and its iron ion complexes against *Helicobacter pylori, Candida albicans* and human immunodeficiency virus (HIV) [17,18].

Eukaryotic and bacterial melanins may exhibit immunomodulatory properties. Cuttlefish melanin is able to activate and polarize dendritic cells, while human neuromelanin activates the nuclear factor kappa B (NF-κB) signaling pathway resulting with the secretion of pro-inflammatory cytokines: tumor necrosis factor alpha (TNF-α) and interleukin (IL)-6 [19]. The role of melanin in the regulation of innate immunity in humans is potentially due to the modulation of phagocytes and complement activity, reduction in the mRNA translation of proinflammatory cytokines and interaction with effector molecules [20]. Melanization is an important part of the cuticular wound healing process in arthropods and functions as part of the innate immune system in isopods by encapsulating parasites with melanin [21]. However, negatively charged melanin isolated from *Cryptococcus neoformans* diminished fungal cell susceptibility to cationic antimicrobial peptides of phagocytes, which was correlated with the downregulation of phagocytosis and diminished secretion of TNF-α, IL-1β, IL-6 and IL-12 as well as modulation of complement activities [22]. The potential wide range of microbial melanin applications drives studies on natural, synthetic and recombinant melanins [23].

Hypothetically, natural bacterial or synthetic formulations of pyomelanin may vary in their biological activity. Before studying different applications of pyomelanin formulations, it seems valuable to verify whether there are differences in physicochemical properties, cytocompatibility, the ability to stimulate the activation of the NF-κB pathway and in vivo toxicity between bacterial pyomelanin and its synthetic form.

Considering the potential application of pyomelanin in medicine, the aim of this study was to characterize the biochemical and physiological properties of natural water-soluble and water insoluble pyomelanin (Pyo_{sol} and Pyo_{insol}) isolated from a culture of *P. aeruginosa* and synthetic pyomelanin (sPyo), obtained by us under laboratory conditions, for further biomedical applications. We optimized *P. aeruginosa* growth conditions for the effective production of pyomelanin, developed an isolation procedure for natural water-soluble pyomelanin and a procedure for pyomelanin synthesis. The chemical structure of the bacterial and synthetic pyomelanin was investigated by Fourier transform infrared (FTIR) spectroscopy and the thermal properties were characterized by means of differential scanning calorimetry (DSC) and thermogravimetry (TGA). The cytocompatibility of pyomelanin from both sources was determined using an in vitro model on reference L-929 fibroblasts and THP-1 human monocytes. Noncytotoxic concentrations were established in the reference MTT reduction assay. In our in vivo biosafety study, we used the *Galleria melonella* insect model as an alternative to rodents. Larvae *of G. mellonella* are widely used in the toxicity assessment of new biopharmaceuticals, bioactive substances and newly synthesized chemicals [24]. Due to the ability to maintain larvae in a temperature range from 20 °C to 42 °C, this insect model mimics the physiological conditions of mammals during in vivo toxicity assessment [25]. In addition, *G. melonella* larvae provide an inexpensive and convenient way, free from legal or ethical restrictions, to assess the safety of new biomolecules, and the results generated from this model show a strong correlation with those obtained from mammalian systems. The use of *G. mellonella* larvae provides a more accurate representation of the biocompound's interactions with the host organism than studies on cell lines, which allows for the precise selection of concentrations before animal testing [26]. Furthermore, the ability of pyomelanin to activate the nuclear factor kappa B (NF-κB) signaling pathway was determined using the human recombinant monocyte model THP1-Blue™ NF-κB.

2. Results

2.1. The Efficiency of the Pyomelanin Biosynthesis and Chemical Synthesis

To determine the application potential of different pyomelanin variants, we first analyzed the extraction and synthesis efficiencies. We extracted two variants of natural pyomelanin produced by *P. aeruginosa* during growth on pyomelanin minimal medium (PMM) dedicated to these bacteria: PMM I, PMM II and pyomelanin synthesized from HGA. The efficiency of the extraction of natural pyomelanin formulations vs. the efficiency of pyomelanin synthesis is shown in Table 1. Both variants of natural pyomelanin, Pyo_{insol} and Pyo_{sol}, were produced more efficiently by *P. aeruginosa* after 7 days of growth of this bacterium on PMM II than by PMM I medium (1.79 ± 0.18 g/L and 1.22 ± 0.10, respectively, $p < 0.05$) (Table 1). Compared to the high-efficiency bacterial biosynthesis of pyomelanin, the yield of sPyo production from HGA was equal to 0.077 ± 0.01 g/1 g HGA.

Table 1. Efficiency of the Extraction of *P. aeruginosa* Pyomelanin from Minimal Growth Media and the Synthesis of Pyomelanin from HGA.

P. aeruginosa Growth Medium	Pyo_{sol} [g/L]	Pyo_{insol} [g/L]	sPyo [g/1 g of HGA]
PMM I	1.13 ± 0.12	0.71 ± 0.04	0.077 ± 0.01
PMM II	1.79 ± 0.18	1.22 ± 0.10	

2.2. Identification of the Functional Groups and Linkages in Pyomelanin Molecules

The structure of the pigments was investigated by FTIR spectroscopy, a technique in which a sample's absorbance of infrared light at various wavelengths is measured to determine the structure of molecules. The infrared spectrum includes absorbance bands corresponding with the various vibrations of the sample's atoms. Each chemical molecule will produce a unique infrared spectrum. It should be mentioned that structural characterization of melanin is elusive due to its complexity and metabolic residues, such as proteins, amino acids and carbohydrates, in cases of microbial origin. The FTIR spectra of microbial Pyo_{sol} and Pyo_{insol} and synthetic sPyo are presented in Figure 1. The FTIR spectra of bacterial Pyo_{sol} and Pyo_{insol} were similar at wavenumbers higher than 1600 cm^{-1}. Between 3700 and 3000 cm^{-1}, a large absorption band resulted from overlapping -OH groups and unsaturated carbon or aromatic rings.

At lower wavenumbers, the bands with three maxima assigned to the stretching vibrations of aliphatic C–H were at 2953, 2925, and 2855 cm^{-1} in Pyo_{insol} and 2960, 2944, and 2873 cm^{-1} in Pyo_{sol}. The band at 1700 cm^{-1} corresponding to carbonyl stretching (C=O) of the COOH groups was visible in Pyo_{insol}, whereas this band was absent in Pyo_{sol}. The bands at 1631 and 1613 cm^{-1} in Pyo_{insol} and at 1601 cm^{-1} in Pyo_{sol} were typical for C=C bonds conjugated with C=O groups (quinones). At wavelengths shorter than 1600 cm^{-1}, the differences between the spectra of Pyo_{sol} and Pyo_{insol} were more considerable. The bands at 1515 cm^{-1} ascribed to aromatic C_{ar}–H bonds, and at 1441 cm^{-1} from in-plane aromatic skeletal vibrations of C=C, were only visible in Pyo_{insol}. Similarly, the band of the phenolic–OH links at 1216 cm^{-1} appeared only in the Pyo_{insol} FTIR spectrum. The band associated with the O–H bonds of hydroxyl groups attached to the ring was strong at 1402 cm^{-1} in Pyo_{sol} and weak at 1385 cm^{-1} in Pyo_{insol}. The strong band at 1082 cm^{-1} in Pyo_{sol} and the shoulder band at 1065 cm^{-1} in Pyo_{insol} could be related to the stretching of a C–O band of a primary alcohol group (phenolic groups). The strong band at 857 cm^{-1} in Pyo_{sol} and weak band at 828 cm^{-1} in Pyo_{insol} were related to out-of-plane deformation vibrations of aromatic C_{ar}–H bonds.

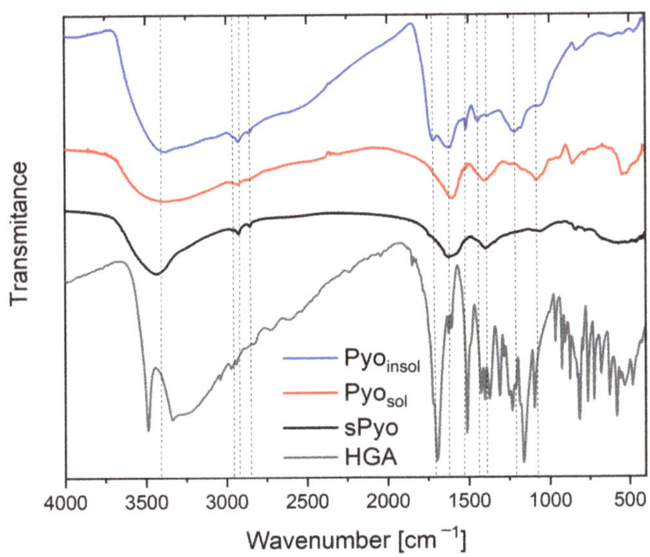

Figure 1. Fourier transform infrared (FTIR) spectra of the water-soluble pyomelanin (Pyo_{sol}), water-insoluble pyomelanin (Pyo_{insol}), synthetic pyomelanin (sPyo) and homogentisic acid (HGA).

The FTIR spectrum of sPyo in the range of 4000–1600 cm^{-1} was similar to the spectra of both types of bacterial pyomelanin, except that the band assigned to the O–H stretch had a narrower range of 3700–3300 cm^{-1}. The bands corresponding to aliphatic C–H bonds were at 2957, 2924 and 2852 cm^{-1}. The sPyo spectrum also presented a weak shoulder peak at 1738 cm^{-1} from COOH groups. The bands at 1515 and 1216 cm^{-1} visible for Pyo_{insol} were absent, similar to that of Pyo_{sol}. There were bands at 1395 and 1062 cm^{-1} from O–H and C_{ar}–O vibrations of hydroxyl groups attached to the ring, and bands from aromatic C–H bonds at 835 and 780 cm^{-1}. In the polymer chain of sPyo, there were more C_{ar}–C_{ar} linkages between rings, as indicated by the absence of a band at 1515 cm^{-1}.

2.3. *The Thermal Stability and Thermal Properties of Pyomelanins*

To determine the thermal stability of the pigments, thermogravimetric measurements in an inert atmosphere were performed. Thermal stability is a key parameter determining the possibility of processing pyomelanin by thermal processing methods used for the fabrication of composites and blends with other thermoplastic biopolymers. The TGA curves and the first derivative of the mass with respect to time (dm/dt = f(T)) curves of the dyes are presented in Figure 2, and the estimated characteristic parameters of thermal degradation are collected in Table 2. Based on the shape of the TGA and first derivative curves, it can be concluded that the dyes differed in degradation mechanism. The common feature of the dyes was the presence of the first stage of mass loss occurring up to 125 °C.

The maximum rate of the first stage of mass loss, set as the maximum of the first peak on dm/dt = f(T) curves, was located at a similar temperature denoted as the $T^{1st\ peak}$. This stage of mass loss could be assigned to the loss of volatile low molar mass species. The highest mass loss up to 125 °C was exhibited by Pyo_{sol}. Only at a temperature range of mass loss up to 125 °C did the endothermic effect occur for Pyo_{sol}, as indicated by the endothermic peak on the DSC curve of Pyo_{sol} (Figure 3). The temperature range of the endothermic effect, the corresponding highest mass loss in the case of Pyo_{sol} and the fact that Pyo_{sol} was isolated from the water phase, allows the assuming of the first stage of mass loss to water loss (Figure 2). The second stage of mass loss could be assigned to the beginning of pyrolysis of pyomelanin. The temperature of onset of the second mass loss can be taken as an upper limit of the thermal stability of pyomelanin (T_{deg}^{onset}). The highest T_{deg}^{onset} was observed for Pyo_{insol} (196.4 °C); for other pyomelanins, T_{deg}^{onset} was lower and was located at 173.0 °C and 158.0 °C for Pyo_{sol} and sPyo, respectively. Moreover, for sPyo, above 125 °C, continuous mass loss occurred up to the onset of the main degradation. Thus, Pyo_{insol} and Pyo_{sol} showed superior thermal stability to that of sPyo. Pyo_{sol} had a significantly greater residue at 800 °C (75%) than Pyo_{insol} and sPyo (ca. 40%), indicating a greater proportion of aromatic moieties or other conjugated unsaturated bonds in polymer chains.

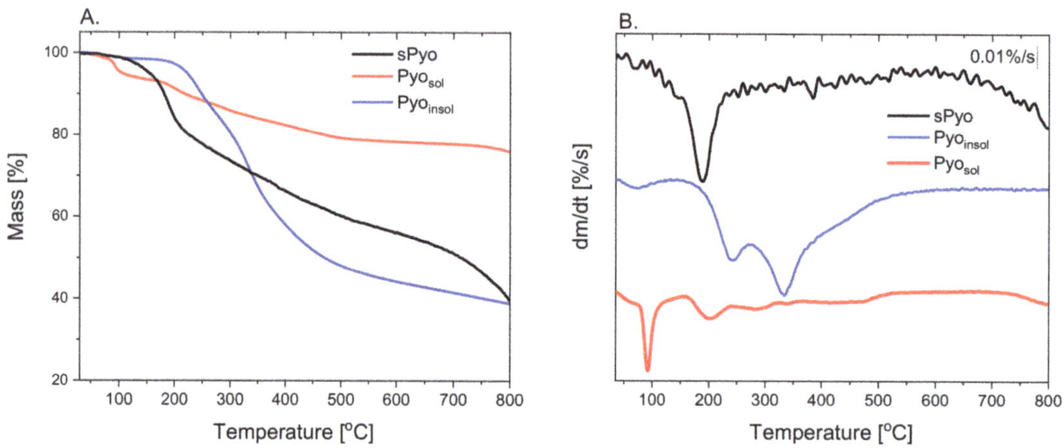

Figure 2. The thermogravimetry (TGA) (**A**) and the first derivative dm/dt (**B**) curves of the pyomelanins.

Table 2. Thermal Stability Parameters of the Polymers Estimated from Thermogravimetry and Thermal Properties from Differential Scanning Calorimetry.

Form of Pyomelanin	$T^{1st\ peak}$ [°C]	Mass Loss Up to 125 °C [%]	T_{deg}^{onset} [°C]	T_{deg}^{peak} [°C]	Residue at 800 °C [%]	T_1 [°C]	ΔH_1 [J/g]
sPyo	64.2	2.90	158.0	176.3	39.55	—	—
Pyo_{insol}	72.0	1.60	196.4	233.3	38.85	—	—
Pyo_{sol}	85.6	7.09	173.0	193.7	75.91	78.5	19.8

The TGA results confirmed that the procedures of isolation and purification of the pigments affect the thermal properties due to differences in the structure and composition of the final products, as indicated by FTIR analysis.

The first heating DSC curves of pyomelanin are presented in Figure 3, and the estimated thermal parameters are detailed in Table 2. On the DSC curves of the pigments, an endothermic peak corresponding to degradation was visible. The localization of the degradation peak maximum (T_{deg}^{peak}) was similar for sPyo and Pyo_{sol}, and that for Pyo_{insol}

started at a lower temperature. For Pyo_{sol}, a weak endothermic peak with a maximum at 78.5 °C was visible. The assignment of the endothermic effect to water loss agreed with the TGA results. At this temperature range, mass loss occurred for Pyo_{sol}, with the highest rate at 85.6 °C, as estimated from the first derivative dm/dt(T) curve. We also observed that Pyo_{sol} exhibited hygroscopic properties.

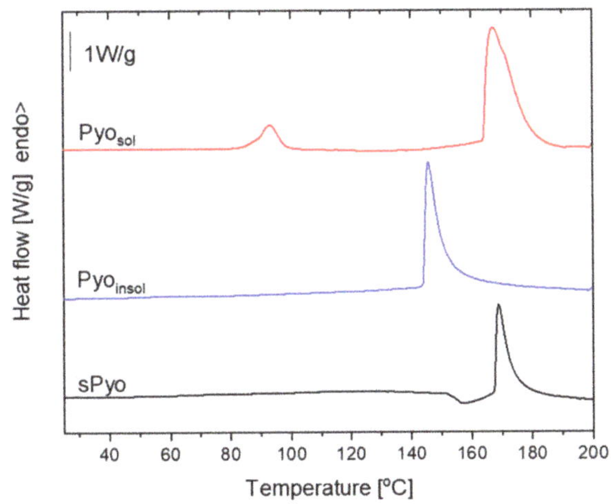

Figure 3. The first heating differential scanning calorimetry (DSC) curves of pyomelanin.

2.4. Biocompatibility

2.4.1. Bacterial Pyomelanins Show Higher Cytocompatibility Than the Synthetic Pyomelanin

Looking for new biomedical applications of bacterial polymers, including pyomelanin, requires characterizing their safety at the in vitro level and determining doses that do not show toxic effects. Moreover, differences in structure and thermal stability may translate into interactions of pyomelanin with cells through different ranges of safe concentrations. The influence of each form of pyomelanin on the viability of eukaryotic cells was assessed in the MTT reduction assay using the reference L-929 mouse fibroblasts and human THP-1 monocytes to exclude the cytotoxic effects toward immune cells. The ranges of safe concentrations of different pyomelanin formulations for eukaryotic cells are shown in Figure 4. Pyo_{sol} and Pyo_{insol}, in the full range of tested concentrations (1–1024 µg/mL), did not decrease the number of viable target cells, which were able to reduce MTT (both L-929 and THP-1-cell lines), below the 70% cell level, as required by the ISO norm. The unfavorable effect of reduced cell viability was observed only in the case of sPyo. Less than 70% of mouse fibroblasts or human monocytes were able to reduce MTT in the milieu of sPyo used in the concentration range 64–1024 µg/mL ($p < 0.05$) and 32–1024 µg/mL ($p < 0.05$), respectively (Figure 4).

2.4.2. Water-Soluble Pyomelanin Induces NF-κB Pathway Activation

In this study, THP1-Blue™ NF-κB human monocytes were used as biosensors of the NF-κB-driven signaling pathway in innate immune cells. NF-κB induction in these transformed cells results in the SEAP secretion. The amount of SEAP is proportional to cell activation. The levels of induction and activation of NF-κB in THP1-Blue™ NF-κB monocytes in response to the tested variants of pyomelanin are expressed as absorbance in Figure 5.

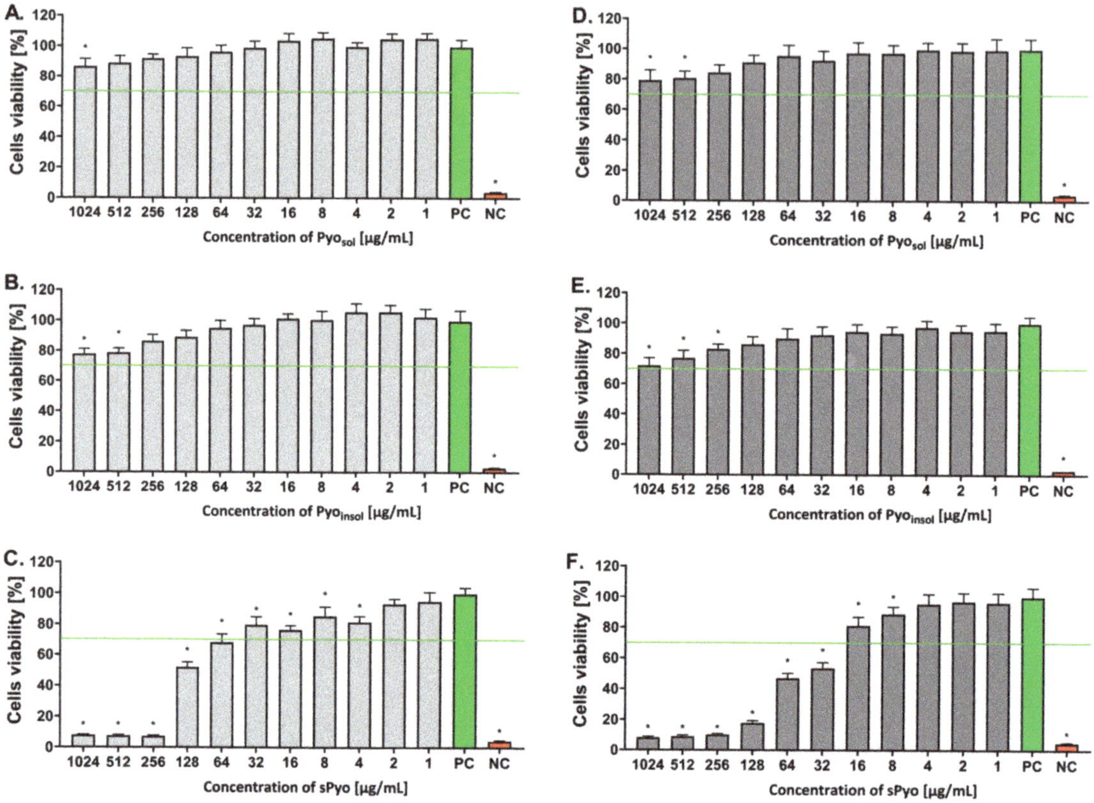

Figure 4. The percentage of viable cells after exposure to different concentrations of the tested pyomelanin variants. Viability of murine fibroblasts L-929 (**A–C**) and human monocytes (**D–F**) incubated for 24 h with water-soluble pyomelanin (Pyo$_{sol}$) (**A**,**D**), water-insoluble pyomelanin (Pyo$_{insol}$) (**B**,**E**) or synthetic pyomelanin (sPyo) (**C**,**F**), evaluated in the 3-(4,5-dimethylthiazol-2-yl)-2,5-diphenyltetrazolium bromide (MTT) reduction assay according to ISO-10993-5:2009. Cells incubated in the cell culture medium alone, without pyomelanin, served as a positive control (PC) of cell viability (100%). Cells treated with 3% H_2O_2 were a negative control (NC) (no viable cells). Data are presented as the mean ± standard deviation (SD) of five separate experiments (six replicates for each experimental variant). The green line indicates the minimum level (70%) of viable cells, which are able to reduce MTT according to the ISO norm. Statistical significance was calculated using ANOVA analysis, followed by Dunnett's post hoc test. Significant difference *—$p < 0.05$): cells exposed to tested pyomelanins vs. cells in culture medium alone. Statistical significance between Pyo concentrations was calculated using ANOVA analysis, followed by Tukey's post hoc test. The cytotoxicity of all forms of Pyo was dose-dependent in the concentration range 1024–256 µg/mL ($p < 0.05$).

Pyo$_{sol}$ activated NF-κB in the concentration range of 1–1024 µg/mL (the absorbance ranged from 0.70 to 1.43, $p < 0.05$) (Figure 5A). In cell cultures treated with Pyo$_{insol}$, a significant induction of NF-κB was observed in the concentration range of 64–1024 µg/mL ($p < 0.05$) (Figure 5B). However, the level of activation was lower than that in cell cultures exposed to Pyo$_{sol}$ (absorbance range: 0.30–0.53). Moreover, sPyo induced NF-κB in the concentration range of 1–64 µg/mL; however, there were no differences between induction levels in response to different pyomelanin concentrations within this range (the absorbance range 0.29–0.41) (Figure 5C). The level of NF-κB activation in response to *E. coli* LPS shown as absorbance was equal to 2.07 ($p < 0.05$) (Figure 5A–C).

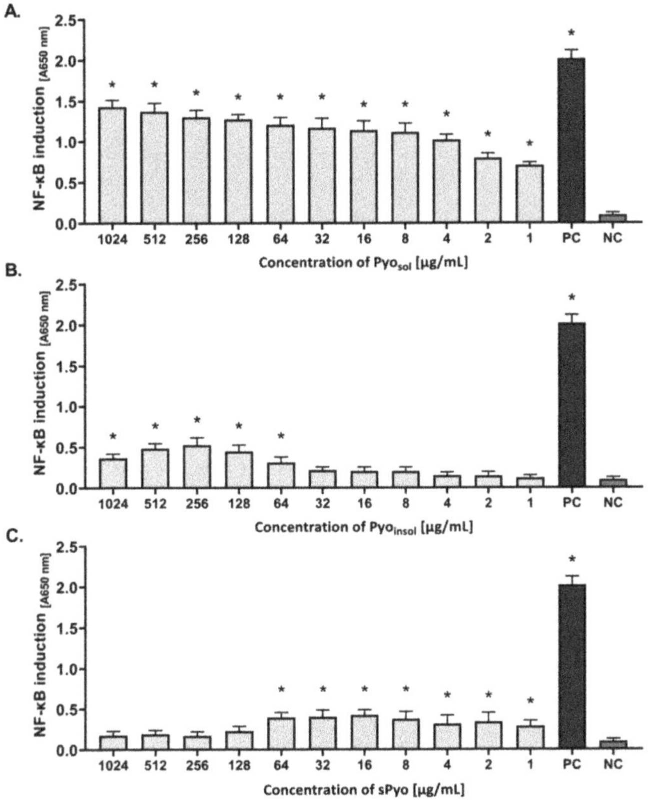

Figure 5. The level of activation of THP1-Blue™ NF-κB monocytes in response to the tested variants of pyomelanin. Cells were incubated for 24 h with (**A**) water-soluble pyomelanin (Pyo_{sol}), (**B**) water-insoluble pyomelanin (Pyo_{insol}), (**C**) synthetic pyomelanin (sPyo), or lipopolysaccharide (LPS) of *Escherichia coli* as a positive control (PC). Cells in culture medium alone served as the negative control (NC). The secreted embryonic alkaline phosphatase, which was used as an indicator of nuclear factor kappa B (NF-κB) activation, was quantified spectrophotometrically (OD = 650 nm) after enzymatic substrate conversion. Data are presented as the mean ± standard deviation (SD) of five separate experiments (six replicates of each experimental variant). Statistical significance was calculated using ANOVA analysis, followed by Dunnett's post hoc test. Statistical significance between Pyo concentrations was calculated using ANOVA analysis, followed by Tukey's post hoc test. Significant difference *, $p < 0.05$. A (absorbance)—optical density 650 nm.

2.4.3. In Vivo Toxicity of Pyomelanin

In the current study, *G. mellonella* larvae were used as a nonmammalian insect model reflecting the biological complexity of live organisms, which is an ethically accepted alternative for the examination of new formulation safety in vivo. By assessing the four physiological functions of the wax moth larvae after injection with the tested variants of pyomelanin, the HISS of the insects was evaluated and presented as a heatmap (Figure 6). Larvae with a total score in the range of 8.5–10.0 points were regarded as healthy, and the substances tested were considered nontoxic in this in vivo larval model.

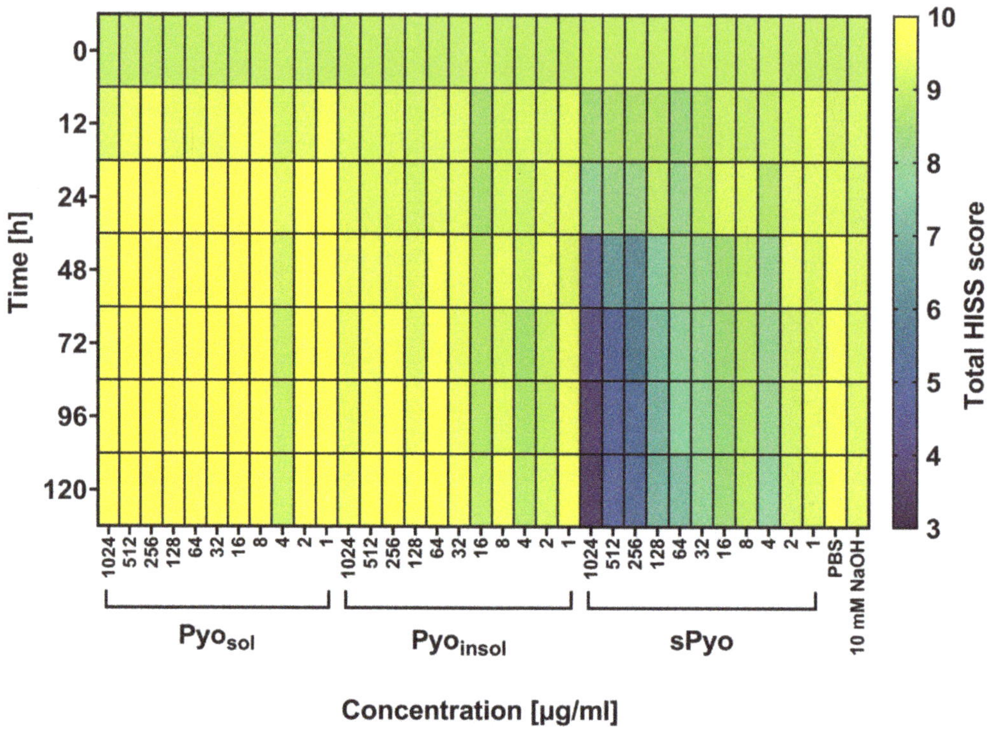

Figure 6. Total Health Index Scoring System (HISS) heatmap for *G. mellonella* larvae treated with tested pyomelanin variants. Larvae were injected with water-soluble (Pyo$_{sol}$), water-insoluble (Pyo$_{insol}$) or synthetic pyomelanin (sPyo) in the concentration range of 1–1024 µg/mL or control solvent solutions: phosphate-buffered saline (PBS) or 50 mM NaOH. At 0, 12, 24, 48, 72, 96, and 120 h after the injection of pyomelanin, the health of the larvae was evaluated using HISS and expressed as the total HISS score. Statistical significance was calculated using ANOVA analysis, followed by Dunnett's post hoc test. Statistical significance between Pyo concentrations was calculated using ANOVA analysis, followed by Tukey's post hoc test. sPyo toxicity to *G. mellonella* was dose-dependent in concentration range 1024–64 µg/mL ($p < 0.05$).

As shown in Figure 6, no deleterious effects of the two variants of bacterial pyomelanin (Pyo$_{sol}$ and Pyo$_{insol}$) were seen in the *G. melonella* in vivo model at 0, 12, 24, 48, 72, 96 and 120 h after injection of insect larvae with pyomelanin. The total HISS scores for larvae treated with *P. aeruginosa* pyomelanins were close to the baseline HISS score, which was equal to 9.0. The total HISS scores in larvae injected with sPyo in the range of 4–1024 µg/mL were lower (7.79–3.58) than the total HISS scores in larvae inoculated with natural *P. aeruginosa* pyomelanin variants. Pyo$_{sol}$ and Pyo$_{insol}$ did not affect the viability of insect larvae, while sPyo used in the range of 64–1024 µg/mL significantly decreased the number of live larvae (42–83%, $p < 0.05$) compared to control larvae not injected with pyomelanin (Supplementary Figure S2).

3. Discussion

In our study, we described for the first time the method of culturing *Pseudomonas aeruginosa* on the PMM II medium and the method of obtaining water-soluble pyomelanin (Pyo$_{sol}$). In addition, we compared the physicochemical and biological properties of bacterial Pyo$_{sol}$ to its insoluble form (Pyo$_{insol}$) and synthetic pyomelanin (sPyo).

We compared our results with several reports on the efficiency of the production of microbial melanins, which depends on the metabolic abilities of bacteria and culture conditions. Madhusudhan et al. reported a production efficiency of extracellular water-soluble melanin of *Streptomyces lusitanus* at the levels of 0.264 g/L and 5.29 g/L [27]. Lagunas-Muñoz et al. showed that recombinant *E. coli* expressing the tyrosinase coding gene from *Rhizobium* produced 6.0 g/L melanin [28]. The yield of water-insoluble bacterial melanins showed a wide range for different species of bacteria cultured in bioreactors: 13.7 g/L for *Streptomyces kathirae*, 3.76 g/L for *Flavobacterium kingsejongi* and 0.125 g/L for *Streptomyces glaucens* [23,29,30]. Significant differences in melanin production efficiency were also observed within the *Pseudomonas* species, with 6.7 g/L water-insoluble melanin obtained from *Pseudomonas stutzeri* and 0.35 g/L obtained from ***Pseudomonas putida*** [31,32]. A high efficiency of bacterial pyomelanin production with limited secretion of undesirable substances is key to obtaining a bacterial pigment for further physicochemical and biological studies.

The FTIR spectra of microbial Pyo_{sol}, Pyo_{insol} and sPyo exhibit the characteristic bands of the pyomelanins described in the literature [33,34]. Based on the relatively high degree of similarity of the spectra of sPyo and Pyo_{sol}, it can be concluded that the structure of sPyo is more similar to that of Pyo_{sol} than that of Pyo_{insol}. The carbonyl stretching (C=O) of the COOH groups was only visible in Pyo_{insol}. The lack of this band can be similarly found in the literature for microbial melanin isolated in acid precipitation [4,35,36]. The FTIR spectrum of Pyo_{insol} was similar to the FTIR spectrum of melanin produced from a deep-sea sponge-associated *Pseudomonas* strain [35] and pyomelanin from a culture of *Halomonas titanicae* which was produced through the 4-hydroxyphenylacetic acid-1-hydroxylase route [6]. Lorquin et al. concluded that the presence of the band ascribed to aromatic C_{ar}–H has a significant meaning in terms of the type of ring linkages. This group suggested that a pyomelanin that does not have this band has fewer free C_{ar}–H locations and more Car-Car connections between rings in the chain structure [6]. It may suggest that Pyo_{sol} contained more Car-Car linkages than Pyo_{insol}. Moreover, the comparison of the FTIR spectra of sPyo and HGA showed that the sPyo did not contain HGA residue detectable by FTIR measurements. Differences in chemical structure may influence the solubility of various forms of pyomelanins, their thermal stability and biological activity; however, further studies are required.

The TGA and DSC results confirmed that the procedures of isolation and purification of the pigments affect the thermal properties due to differences in the structure and composition of the final products, as indicated by FTIR analysis. Pyo_{sol} showed a significantly greater residue at 800 °C than Pyo_{insol} and sPyo, indicating a greater proportion of aromatic moieties or other conjugated unsaturated bonds in polymer chains [37]. This result was in agreement with the FTIR results revealing a higher content of C_{ar}–C_{ar} linkages between rings in Pyo_{sol} than in the other pyomelanin samples. For Pyo_{sol}, we identified a weak endothermic peak with a maximum at 78.5 °C which is associated with water loss [38]. The presence of this peak only for Pyo_{sol} was a consequence of the isolation of pyomelanin from the aqueous phase. Similar to our DSC results for Pyo_{sol}, the showing of two endothermic peaks was reported for microbial melanin by Kiran et al. [35]. We also showed that Pyo_{sol} exhibited hygroscopic properties. Melanin is known for its hygroscopic character and strong association with water [39].

The use of bacterial-derived pyomelanin and synthetic pyomelanin as potential immunomodulators and bioactive substances for further targeted biomedical applications requires determining the range of cytocompatible concentrations to avoid negative effects on cell metabolism and viability. We have shown that bacterial pyomelanins are characterized by high in vitro safety for L-929 fibroblasts and THP-1 monocytes compared to the synthetic form of this pigment. The high level of Pyo_{sol} and Pyo_{insol} cytocompatibility resulted from the effective removal of lipopolysaccharide and other bacterial metabolites that are cytotoxic. The lower cytocompatibility of sPyo compared to Pyosol or Pyo_{sol} against L-929 fibroblasts and human monocytes, as shown in this study, might be due to polymerization and structural differences between sPyo and Pyo_{insol} or Pyo_{sol} (a lower ability to create hydrogen bonds). Interestingly, no residual HGA was observed in the sPyo samples, which could adversely affect cell viability. However, further studies are needed to determine the components influencing the biological activity of the studied pyomelanins. Potentially, the biological activity depends on the complexity of interactions between various functional groups within each variant of pyomelanin.

Several studies have demonstrated the biosafety ranges of microbial melanins in vitro. Oh et al. reported that melanin from *Amorphotheca resinae* in the range of 200–4000 μg/mL did not affect the viability of human keratinocytes HaCaT after 24 h of exposure of cells [40]. Lorquin et al. showed that pyomelanin from *H. titanicae* and its synthetic form resulting from HGA polymerization in the milieu of Mn^{2+} were noncytotoxic to human epidermal keratinocytes [6]. Melanin from *Dietzia schimae*, which possesses photoprotective activity, was safe for human fibroblast hFB at concentrations below ≤500 μg/mL [41]. Pyomelanin produced by various *Pseudomonas* species may differ in cytotoxicity toward eukaryotic cells. In the study by Kurian and Bhat, the highest concentration of melanin from *Pseudomonas stuteri*, which was safe for L-929 fibroblasts, was 100 μg/mL [42]. The cytotoxicity of pyomelanin from *P. putida* against A-375, HeLa Kyoto, HEPG2 or Caco2 cell lines was examined by Ferraz et al. and expressed as the cytotoxicity index IC_{50}, with values of 1770 μg/mL, 2510 μg/mL, 890 μg/mL and 1080 μg/mL, respectively [43].

Monocytes play a key role in the development of inflammatory and immune responses, which determine the elimination of infectious agents, induction of antigen-specific adaptive immunity and tissue regeneration; thus, testing new components with medical potential in humans regarding the effectiveness of monocyte activation is needed [44]. The level of activation may vary depending on the cell type, the chemical structure of biocomponents and the cell milieu. In response to tissue damage, monocytes and macrophages deliver proinflammatory cytokines, including chemokines, which facilitate the recruitment of immunocompetent cells, and the removal of debris, which is a prerequisite for successful healing. In subsequent stages of healing, macrophages can reduce inflammation, through the secretion of anti-inflammatory cytokines, can control the differentiation of stem cells, and can regulate angiogenesis [45]. It has been revealed that acute inflammation or low doses of proinflammatory cytokines are necessary for the reconstruction of bone tissue; therefore, modulation of the NF-κB pathway by pyomelanin may influence bone remodeling [46].

In this study, we have shown that Pyo_{sol} is non-cytotoxic in the widest range of concentrations, which is compatible with NF-κB induction compared to the activity of sPyo or Pyo_{insol}. However, the poor activation of NF-κB by water-insoluble and synthetic forms of pyomelanin does not rule them out from further biological studies. Depending on the application context, the proinflammatory or anti-inflammatory activity of pyomelanin can be considered. Proinflammatory activity is desirable in fighting against infection and in the early stages of tissue regeneration, whereas to prevent chronic inflammation, anti-inflammatory properties of biocomponents are needed. Allam et al. reported that *Streptomyces longisporoflavus* melanin improves immune defense against *Escherichia coli* infection [47]. The severity of oxidative stress is the result of a strong inflammatory response due to the activation of monocytes. Langhfelder et al. reported on the ability of fungal melanins to neutralize such stress [11]. On the other hand, the initiation of regeneration processes requires the stimulation of monocytes to secrete proinflammatory cytokines, which was recently reported [19]. The Pyo_{sol} used in this study seems to meet the highest biological safety, combined with the activation of monocytes, and due to this may be further tested for immunomodulatory and pro-regenerative properties in different cell models.

The results from the in vivo model of *G. mellonella* obtained in this study correspond to the observations on the safety of various forms of pyomelanin in the in vitro model. These results are very useful in selecting potential pyomelanin applications. In particular, Pyo_{sol} and Pyo_{insol} seem to meet the requirements for further studies on potential medical applications. In contrast, sPyo can be further investigated for nonmedical applications.

4. Materials and Methods

4.1. Culture of Pseudomonas aeruginosa

To obtain pyomelanin with a limited content of undesirable substances (e.g., alginate, excess protein products), a new minimal liquid medium for *P. aeruginosa* cultivation was developed. Two versions of pyomelanin minimal medium (PMM) were prepared. The first PMM version (PMM I) contained 2.0 g of KH_2PO_4, 5.0 g of NaCl, 0.1 g of $MgSO_4$, 2.0 g of L-tyrosine and 2.0 g of glucose per 1000 mL of distilled water. The second version (PMM II) was additionally supplemented with 1.5 g of arabinose and 1.35 g of malic acid. All chemicals were purchased from PolAura, Dywity, Poland. After dissolving the substrates, the pH of both media was adjusted to 7.0 with 0.5 M NaOH. PMM I and PMM II were autoclaved at 121 °C, 2.5 Ba. Luria–Bertani (LB) broth medium was inoculated with the *P. aeruginosa* Mel^+ strain deposited in the collection of the Department of Immunology and Infectious Biology UŁ, Poland, and cultured (37 °C, 18 h) to obtain an initial bacterial suspension. After incubation, 300 mL of PMM II was inoculated with 1.0 mL of a 1.0 McFarland bacterial suspension and grown for 5 days (37 °C, shaking at 120 rpm) when the culture medium changed to a deep black-brown color. To increase the production of pyomelanin, after cultivation, the bacterial cultures were exposed to sunlight for 2 days at room temperature.

4.2. Isolation of Pyo_{insol}

To isolate Pyo_{insol} from the cell-free supernatant, 300 mL of bacterial culture was centrifuged at 3300× *g* for 15 min, and then the supernatant was acidified with 6.0 M HCl (PolAura, Dywity, Poland) to pH 2.0 and stored, protected from light, at room temperature for 5 days. Thereafter, the supernatant was boiled for 45 min to avoid the formation of melanoidins after cooling, and the supernatant was centrifuged at 3300× *g* for 25 min. The pellet of Pyo_{insol} was washed three times with 25 mL of 0.1 M HCl and then three times with double distilled water. Afterward, 10 mL of ethanol (99.9%) (Chempur, Piekary Śląskie, Poland) was added to the pigment pellet and placed in a water bath (95 °C, 30 min). After storage in an incubator (50 °C, overnight), for complete evaporation of alcohol, the pyomelanin was washed twice with ethanol and air dried.

4.3. Isolation of Pyo_{sol}

To obtain Pyo_{sol}, the postculture bacterial cell-free supernatant was incubated with chloroform in a 1:1 ratio under shaking conditions for 24 h (room temperature, shaking at 120 rpm). The aqueous phase containing pyomelanin was then separated from the chloroform and protein phases using a separating funnel. To remove residual protein contaminants, the aqueous layer was centrifuged at 5300× *g* for 30 min. Pyo_{sol} was concentrated and purified from low-molecular-weight soluble substances by centrifuging the supernatant on an ultrafiltration unit (3300× *g*, 60 min., MWCO 30 kDa) (Sartorius, Göttingen, Germany). The bacterial pigment was dried at 50 °C overnight.

4.4. Synthesis of Pyomelanin

HGA (TCI, Eschborn, Germany), which is the main precursor for pyomelanin in *P. aeruginosa*, was used to prepare sPyo. HGA (1.0 g) was dissolved in 400 mL of distilled water (heated to 50 °C), and then a solution of 4.0 M NaOH was added to achieve pH 10.5. Autoxidation of HGA to sPyo was carried out for 10 days at 37 °C in the absence of light. The tube with the HGA solution was opened once a day to provide a new portion of oxygen. When a dark brown pigment was observed in the tube, sPyo was precipitated with 10.0 M HCl (to pH 6.0), left to sediment for 24 h and centrifuged (3300× *g*, 25 min). The pyomelanin pellet was suspended in 2.2 M HCl and left for 2 days to stabilize the pigment. Then, sPyo was centrifuged (6600× *g*, 10 min) and washed three times with 0.1 M HCl and double-distilled water.

4.5. Purification of Bacterial Pyomelanins

Lipopolysaccharides (LPSs) were removed from *P. aeruginosa* pyomelanin by affinity chromatography using Pierce™ High Capacity Endotoxin Removal Spin Columns (Thermo Scientific, Waltham, MA, USA). The resin and column were prepared and equilibrated according to the manufacturer's protocol. Samples of Pyo_{sol} and Pyo_{insol} (5 mg/mL) were applied to the columns, incubated for 3 h with gentle mixing, centrifuged at 500× *g*, collected into new tubes and dried at 50 °C overnight. The pyomelanin pellet was washed with chloroform, ethyl acetate, ethanol and water. For further experiments, the pyomelanin was stored in a dark and dry place at 4 °C. The methodology for the isolation and purification of bacterial pyomelanins is shown in Figure 7.

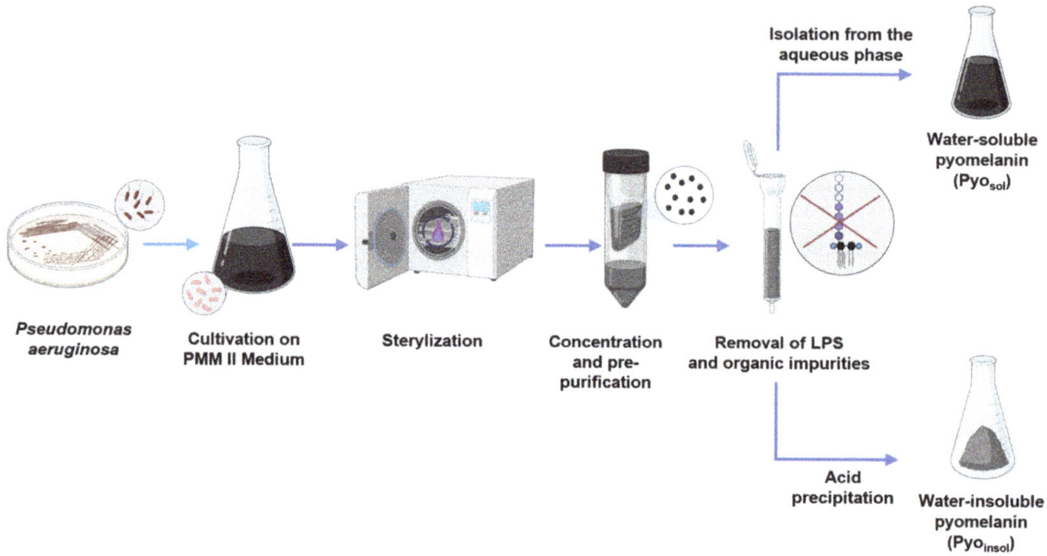

Figure 7. Schematic representation of isolation and purification of bacterial pyomelanins. Abbreviations: water-soluble pyomelanin (Pyo_{sol}), water-insoluble pyomelanin (Pyo_{insol}), synthetic pyomelanin (sPyo), lipopolysaccharide (LPS).

4.6. Fourier Transform infrared (FTIR) Spectroscopy

FTIR spectra in transmission mode were collected from 4000 to 400 cm^{-1} using the KBr pellet technique on a Thermo Nicolet Nexus FTIR spectrometer (Thermo Fisher Scientific, Waltham, MA, USA) and analyzed with Thermo Scientific Omnic Software ver. 8.3.

4.7. Thermogravimetric Analysis (TGA)

TGA measurements were performed using a TGA/DSC1 Mettler Toledo system (Mettler Toledo, Greifenesee, Switzerland) [48]. Samples were heated from 25 °C to 800 °C at a rate of 10 °C/min under 60 mL/min of nitrogen flow. The evaluation of the TGA curves was performed using STARe ver. 16.20c software (Mettler Toledo, Greifenesee, Switzerland). The first derivative of mass over time was calculated with OriginPro ver. 2021 (OriginLab Corporation, Northampton, MA, USA) and plotted against temperature. The Savitzky-Golay smoothing algorithm was implemented. A 20 point window and a second-order polynomial was used.

4.8. Differential Scanning Calorimetry (DSC)

DSC measurements were performed using a Mettler Toledo DSC1 system (Mettler Toledo, Greifenesee, Switzerland) coupled with a Huber TC 100 intracooler (Huber USA, Inc., Raleigh, USA) [49]. The instrument was calibrated using indium (T_m = 156.6 °C,

ΔH_m = 28.45 J/g) and zinc (T_m= 419.7 °C, ΔH_m = 107.00 J/g) standards. Samples (~3.5 mg) were measured in 40 μL aluminum pans under a constant nitrogen purge (60 mL/min) from 0 °C to 200 °C. The heating and cooling rates were set to 10 °C/min. The recorded DSC curves were normalized to the sample mass. The evaluation of the DSC curves was performed using STAR^e ver. 16.20c software (Mettler Toledo, Greifenesee, Switzerland).

4.9. *Assessment of Pyomelanin Biocompatibility*

4.9.1. Cell Cultures

The biocompatibility of pyomelanin was assessed in vitro according to ISO 10993-5:2009 (Biological evaluation of medical devices—Part 5: Tests for in vitro cytotoxicity) using two cell lines: the reference L-929 (CCL-1™) mouse fibroblasts and human monocytes THP-1 (TIB-202™), which were obtained from the American Type Culture Collection (ATCC, Manassas, VA, USA). Prior to experiments, cells were cultured in Roswell Park Memorial Institute (RPMI)-1640 medium supplemented with 10% heat-inactivated fetal calf serum (FCS; HyClone Cytiva, Marlborough, MA, USA) and the antibiotics penicillin (100 U/mL) and streptomycin (100 μg/mL) (Sigma-Aldrich, Darmstadt, Germany). Mouse fibroblasts and human monocytes were incubated at 37 °C in a humidified atmosphere containing 5% CO_2 until the formation of the cell monolayer. Before being used in the experiments, the cell viability and cell density were assessed by trypan blue exclusion assay using a counting Bürker chamber (Blaubrand, Wertheim, Germany). The cells were used in the experiments only if cell viability was higher than 95%.

4.9.2. MTT Reduction Assay

The biocompatibility of pyomelanins was assessed in vitro in cell cultures using a 3-(4,5-dimethylthiazol-2-yl)-2,5-diphenyltetrazolium bromide (MTT, Sigma-Aldrich, Darmstadt, Germany) reduction assay as previously described [50] and as recommended by the Food and Drug Administration and ISO norm 109935 [ISO 10993-10995:2009. Biological evaluation of medical devices—Part 5: Tests for in vitro cytotoxicity]. L-929 fibroblasts or THP-1 monocytes adjusted to a density of 2×10^5 cells/mL were seeded (20,000 cells per well) in 96 well culture plates (Nunclon Delta Surface, Nunc, Rochester, NY, USA) and incubated overnight prior to stimulation with pyomelanin. Cell morphology and confluency were controlled using an inverted contrast phase microscope (Motic AE2000, Xiamen, China). Stock solutions of Pyo_{insol} or sPyo, 10 mg/mL in 50 mM NaOH (PolAura, Dywity, Poland) in complete RPMI-1640 (cRPMI-1640) were diluted with medium to concentrations of 1024, 512, 256, 128, 64, 32, 16, 8, 4, 2 and 1 μg/mL. An identical series of dilutions was prepared for Pyo_{sol} starting from a stock solution at a concentration of 2 mg/mL initially dissolved in cRPMI-1640. The pyomelanin solutions were sterilized by filtration using filters with a 0.22 μm pore diameter (Sartorius, Göttingen, Germany). Suspensions of pyomelanins were distributed to the wells of culture plates (6 replicates for each experimental variant) containing cell monolayers. After 24 h of incubation, the condition of the cell monolayers was verified under an inverted contrast phase microscope. The cell cultures in medium without pyomelanin were used as a positive control (PC) of cell metabolic activity, whereas the cell cultures in 3% H_2O_2 served as a negative control (NC). To quantify the metabolic activity of cells, 20 μL of MTT was added to each well, and incubation was carried out for the next 4 h. The plates were centrifuged at 450× *g* for 10 min, the supernatant was removed, and the formazan crystals were dissolved with 100 μL of dimethyl sulfoxide (Sigma Aldrich, Seelze, Germany). The absorbance was determined spectrophotometrically using a Multiskan EX reader (Thermo Scientific, Waltham, MA, USA) at 570 nm. The effectiveness of MTT reduction was calculated based on the following formula: MTT reduction relative to untreated cells (%) = (absorbance of treated cells/absorbance of untreated cells × 100%) − 100%.

4.9.3. Activation of Monocytes

THP1-Blue™ NF-κB monocytes (InvivoGen, San Diego, CA, USA), derived from human THP-1 monocytes, were used to determine the activation of the NF-κB signal transduction pathway, as previously described [51], in response to exposure of cells to sPyo, Pyo_{insol} or Pyo_{sol}. THP1-Blue™ NF-κB cells are specific biosensors of the NF-κB pathway, which is typical for innate immune cells [52,53]. The induction of NF-κB results in secretion of embryonic alkaline phosphatase (SEAP) by these cells. Cell suspensions, 2×10^6 cells/mL in selective RPMI-1640 supplemented with heat-inactivated 10% FCS (HyClone, Cytiva, Marlborough, MA, USA), 25 mM 4-(2-hydroxyethyl)-1-piperazineethanesulfonic acid (HEPES), 100 U/mL penicillin, 100 μg/mL streptomycin, 2 mM glutamine and selective antibiotics (100 μg/mL normocin and 10 μg/mL blasticidin) (InvivoGen, San Diego, CA, USA), at a density below 2×10^6 cells/mL, were cultured for 5 days in a humidified 5% CO_2 atmosphere at 37 °C. Freshly prepared suspensions of monocytes in culture medium were distributed to the wells of culture plates (1×10^5 cells/well; 180 μL). Then, 20 μL of tenfold-concentrated pyomelanin solution was added to selected wells (in six replicates) to a final concentration of 1024, 512, 256, 128, 64, 32, 16, 8, 4, 2 and 1 μg/mL. Cells were incubated for 24 h in an incubator. Monocytes in selective RPMI-1640 alone served as an negative control (NC), whereas monocytes stimulated with 10 ng/mL LPS from *Escherichia coli* O55:B5 (Sigma-Aldrich, Darmstadt, Germany) were used as a positive control (PC) for NF-κB activation. The level of SEAP secretion was determined in the cell culture supernatants. Cell-free supernatant (20 μL) was mixed with 180 μL QUANTI-Blue™ (InvivoGen, San Diego, CA, USA) and incubated at 37 °C for 4 h. Absorbance was measured at 650 nm using a Multiskan EX reader (Thermo Scientific, Waltham, MA, USA). The results are expressed as the mean and standard deviation (SD) of five experiments performed in six replicates for each experimental variant.

4.9.4. In Vivo Toxicity Assay

Different formulations of pyomelanin were examined for their toxicity in vivo using the model of last instar *Galleria mellonella* larvae, which enables real-time cytotoxicity testing [53–55]. In the assessment of the last instar, the size of the larva, width and degree of scleritization of the head capsule were analyzed, and the ecdysial line along the middle of the dorsal side was observed. Before the experiment, the last instar of larvae (purchased from a local vendor), confirmed by the specialist from the Department of Ecology and Vertebrate Zoology, Faculty of Biology and Environmental Protection, University of Łódź, Poland, were stored in the dark in a refrigerator at 15 °C to minimize transformation into adult form. Prior to conducting the assay, each larva was sterilized with a cotton swab dipped in 70% ethanol. The bioassay was performed in glass Petri dishes. Twelve larvae (200–300 mg weight) were injected with 10 μL of Pyo_{insol}, Pyo_{sol} or sPyo into the hemocoel via the intersegmental membrane near the last left proleg using a microsyringe (Sigma Aldrich, Darmstadt, Germany). Control larvae were injected with phosphate-buffered saline (PBS) or 10 mM NaOH (control solvent for Pyo_{sol} and sPyo). The physiological activities of insects or the number of dead insects were recorded at 0, 12, 24, 48, 72, 96 and 120 h after the injection. The larval health status was evaluated using the Health Index Scoring System (HISS), which is based on the following symptoms: larval mobility, cocoon formation, melanization of the body integuments and survival [55] (Table 3). In this system, the total melanization of the larvae (black larvae) and loss of larval motility correlate with the death of the larvae [56]. Representative pictures of morphological changes in larvae injected with Pyo_{sol}, Pyo_{insol} or sPyo are shown in Supplementary Figure S1.

Table 3. The Health Index Scoring System (HISS) for *G. mellonella* Larvae.

Category	Description	Score
Mobility	no movement	0
	minimal movement on stimulation	1
	movement when stimulated	2
	movement without stimulation	3
Cocoon formation	no cocoon	0
	partial cocoon	0.5
	full cocoon	1
Melanization	black larvae	0
	black spots on brown larvae	1
	≥3 spots on beige larvae	2
	<3 spots on beige larvae	3
	no melanization	4
Survival	dead	0
	alive	2

4.10. Statistical Analysis

The Kolmogorov–Smirnov test was used to test the normality of the data. Intergroup outcomes were compared for statistical significance using ANOVA (analysis of variance) followed by Dunnett's post hoc test. Statistical significance between pyomelanin concentrations was calculated using ANOVA analysis, followed by Tukey's post hoc test. In all cases, significance was accepted at $p < 0.05$. All analyses were performed using GraphPad Prism 9 software (GraphPad Software, San Diego, CA, USA).

5. Conclusions

Taking into account the need to search for biocomponents, including those of bacterial origin with multidirectional biological activity (antimicrobial, immunomodulatory and proregenerative), the aim of this study was to obtain natural bacterial pyomelanin from *Pseudomonas aeruginosa*, namely, Pyo_{sol} and Pyo_{insol}, and synthetic pyomelanin (sPyo). Furthermore, we characterized these three variants of pyomelanin in terms of chemical structure and biosafety for further targeted biomedical research. The culture medium increasing the production of pyomelanin was developed as well as the conditions for the isolation and synthesis of pyomelanin from HGA. FTIR analysis showed that the most important difference between variants of pyomelanin concerns the connections of aromatic rings in polymer chains. In the case of Pyo_{sol} and sPyo, the C_{ar}-C_{ar} connections between rings dominate the chain structure, whereas Pyo_{insol} showed fewer C_{ar}-C_{ar} links between rings. For Pyo_{sol}, the wide band at 3700–3000 cm^{-1} indicated the presence of condensed double bonds as well as the presence of OH groups involved in hydrogen interactions, e.g., with water. We observed that Pyo_{sol} exhibited hygroscopic properties (atmospheric moisture caused clumping). TGA confirmed the highest water content in Pyo_{sol}. Further chemical research is required to fully define the structural differences in the molecules of different variants of pyomelanin. A high level of biological safety allows for the recommendation of Pyo_{sol} and Pyo_{insol} for further biomedical studies, including research on their antimicrobial, immunomodulatory, and proregenerative activities. Investigating the biosafety and modulation of the physiological activity of other cell lines will guide research on pyomelanin towards the development of pyomelanin bioactive preparations.

Supplementary Materials: The following supporting information can be downloaded at: https://www.mdpi.com/article/10.3390/ijms24097846/s1.

Author Contributions: Conceptualization, M.M.U. and M.C.; methodology, M.M.U., M.N. and M.G.; validation, M.M.U., M.N. and M.G.; formal analysis, M.M.U., K.R. and P.P.; investigation, M.M.U., M.N. and M.G.; resources, M.M.U., M.N. and M.G.; data curation, M.M.U., M.N. and M.G.; writing—original draft preparation, M.M.U., P.P. and M.C.; writing—review and editing, K.R., M.M.U., M.G.

and M.N.; supervision, M.C. and K.R.; project administration, K.R.; funding acquisition, K.R. All authors have read and agreed to the published version of the manuscript.

Funding: This research was funded by the Foundation for Polish Science through the European Union under the European Regional Development Fund, within the TEAM-NET program entitled "Multifunctional biologically active composite for application in bone regenerative medicine". (POIR.04.04.00-00-16D7/18-00).

Institutional Review Board Statement: Not applicable.

Informed Consent Statement: Not applicable.

Data Availability Statement: The data generated during this study are available at University of Łódź, Faculty of Biology and Environmental Protection, Department of Immunology and Infectious Biology, Łódź, 90-237, Poland, and are available from the corresponding authors upon request.

Acknowledgments: The authors are thankful to Mariusz Tszydel, from the Department of Ecology and Vertebrate Zoology, Faculty of Biology and Environmental Protection, University of Łódź, for their expert evaluation and confirmation of the *Galleria mellonella* larval stage used in the in vivo toxicity assessment.

Conflicts of Interest: The authors declare no conflict of interest. The funders had no role in the design of the study; in the collection, analyses, or interpretation of data; in the writing of the manuscript; or in the decision to publish the results.

References

1. Behzadi, P.; Baráth, Z.; Gajdács, M. It's Not Easy Being Green: A Narrative Review on the Microbiology, Virulence and Therapeutic Prospects of Multidrug-Resistant *Pseudomonas aeruginosa*. *Antibiotics* **2021**, *10*, 42. [CrossRef]
2. Liebgott, P.P.; Labat, M.; Amouric, A.; Tholozan, J.L.; Lorquin, J. Tyrosol Degradation Via the Homogentisic Acid Pathway in a Newly Isolated Halomonas Strain from Olive Processing Effluents. *J. Appl. Microbiol.* **2008**, *105*, 2084–2095. [CrossRef]
3. Ketelboeter, L.M.; Potharla, V.Y.; Bardy, S.L. NTBC Treatment of the Pyomelanogenic *Pseudomonas aeruginosa* Clinical Isolate PA1111 Inhibits Pigment Production and Increases Sensitivity to Oxidative Stress. *Curr. Microbiol.* **2014**, *69*, 343–348. [CrossRef]
4. David, C.; Daro, A.; Szalai, E.; Atarhouch, T.; Mergeay, M. Formation of Polymeric Pigments in the Presence of Bacteria and Comparison with Chemical Oxidative Coupling—II. Catabolism of Tyrosine and Hydroxyphenylacetic Acid by Alcaligenes eutrophus CH34 and Mutants. *Eur. Polym. J.* **1996**, *32*, 669–679. [CrossRef]
5. Roy, S.; Rhim, J.W. New Insight into Melanin for Food Packaging and Biotechnology Applications. *Crit. Rev. Food Sci. Nutr.* **2021**, *62*, 4629–4655. [CrossRef]
6. Lorquin, F.; Ziarelli, F.; Amouric, A.; Di Giorgio, C.; Robin, M.; Piccerelle, P.; Lorquin, J. Production and Properties of Non-Cytotoxic Pyomelanin by Laccase and Comparison to Bacterial and Synthetic Pigments. *Sci. Rep.* **2021**, *11*, 8538. [CrossRef] [PubMed]
7. Fonseca, É.; Freitas, F.; Caldart, R.; Morgado, S.; Vicente, A.C. Pyomelanin Biosynthetic Pathway in Pigment-Producer Strains from the Pandemic *Acinetobacter baumannii* IC-5. Mem. *Inst. Oswaldo Cruz.* **2020**, *115*, e200371. [CrossRef] [PubMed]
8. Zeng, Z.; Cai, X.; Wang, P.; Guo, Y.; Liu, X.; Li, B.; Wang, X. Biofilm Formation and Heat Stress Induce Pyomelanin Production in Deep-Sea *Pseudoalteromonas* sp. SM9913. *Front. Microbiol.* **2017**, *8*, 1822. [CrossRef] [PubMed]
9. Rodríguez-Rojas, A.; Mena, A.; Martín, S.; Borrell, N.; Oliver, A.; Blázquez, J. Inactivation of the hmgA Gene of *Pseudomonas aeruginosa* Leads to Pyomelanin Hyperproduction, Stress Resistance and Increased Persistence in Chronic Lung Infection. *Microbiology* **2009**, *155*, 1050–1057. [CrossRef]
10. Baishya, J.; Wakeman, C.A. Selective Pressures during Chronic Infection Drive Microbial Competition and Cooperation. *NPJ Biofilms Microbiomes* **2019**, *5*, 16. [CrossRef]
11. Langfelder, K.; Streibel, M.; Jahn, B.; Haase, G.; Brakhage, A.A. Biosynthesis of Fungal Melanins and Their Importance for Human Pathogenic Fungi. *Fungal Genet. Biol.* **2003**, *38*, 143–158. [CrossRef] [PubMed]
12. Zeng, Z.; Guo, X.P.; Cai, X.; Wang, P.; Li, B.; Yang, J.L.; Wang, X. Pyomelanin from *Pseudoalteromonas lipolytica* Reduces Biofouling. *Microb. Biotechnol.* **2017**, *10*, 1718–1731. [CrossRef] [PubMed]
13. Weidenfeld, I.; Zakian, C.; Duewell, P.; Chmyrov, A.; Klemm, U.; Aguirre, J.; Ntziachristos, V.; Stiel, A.C. Homogentisic Acid-Derived Pigment as a Biocompatible Label for Optoacoustic Imaging of Macrophages. *Nat. Commun.* **2019**, *10*, 5056. [CrossRef] [PubMed]
14. Cheng, Z.; Valença, W.O.; Dias, G.G.; Scott, J.; Barth, N.D.; de Moliner, F.; Souza, G.B.P.; Mellanby, R.J.; Vendrell, M.; da Silva Júnior, E.N. Natural Product-inspired Profluorophores for Imaging NQO1 Activity in Tumour Tissues. *Bioorg. Med. Chem.* **2019**, *27*, 3938–3946. [CrossRef]
15. Mavridi-Printezi, A.; Guernelli, M.; Menichetti, A.; Montalti, M. Bio-Applications of Multifunctional Melanin Nanoparticles: From Nanomedicine to Nanocosmetics. *Nanomaterials* **2020**, *10*, 2276. [CrossRef]

16. Li, H.; Zhou, X.; Huang, Y.; Liao, B.; Cheng, L.; Ren, B. Corrigendum: Reactive Oxygen Species in Pathogen Clearance: The Killing Mechanisms, the Adaption Response, and the Side Effects. *Front. Microbiol.* **2021**, *12*, 685133. [CrossRef]
17. Tapia, C.V.; Falconer, M.; Tempio, F.; Falcón, F.; López, M.; Fuentes, M.; Alburquenque, C.; Amaro, J.; Bucarey, S.A.; Di Nardo, A. Melanocytes and Melanin Represent a First Line of Innate Immunity against *Candida albicans*. *Med. Mycol.* **2014**, *52*, 445–454. [CrossRef]
18. Elobeid, A.S.; Kamal-Eldin, A.; Abdelhalim, M.A.K.; Haseeb, A.M. Pharmacological Properties of Melanin and its Function in Health. *Basic Clin. Pharmacol. Toxicol.* **2017**, *120*, 515–522. [CrossRef]
19. Cuzzubbo, S.; Carpentier, A.F. Applications of Melanin and Melanin-Like Nanoparticles in Cancer Therapy: A Review of Recent Advances. *Cancers* **2021**, *13*, 1463. [CrossRef] [PubMed]
20. Koike, S.; Yamasaki, K. Melanogenesis Connection with Innate Immunity and Toll-like Receptors. *Int. J. Mol. Sci.* **2020**, *21*, 9769. [CrossRef]
21. Bilandžija, H.; Laslo, M.; Porter, M.L. Melanization in Response to Wounding is Ancestral in Arthropods and Conserved in Albino Cave Species. *Sci. Rep.* **2017**, *7*, 17148. [CrossRef]
22. Liu, S.; Youngchim, S.; Zamith-Miranda, D.; Nosanchuk, J.D. Fungal Melanin and the Mammalian Immune System. *J. Fungi* **2021**, *7*, 264. [CrossRef]
23. Lee, H.S.; Choi, J.Y.; Kwon, S.J.; Park, E.S.; Oh, B.M.; Kim, J.H.; Lee, P.C. Melanin Biopolymer Synthesis Using a New Melanogenic Strain of *Flavobacterium kingsejongi* and a Recombinant Strain of *Escherichia coli* Expressing 4-Hydroxyphenylpyruvate Dioxygenase from *F. kingsejongi*. *Microb. Cell Fact.* **2022**, *21*, 75. [CrossRef] [PubMed]
24. Coates, C.J.; Lim, J.; Harman, K.; Rowley, A.F.; Griffiths, D.J.; Emery, H.; Layton, W. The Insect, *Galleria mellonella*, is a Compatible Model for Evaluating the Toxicology of Okadaic Acid. *Cell Biol. Toxicol.* **2019**, *3*, 219–232. [CrossRef] [PubMed]
25. Firacative, C.; Khan, A.; Duan, S.; Ferreira-Paim, K.; Leemon, D.; Meyer, W. Rearing and Maintenance of *Galleria mellonella* and it's a pplication to study fungal virulence. *J. Fungi* **2020**, *3*, 130. [CrossRef] [PubMed]
26. Piatek, M.; Sheehan, G.; Kavanagh, K. *Galleria mellonella*: The versatile host for drug discovery, In vivo toxicity testing and characterising host-pathogen interactions. *Antibiotics* **2021**, *12*, 1545. [CrossRef]
27. Madhusudhan, D.N.; Mazhari, B.B.Z.; Dastager, S.G.; Agsar, D. Production and Cytotoxicity of Extracellular Insoluble and Droplets of Soluble Melanin by *Streptomyces lusitanus* DMZ-3. *Biol. Med. Res. Int.* **2014**, *2014*, 306895.
28. Lagunas-Muñoz, V.H.; Cabrera-Valladares, N.; Bolívar, F.; Gosset, G.; Martínez, A. Optimum Melanin Production Using Recombinant *Escherichia coli*. *J. Appl. Microbiol.* **2006**, *101*, 1002–1008. [CrossRef]
29. Guo, J.; Rao, Z.; Yang, T.; Man, Z.; Xu, M.; Zhang, X. High-Level Production of Melanin by a Novel Isolate of *Streptomyces kathirae*. *FEMS Microbiol. Lett.* **2014**, *357*, 85–91. [CrossRef]
30. Ahn, S.Y.; Jang, S.; Sudheer, P.D.V.N.; Choi, K.Y. Microbial Production of Melanin Pigments from Caffeic Acid and L-Tyrosine Using *Streptomyces glaucescens* and FCS-ECH-Expressing *Escherichia coli*. *Int. J. Mol. Sci.* **2021**, *22*, 2413. [CrossRef]
31. Kumar, C.G.; Sahu, N.; Reddy, G.N.; Prasad, R.B.N.; Nagesh, N.; Kamal, A. Production of Melanin Pigment from *Pseudomonas stutzeri* Isolated from Red Seaweed *Hypnea musciformis*. *Lett. Appl. Microbiol.* **2013**, *57*, 295–302. [CrossRef]
32. Martínez, L.M.; Martinez, A.; Gosset, G. Production of Melanins with Recombinant Microorganisms. *Front. Bioeng. Biotechnol.* **2019**, *7*, 285. [CrossRef] [PubMed]
33. Schmaler-Ripcke, J.; Sugareva, V.; Gebhardt, P.; Winkler, R.; Kniemeyer, O.; Heinekamp, T.; Brakhage, A.A. Production of Pyomelanin, a Second Type of Melanin, Via the Tyrosine Degradation Pathway in *Aspergillus fumigatus*. *Appl. Environ. Microbiol.* **2009**, *75*, 493–503. [CrossRef]
34. Tahar, I.B.; Kus-Liśkiewicz, M.; Lara, Y.; Javaux, E.; Fickers, P. Characterization of a Nontoxic Pyomelanin Pigment Produced by the Yeast *Yarrowia lipolytica*. *Biotechnol. Prog.* **2020**, *36*, e2912. [PubMed]
35. Kiran, G.S.; Jackson, S.A.; Priyadharsini, S.; Dobson, A.D.W.; Selvin, J. Synthesis of Nm-PHB (Nanomelanin-Polyhydroxy Butyrate) Nanocomposite Film and Its Protective Effect against Biofilm-Forming Multi Drug Resistant *Staphylococcus aureus*. *Sci. Rep.* **2017**, *7*, 9167. [CrossRef]
36. Furtado, G.P.; Lourenzoni, M.R.; Fuzo, C.A.; Fonseca-Maldonado, R.; Guazzaroni, M.E.; Ribeiro, L.F.; Ward, R.J. Engineering the Affinity of a Family 11 Carbohydrate Binding Module to Improve Binding of Branched Over Unbranched Polysaccharides. *Int. J. Biol. Macromol.* **2018**, *120*, 2509–2516. [CrossRef]
37. Bair, H.E. *Thermal Characterization of Polymeric Materials*, 2nd ed.; Turi, E.A., Ed.; Academic Press: San Diego, CA, USA, 1997; pp. 2263–2420.
38. Lu, J. Characterization and Pseudopolymorphism of l-Phenylalanine Anhydrous and Monohydrate Forms. *Afr. J. Pharm. Pharmacol.* **2012**, *6*, 269–277. [CrossRef]
39. Cordero, R.J.; Casadevall, A. Functions of Fungal Melanin Beyond Virulence. *Fungal Biol. Rev.* **2017**, *31*, 99–112. [CrossRef] [PubMed]
40. Oh, J.J.; Kim, J.Y.; Son, S.H.; Jung, W.J.; Kim, D.H.; Seo, J.W.; Kim, G.H. Fungal Melanin as a Biocompatible Broad-Spectrum Sunscreen with High Antioxidant Activity. *RSC Adv.* **2021**, *11*, 19682–19689. [CrossRef]
41. Eskandari, S.; Etemadifar, Z. Biocompatibility and Radioprotection by Newly Characterized Melanin Pigment and its Production from *Dietzia schimae* NM3 in Optimized Whey Medium by Response Surface Methodology. *Ann. Microbiol.* **2021**, *71*, 17. [CrossRef]
42. Kurian, N.K.; Bhat, S.G. Data on the Characterization of Non-Cytotoxic Pyomelanin Produced by Marine *Pseudomonas stutzeri* BTCZ10 with Cosmetological Importance. *Data Brief* **2018**, *18*, 1889–1894. [CrossRef] [PubMed]

43. Ferraz, A.R.; Pacheco, R.; Vaz, P.D.; Pintado, C.S.; Ascensão, L.; Serralheiro, M.L. Melanin: Production from Cheese Bacteria, Chemical Characterization, and Biological Activities. *Int. J. Environ. Res. Public Health* **2021**, *18*, 10562. [CrossRef] [PubMed]
44. Ogle, M.E.; Segar, C.E.; Sridhar, S.; Botchwey, E.A. Monocytes and Macrophages in Tissue Repair: Implications for Immunoregenerative Biomaterial Design. *Exp. Biol. Med.* **2016**, *10*, 1084–1097. [CrossRef]
45. Oishi, Y.; Manabe, I. Macrophages in Inflammation, Repair and Regeneration. *Int. Immunol.* **2018**, *30*, 511–528. [CrossRef] [PubMed]
46. Lin, T.H.; Pajarinen, J.; Lu, L.; Nabeshima, A.; Cordova, L.A.; Yao, Z.; Goodman, S.B. NF-κB as a Therapeutic Target in Inflammatory-Associated Bone Diseases. *Adv. Protein Chem. Struct. Biol.* **2017**, *107*, 117–154. [PubMed]
47. Prime, B.R.; Bair, H.E.; Vyazovkin, S.; Gallagher, P.K.; Riga, A. Thermogravimetric Analysis (TGA). In *Thermal Analysis of Polymers, Fundamentals and Applications;* Menczel, J.D., Prime, B.R., Eds.; John Wiley & Sons, Inc.: Hoboken, NJ, USA, 2009; pp. 241–317.
48. Menczel, J.D.; Judovits, L.; Prime, B.R.; Bair, H.E.; Reading, M.; Swier, S. Differential Scanning Calorimetry (DSC). In *Thermal Analysis of Polymers, Fundamentals and Applications;* Menczel, J.D., Prime, B.R., Eds.; John Wiley & Sons, Inc.: Hoboken, NJ, USA, 2009; pp. 7–239.
49. Allam, N.; El-Zaher, E.A.; Allam, A. Immunological Efficiency of Microbial Melanin on Bacterial Pathogenicity. *Int. J. Cancer Biomed. Res.* **2020**, *4*, 243–258. [CrossRef]
50. Mnich, E.; Kowalewicz-Kulbat, M.; Sicińska, P.; Hinc, K.; Obuchowski, M.; Gajewski, A.; Moran, A.P.; Chmiela, M. Impact of *Helicobacter pylori* on the Healing Process of the Gastric Barrier. *World J. Gastroenterol.* **2016**, *33*, 7536–7558. [CrossRef] [PubMed]
51. Biernat, M.; Szwed-Georgiou, A.; Rudnicka, K.; Płociński, P.; Pagacz, J.; Tymowicz-Grzyb, P.; Woźniak, A.; Włodarczyk, M.; Urbaniak, M.M.; Krupa, A.; et al. Dual Modification of Porous Ca-P/PLA Composites with APTES and Alendronate Improves their Mechanical Strength and Cytobiocompatibility towards Human Osteoblasts. *Int. J. Mol. Sci.* **2022**, *23*, 14315. [CrossRef]
52. Liu, T.; Zhang, L.; Joo, D.; Sun, S.C. NF-κB Signaling in Inflammation. *Signal Transduct. Target. Ther.* **2017**, *2*, 17023. [CrossRef] [PubMed]
53. Zinatizadeh, M.R.; Schock, B.; Chalbatani, G.M.; Zarandi, P.K.; Jalali, S.A.; Miri, S.R. The Nuclear Factor Kappa B (NF-kB) Signaling in Cancer Development and Immune Diseases. *Genes Dis.* **2021**, *8*, 287–297. [CrossRef] [PubMed]
54. Wu, G.; Yi, Y. Haemocoel Injection of PirA1B1 to *Galleria mellonella* Larvae Leads to Disruption of the Haemocyte Immune Functions. *Sci. Rep.* **2016**, *6*, 34996. [CrossRef] [PubMed]
55. Tsai, C.J.Y.; Loh, J.M.S.; Proft, T. *Galleria mellonella* Infection Models for the Study of Bacterial Diseases and for Antimicrobial Drug Testing. *Virulence* **2016**, *7*, 214–229. [CrossRef] [PubMed]
56. Allegra, E.; Titball, R.W.; Carter, J.; Champion, O.L. *Galleria mellonella* Larvae Allow the Discrimination of Toxic and Non-Toxic Chemicals. *Chemosphere* **2018**, *198*, 469–472. [CrossRef] [PubMed]

Disclaimer/Publisher's Note: The statements, opinions and data contained in all publications are solely those of the individual author(s) and contributor(s) and not of MDPI and/or the editor(s). MDPI and/or the editor(s) disclaim responsibility for any injury to people or property resulting from any ideas, methods, instructions or products referred to in the content.

International Journal of *Molecular Sciences*

Article

Fundamental Properties and Clinical Application of 3D-Printed Bioglass Porcelain Fused to Metal Dental Restoration

Yangan Yun [1], Hyeon Kang [1,2], Eun-Chae Kim [3], Sangwon Park [1], Yong-Seok Lee [4,*] and Kwidug Yun [1,*]

1 Department of Prosthodontics, School of Dentistry, Chonnam National University, 33 Yongbongro, Bukgu, Gwangju 61186, Republic of Korea
2 Ace Dental Clinic, 56, Namak 3-ro, Samhyang-eup, Muan-gun 58567, Republic of Korea
3 Department of Dental Laboratory Technology, Gwangju Health University, 73, Bungmun-daero 419beon-gil, Gwangsan-gu, Gwangju 62287, Republic of Korea
4 Department of Mechanical Engineering, Myeongji University, Yongin 17058, Republic of Korea
* Correspondence: yslee23@mju.ac.kr (Y.-S.L.); ykd@jnu.ac.kr (K.Y.); Tel.: +82-31-330-6423 (Y.-S.L.); +82-62-530-5631 (K.Y.)

Abstract: The purpose of this study is to evaluate the mechanical properties and clinical fitness of 3D-printed bioglass porcelain fused to metal (PFM) dental crowns. To evaluate the mechanical properties, tensile strength, Vickers microhardness, shear bond strength, and surface roughness tests of the SLM printed Co-Cr alloy was conducted. A right mandibular 1st molar tooth was prepared for a single dental crown ($n = 10$). For a three-unit metal crown and bridge, the right mandibular first premolar and first molar were prepared. Bioglass porcelain was fired to fabricate PFM dental restorations. A clinical gap was observed and measured during each of the four times porcelain was fired. A statistical analysis was conducted. The SLM technique showed the largest statistically significant tensile strength and a 0.2% yield strength value. The milling technique had the lowest statistically significant compressive strength value. The shear bond strength and surface roughness showed no statistically significant difference between the fabricated method. There was a statistically significant change in marginal discrepancy according to the porcelain firing step. The casting technique showed the greatest statistically significant margin discrepancy value. The SLM method showed better fitness than the traditional casting method and showed better mechanical properties as a dental material.

Keywords: 3D printing; SLM; margin fitness; dental restoration

check for updates

Citation: Yun, Y.; Kang, H.; Kim, E.-C.; Park, S.; Lee, Y.-S.; Yun, K. Fundamental Properties and Clinical Application of 3D-Printed Bioglass Porcelain Fused to Metal Dental Restoration. *Int. J. Mol. Sci.* **2023**, *24*, 7203. https://doi.org/10.3390/ijms24087203

Academic Editor: Mary Anne Melo

Received: 28 February 2023
Revised: 29 March 2023
Accepted: 11 April 2023
Published: 13 April 2023

Copyright: © 2023 by the authors. Licensee MDPI, Basel, Switzerland. This article is an open access article distributed under the terms and conditions of the Creative Commons Attribution (CC BY) license (https://creativecommons.org/licenses/by/4.0/).

1. Introduction

Porcelain fused to metal (PFM) restoration refers to porcelain restoration supported by a metal substructure, and it is one of the most commonly used restorations in dentistry [1]. PFM restoration is made of non-precious metals, such as Ni-Cr or Co-Cr alloys, instead of noble metal alloys due to their high cost. These non-noble metal alloys are affordable, exhibit high strength, modulus of elasticity and corrosion resistance, and therefore represent perfect candidates for metal substructures [1]. Especially, Co-Cr alloy has been used primarily as a metal substructure in removable partial dentures in the past. Currently, its implementation as a metal substructure in PFM restoration is increasing due to fewer side effects and higher internal corrosion resistance than Ni-Cr alloy [2,3]. Although metal coping structures using Co-Cr alloy have been mostly manufactured with the conventional lost-wax casting technique, the high fusion temperature of Co-Cr alloy is a hindrance to successful fabrication, in addition to inevitable errors associated with serial fabrication procedures.

Computer-Aided Design/Computer-Aided Manufacturing (CAD/CAM) techniques play a major role in the production of biomechanical appliances and dental prosthodontic restorations [4,5]. Currently, dentists face challenges in providing individual restorations and prosthodontic reconstructions customized to each patient, which entails recreating complicated anatomic forms [6]. Although it may be possible to manufacture complex restorative structures using subtractive manufacturing techniques of multi-axial CAD/CAM milling equipment, it is very time-consuming and generates unnecessary material waste; often, the final product lacks precision depending on the size and shape of the milling equipment [7–9]. However, additive manufacturing, first suggested in the 1980s, represents the opposite concept of subtractive manufacturing. Recently, it has been studied in various fields, including aerospace technology and precision part manufacturing. It is expected to lead to phenomenal changes in manufacturing industries. The American Society for Testing and Materials (ASTM) defines additive manufacturing as a set of procedures combining materials to design the desired object, which generally results in the accumulation of specific layers of materials. Additive manufacturing procedures enable the precise production of various complicated restorations and structures in the field of dentistry and surgery [10,11].

Additive manufacturing can be classified according to methods such as material procurement, sources of energy, and accumulation volume. It is also largely categorized by powder bed systems, powder feed systems, and wire feed systems. Selective laser melting (SLM) is one of the metal accumulation techniques that belong to the powder bed system, which generates structures by directly irradiating metal powder with a high-powered laser beam, resulting in fusion between metallic particles. SLM has been used to make metal copings as well as metal substructures of partial dentures in previous studies. This technique yielded satisfactory results, with the final product having superb mechanical and chemical characteristics, such as high density, strength, and corrosion resistance [12–14].

Co-Cr alloys have been manufactured using both subtractive and additive methods via CAD/CAM technology. Co-Cr alloys can be developed easily and rapidly using additive manufacturing without the need for complex procedures, unlike conventional lost-wax casting techniques [4,10,15]. However, the study of Co-Cr alloy using the SLM technique is still in the initial stages. In addition, the disadvantages of current PFM restorations include the high risk of chipping and the fracture of the porcelain veneer, which is an important factor in determining the clinical success of the restoration [16,17]. Since Co-Cr has been mainly used in removable prostheses in the past, there are relatively few studies on fixed prostheses made in combination with ceramic materials [18–23]. Hence, the comparison of mechanical properties and the bonding strength of metal and porcelain of the restorative structure made with conventional manufacturing techniques should be evaluated before the clinical application of Co-Cr alloy using the SLM technique. However, the bonding strength of Co-Cr produced by the SLM method so far is a laboratory experiment that is different from the clinical situation. In addition, porcelain firing is rarely performed in accordance with the clinical situation. Instead of porcelain firing, thermocycling or a firing procedure without porcelain is performed [24]. However, it is very different from the actual clinical situation. Therefore, in this study, single crowns and three-unit bridges were used in a very similar way to clinical situations.

Marginal fit is one of the necessary requirements for the clinical success of dental restorations. Good marginal fit is essential for the long-term success of dental restorations, and the presence of any errors can render harmful effects on the abutment tooth and the supporting periodontal tissue. An inadequate internal fit could cause lead to a lack of resistance of the metal–ceramic restoration, the loss of axial retention, and decreased fracture resistance. According to the definition by Holmes, the internal gap refers to the vertical measurement from the axial wall of the abutment tooth to the internal surface of the restorative structure, and if the same measurement occurs at the marginal area, it is defined as a marginal gap [25]. Mclean confined a marginal gap of 120 μm as a clinically acceptable range [26]. Many studies have been conducted to investigate the marginal fit of conventional cast manufacturing methods of Co-Cr alloy metal–ceramic restorations [27,28]. However, the study determining the marginal fit of Co-Cr alloy restoration obtained by the SLM technique is still limited. Additionally, metal–ceramic restorations go through a series of porcelain firing procedures, which result in distortion of metal coping due to repeated heat treatments, resulting in a poor marginal fit. However, the study of this topic is inconclusive [29–31]. Specifically, there are no papers that have conducted experiments in the form of bridges that are actually used in clinical practice rather than specimens.

Therefore, the purpose of this study was to compare the mechanical properties of Co-Cr alloy generated using SLM, milling, and casting techniques for comparative evaluation of the fitness of the PFM restoration of each of the techniques mentioned above.

2. Results

2.1. Mechanical Properties

2.1.1. Tensile Strength of Metal Alloy

The results of tensile strength according to casting, milling, and SLM manufacturing techniques are presented in Figure 1. The SLM technique showed a higher mean maximum tensile strength compared with the milling and casting techniques, which was statistically meaningful ($p < 0.05$). When comparing the elongation amongst the groups, the data did not show statistically meaningful differences (casting technique: 5.41 ± 1.81%, milling technique: 9.46 ± 3.04%, SLM technique: 8.69 ± 1.69%) ($p > 0.05$) (Figure 2). The 0.2% yield strengths (casting technique: 716.71 ± 50.03 MPa, milling technique: 512.65 ± 86.42 Mpa, SLM technique: 879.30 ± 34.32 Mpa) turned out to be significantly higher in the order of SLM, casting, and milling techniques, respectively ($p < 0.05$) (Figure 2).

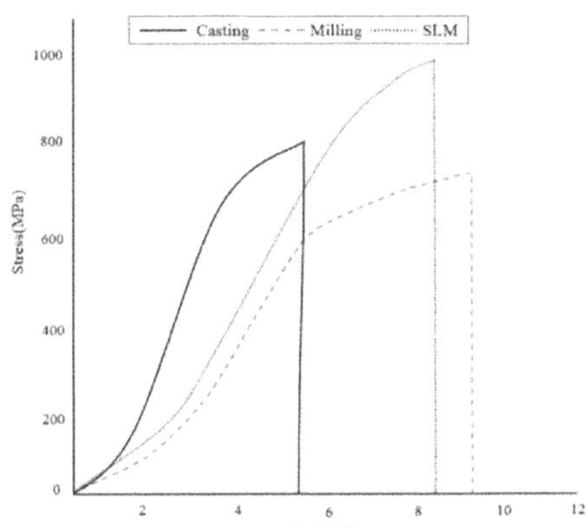

Figure 1. Stress–strain curves of tensile strength test following different manufacturing techniques.

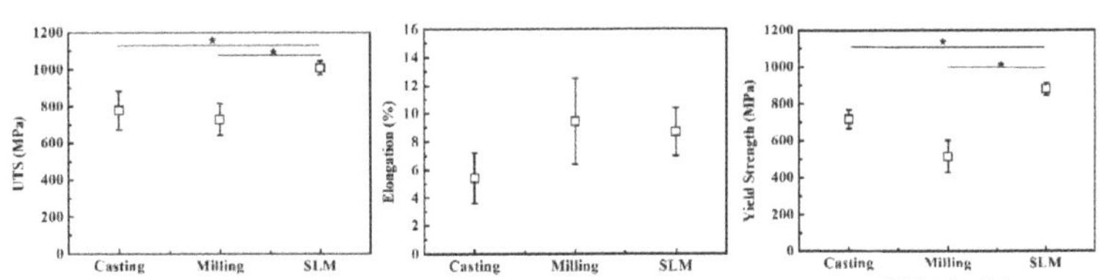

Figure 2. Ultimate tensile strength, elongation and yield strength following different manufacturing techniques. *, significant at $p < 0.05$.

2.1.2. Vickers Microhardness

Figure 3 shows the results of the Vickers microhardness test according to the casting, milling, and SLM techniques (Figure 3). The values of the mean and standard deviation of Vickers microhardness in the SLM, milling, and casting groups were 441.97 ± 16.12 kgf/mm^2, 388.57 ± 19.41 kgf/mm^2, and 431.76 ± 12.85 kgf/mm^2, respectively. The mean value of Vickers microhardness was lower with the milling technique than those of the casting and SLM techniques, which was statistically significant ($p < 0.05$).

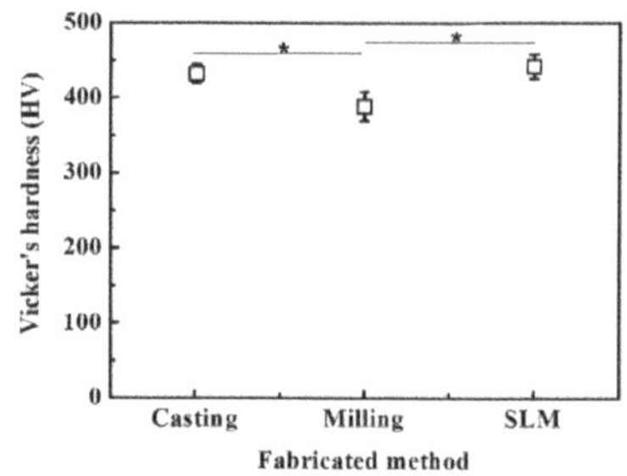

Figure 3. Mean ± SD Vicker's microhardness values following different manufacturing techniques. *, significant at $p < 0.05$.

2.1.3. Shear Bond Strength of Porcelain vs. Metal Alloy

Figure 4 shows the result of the shear bond strength test between Noritake super porcelain Ex-3 porcelain and the Co-Cr alloy specimens made via casting (24.28 ± 2.32 MPa), milling (20.66 ± 1.45 MPa), and SLM (22.55 ± 4.63 MPa) techniques (Figure 4). The difference in the shear bond strength of the three groups was not statistically significant ($p > 0.05$).

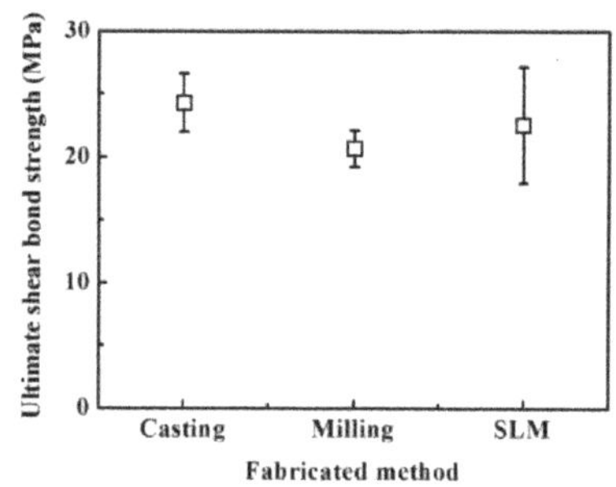

Figure 4. Shear bond strength values following different manufacturing techniques.

Figure 5 illustrates the porcelain attachment fracture surface at ×200 magnification with SEM after the shear bond strength test (Figure 5A–C). All of the specimens showed mixed adhesive failure (between fracture and adhesion surfaces) and cohesive failure (fracture inside the porcelain). The light gray area of the SEM image represents the residual opaque porcelain after the porcelain fracture, and the dark gray area denotes the Co-Cr alloy structure area (Figure 5D). Figure 5E shows the area fraction of adherence porcelain (AFAP) (%) value obtained from Si (silicone) EDS atomic analysis. AFAP (%) values were as follows: casting (74.22 ± 14.90%), SLM (73.85 ± 22.17%), and milling (60.86 ± 22.84%) techniques in the order from high to low but did not appear to be statistically significant.

2.1.4. Metal Surface Roughness

Figure 6 shows the result of the surface roughness test (Ra) according to casting, milling, and SLM techniques measured under four different conditions: (1) raw, (2) raw + sandblasting treatment, (3) polishing, and (4) polishing + sandblasting treatment (Figure 6). The surface roughness measured under raw conditions was as follows: milling (0.37 ± 0.02 μm), SLM-bur (1.73 ± 0.21 μm), casting (1.91 ± 0.11 μm), and SLM (3.13 ± 0.49 μm) in the order of roughness to smoothness. The SLM specimens showed the highest value that was statistically meaningful, and the milling specimens showed the lowest statistically significant value ($p < 0.05$). The surface roughness was still consistent after the sandblasting treatment of the surface under raw conditions with milling (1.03 ± 0.17 μm), SLM-bur (1.95 ± 0.29 μm), casting (2.29 ± 0.24 μm), and SLM (3.31 ± 0.42 μm), which increased in roughness after the specimens were sandblasted. The surface roughness measured under polished conditions was as follows: milling (0.17 ± 0.01 μm), SLM (0.22 ± 0.04 μm), and casting (0.27 ± 0.02 μm). Casting specimens showed a higher surface roughness compared with milling specimens, which was statistically meaningful ($p < 0.05$). The surface roughness measured via sandblasting after polishing was as follows: milling (0.83 ± 0.14 μm), SLM (0.77 ± 0.10 μm), and casting (0.87 ± 0.24 μm), but was not statistically meaningful ($p > 0.05$).

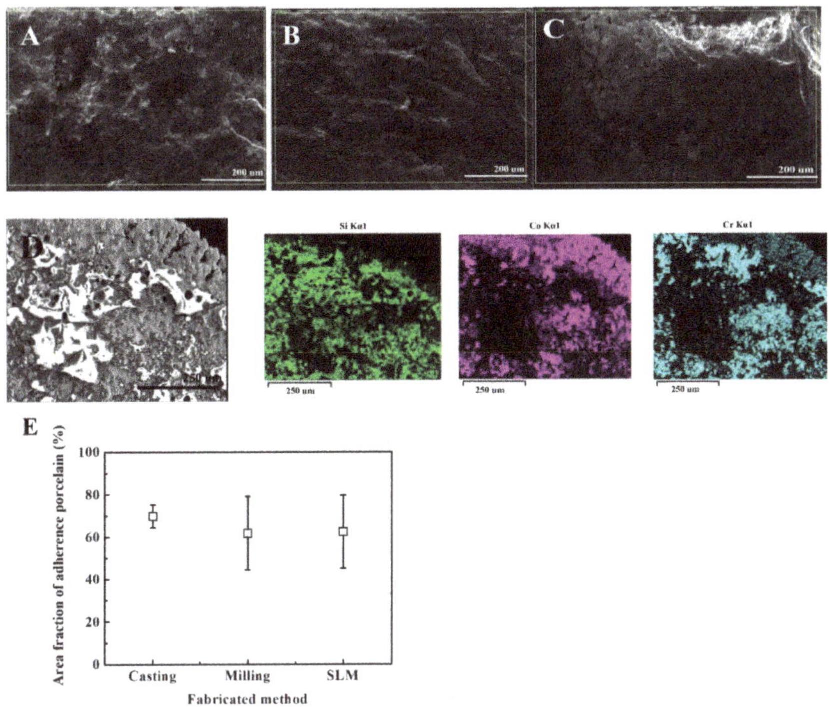

Figure 5. SEM and EDX results (×200 magnification) of casting, milling, SLM specimen metal–ceramic interface after fracture experiment. (**A**) casting, (**B**) milling, (**C**) SLM, (**D**) EDX mapping data on the metal–ceramic interface after fracture experiment. (**E**) area fraction of adherence porcelain values following different manufacturing techniques.

2.1.5. Metallurgic Structure

Figure 7 shows the metal surface structure of Co-Cr alloy specimens manufactured via casting, milling, and SLM techniques. (Figure 7). A significant difference was found on the surface metal structure in terms of the manufacturing techniques. The Co-Cr alloy specimens fabricated via the casting technique showed characteristic dendritic-like shapes. The Co-Cr alloy specimens fabricated with the milling technique showed an island shape. The Co-Cr alloy specimens designed with the SLM technique showed a layered cladding shape at ×100 and ×500 magnification and a fine surface along the inner border with apparent metal fusion at ×2000 magnification (Figure 7).

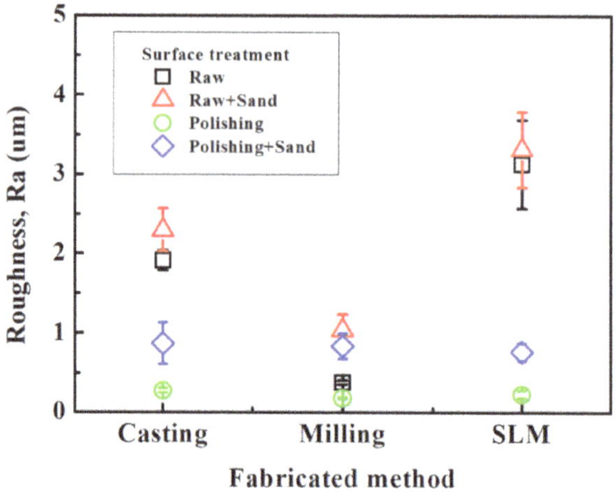

Figure 6. Mean ± SD surface roughness values (Ra) following different manufacturing techniques.

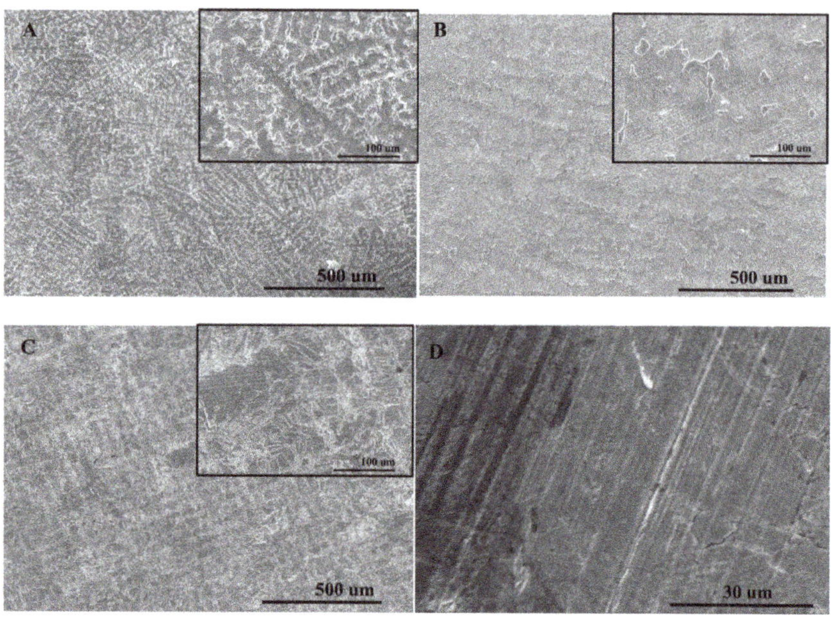

Figure 7. SEM images from polished section of Co-Cr alloys. (**A**) casting (original magnification × 100 and magnification of inserts × 500), (**B**) milling (original magnification × 100 and magnification of inserts × 500), (**C**) SLM (original magnification × 100 and magnification of inserts × 500). (**D**) SLM specimen (original magnification × 2000).

2.2. Dental Restoration Fitness Test

2.2.1. Marginal Fit

The marginal fitness of all groups according to the porcelain firing stages of Co-Cr alloy coping was classified according to firing cycle steps, and the mean and the standard deviation were computed (Table 1).

Table 1. Marginal gap of casting, milling, SLM dental crown.

	Firing Cycle	Casting	Milling	SLM
Single crown	Initial	73.52 ± 13.86 [e]	63.09 ± 17.46 [a]	64.43 ± 13.47 [a]
	Oxidizing	80.39 ± 13.86 [i]	69.58 ± 13.53 [b]	70.52 ± 13.46 [c]
	Opaque	83.23 ± 14.24 [j]	72.20 ± 19.86 [e]	74.44 ± 18.03 [f]
	Dentin	88.53 ± 11.84 [l]	73.90 ± 17.51 [e]	78.67 ± 18.87 [g]
	Glazing	91.65 ± 8.71 [l]	79.70 ± 16.40 [h]	83.83 ± 17.38 [j]
Three-unit bridge	Initial	81.05 ± 20.32 [j]	67.55 ± 9.97 [b]	71.62 ± 16.95 [d]
	Oxidizing	88.04 ± 16.08 [k]	75.68 ± 11.80 [g]	77.63 ± 6.36 [g]
	Opaque	91.65 ± 13.89 [l]	79.80 ± 12.89 [g]	83.76 ± 15.88 [j]
	Dentin	96.09 ± 13.37 [l]	82.05 ± 13.17 [i]	86.45 ± 16.48 [k]
	Glazing	100.07 ± 12.62 [l]	86.59 ± 15.06 [k]	92.64 ± 10.59 [l]

The different letters represent a significant difference at a type-one error rate of 0.05.

The marginal fit showed increasing marginal discrepancy as the porcelain firing stages progressed. Instead, all groups showed a significant difference in the marginal discrepancy between the initial step and after the glazing step ($p < 0.05$) (Table 1).

The marginal fit according to casting, milling and SLM manufacturing techniques after the glazing step suggested that in the case of milling and SLM, the marginal discrepancy values were not statistically meaningful ($p > 0.05$). However, in the case of the casting method, the mean value of the marginal discrepancy was greater when compared with the milling and SLM methods, which was also statistically meaningful ($p < 0.05$) (Table 1).

The marginal discrepancy in the three groups significantly increased as the porcelain firing stages progressed (Table 1). ($p < 0.05$). The mean marginal gap values for the casting group were 81.05 μm, 88.04 μm, 91.65 μm, 96.09 μm, and 100.07 μm after the initial, oxidation, opaque, dentin, and glazing firing steps, respectively. The mean marginal gap values for the milling group were 67.55 μm, 75.68 μm, 79.80 μm, 82.05 μm, and 86.59 μm at the same firing stages. The mean marginal gap values for the SLM group were 71.62 μm, 77.63 μm, 83.76 μm, 86.45 μm, and 92.64 μm at the same firing stages. In the casting, milling, and SLM groups, the marginal discrepancy values increased in the order of the milling, SLM, and casting groups, which was statistically significant ($p < 0.05$). The mean marginal gap value was 91.38 μm for the casting group, 78.33 μm for the milling group, and 82.42 μm for the SLM group. The marginal gap after the completion of the glazing process is the most important value because the metal–ceramic restoration is used after the completion og porcelain firing. The marginal gap values after the completion of the glazing process were 100.07 μm for the casting group, 86.59 μm for the milling group, and 92.64 μm for the SLM group. When comparing the mean value of the marginal discrepancy between the single coping group and the bridge coping group according to the manufacturing techniques, all techniques showed higher marginal discrepancy values in the bridge coping groups than in the single coping groups, which was statistically meaningful ($p < 0.05$).

2.2.2. Internal Fit

The mean and standard deviation of the internal fit were computed in all groups after the porcelain firing process was completed (Table 2). The values were arranged according to the manufacturing techniques, which were analyzed statistically. The axial internal gap was significantly lower in the milling than in the casting and SLM methods ($p < 0.05$), and the occlusal internal gap was significantly higher in the SLM than in the casting and milling methods ($p < 0.05$). When the axial and occlusal internal gaps were combined, the internal gap was higher in the order of the milling, casting, and SLM groups, and the milling and SLM groups showed statistically meaningful differences ($p < 0.05$).

Table 2. Internal gap of casting, milling, SLM single dental crown.

Firing Cycle	Casting	Milling	SLM
Axial	134.9 ± 28.29 [a]	118.35 ± 24.26 [b]	137.16 ± 31.94 [a]
Occlusal	181.84 ± 28.68 [d]	178.94 ± 26.22 [d]	200.86 ± 22.16 [c]
Total	160.98 ± 36.86 [g]	152.01 ± 39.37 [e]	172.55 ± 41.57 [f]

The different letters represent a significant difference at a type-one error rate of 0.05.

For a three-unit bridge, the axial internal gap was lower in the milling group than in the casting and SLM groups, which was statistically meaningful ($p < 0.05$), and the occlusal internal gap was higher in the SLM group than in the casting and milling groups, which was statistically meaningful ($p < 0.05$) (Table 3). With respect to the combined axial and occlusal internal gap, the internal gap increased in the order of the milling, casting, and SLM groups ($p < 0.05$). When comparing the mean internal gap values of the single coping groups and bridge coping groups according to the manufacturing techniques, in all manufacturing techniques, the bridge coping groups showed higher internal discrepancy values than the single coping groups ($p < 0.05$).

Table 3. Internal gaps of a 3-unit bridge in casting, milling, SLM groups after the firing cycle process.

Internal Gap		Casting	Milling	SLM
First premolar	Axial	159.34 ± 10.02 [c]	142.48 ± 12.32 [f]	150.39 ± 9.39 [b]
	Occlusal	179.3 ± 9.65 [c,d,e]	186.42 ± 11.69 [g]	205.07 ± 7.51 [i]
First molar	Axial	159.07 ± 25.31 [c,e]	141.26 ± 26.51 [f]	154.85 ± 14.31 [b]
	occlusal	201.54 ± 33.60 [d]	201.29 ± 27.90 [g]	246.74 ± 18.14 [j]
total		174.82 ± 27.68 [a]	167.86 ± 33.62 [g]	189.26 ± 41.94 [b]

The different letters represent a significant difference at a type-one error rate of 0.05.

3. Discussion

In this study, mechanical properties and restoration fitness tests were conducted with a Co-Cr alloy produced via additive, subtractive, and casting manufacturing techniques. Although the mechanical properties and fitness have been evaluated using additive manufacturing techniques in various studies recently, the study of Co-Cr alloy specimens is still in the beginning stage [18,20–24]. This study investigated the unique traits of metal alloys compared with the mechanical properties of Co-Cr alloy specimens fabricated using SLM, which is an additive manufacturing technique, with specimens made by conventional casting and subtractive manufacturing techniques. In addition, the restoration fit was evaluated by fabricating a PFM crown coping in a similar environment under actual clinical conditions created by executing porcelain build-up.

In the tensile strength test, ultimate tensile strength, elongation, and 2% yield strength were measured and classified according to the groups and specimens meeting the mechanical trait standards of ISO 22674:2006 in this study. The SLM technique showed higher UTS and 2% yield strength than casting and milling techniques, which was statistically meaningful ($p < 0.05$). These results are consistent with the current studies, showing that the Co-Cr alloy manufactured via SLM has improved mechanical traits compared with the conventional manufacturing techniques [8,32,33]. The grain size of the metal alloy is closely correlated with the nucleation rate, which is determined by the degree of supercooling [34]. As the specimens manufactured with the SLM technique exhibit substantially higher differences in temperature compared with those involving the casting technique (casting = 1500 °C, SLM = 1800 °C), it is expected that the grain size is significantly smaller in specimens exposed to the SLM technique. As shown in this study, the SEM images of the Co-Cr alloy specimen showed dendritic and island shapes with casting and milling techniques, respectively, whereas the SLM technique yielded fine grains. Casting technique specimens can be divided into dentritic and interdendritic structures, which separate metal solutes into two domains, resulting in the degradation of mechanical traits. The milling technique also showed a surface metal structure associated with the precipitated island-shaped phase instead of a monophasic structure. Similarly, the Co-Cr alloy specimens fabricated with casting and milling techniques also contained a precipitated phase. Conversely, the Co-Cr alloy specimens obtained via the SLM technique showed a surface structure with fine grains, which may result in more favorable mechanical traits. The distinction in the surface metal structures indicates a difference in solidification and thermo-mechanical procedures, which is consistent with previous studies demonstrating that specimens obtained via the casting and milling techniques carry a dendritic microstructure, whereas specimens prepared via the SLM technique showed a uniform and fine-grain microstructure [34]. The mechanical traits of the Co-Cr alloy are also affected by the composition of metallic phases, in which case, the monophasic composition improves the machinability of the metal alloy. The Co-Cr alloy undergoes a transformation from the FCC phase at high temperature to an HCP phase at low temperature, and in the case of the SLM technique, the specimens are processed via rapid solidification, which results in an alloy consisting mostly of the FCC phase [34]. Therefore, the SLM technique specimens show better machinability than the casting technique specimens.

The SLM technique specimens had the highest Vickers microhardness value at (441.97 kgf/mm^2), which is consistent with recent studies [8]. The Vickers microhardness values of the casting and SLM techniques were higher than that of the milling technique, which was statistically meaningful. Recent studies indicate that the grain size at the time of manufacture is distributed more finely in the SLM technique, which enhances mechanical traits, and residual strength generated during fabrication appears to contribute to higher compressive hardness [8,34].

The results of the surface roughness test showed that the specimens under the raw conditions exhibit a rough surface in the order of milling > casting = SLM-bur > SLM. Metal additive manufacturing contains numerous metal pearls on the surface naturally, which requires finishing with a rotary cutting machine in order to build up an appropriate porcelain superstructure. After the finishing procedure, the surface roughness of the SLM specimens did not show a statistically meaningful difference compared with the casting technique specimens ($p > 0.05$), which coincided with the clinical appearance after the sandblasting processing before porcelain firing. Appropriate surface roughness increases the bonding surface with the porcelain superstructure, which possibly results in increased bond strength. The milling and casting techniques yielded statistically significant differences in surface roughness after polishing ($p < 0.05$). At the same time, the SLM technique specimens processed with surface polishing similarly did not show statistically meaningful differences compared with milling technique specimens ($p > 0.05$). Thus, it can be inferred that the polishing step transforms the macroscopic surface roughness of the SLM technique specimens to that of the milling technique. In addition, when the specimens were processed via sandblasting after polishing, the difference in surface roughness due to the manufacturing technique was not significant ($p > 0.05$). In brief, in the case of the SLM technique, the metal alloy surface under raw conditions contained multiple metal pearls generated during repeated melting and cooling procedures, but after appropriate surface finishing, the specimens showed similar surface roughness as those of the casting and milling techniques, which rendered appropriate surface conditions for the build-up of the porcelain superstructure.

Alloy manufacturing using other techniques increases the differences in the surface shape as well as changes in surface oxide, which affect the strength of the metal–porcelain bond [35]. In this study, the shear bond strength did not show a statistically meaningful difference when porcelain was fused to the Co-Cr alloy specimens via three different techniques ($p > 0.05$). The bond strength between the metal and porcelain is affected by the difference in mechanical and chemical bonding and thermal expansion coefficient [36,37]. According to the results of this study, the lack of difference in the shear bond strength due to the manufacturing techniques means that these complex factors do not contribute to the statistical differences. Likewise, although the result of AFAP showed a residual porcelain attachment rate based on the following order (high to low): casting (74.22 ± 14.90%), SLM (73.85 ± 22.17%), and milling (60.86 ± 22.84%), it was not statistically meaningful ($p > 0.05$). The AFAP of SLM and casting specimens was almost equal, which indicates that it is appropriate for clinical application since it suggests that the metal–porcelain bond strength also shows an equivalent performance.

In this study, the marginal fit was measured during a series of porcelain firing procedures of Co-Cr alloy coping manufactured using casting, milling, and SLM techniques. Accordingly, as the porcelain firing procedure progressed, the marginal gap increased little compared with the previous stage. Comparing the marginal discrepancy of the first stage, the 'initial stage', and the final 'glazing stage', all of the Co-Cr alloy coping specimens manufactured with casting, milling, and SLM techniques showed statistically meaningful differences ($p < 0.05$) due to the accumulation of the marginal distortion of the Co-Cr alloy coping generated during each porcelain firing stage. Although it is suggested that the clinical objective is to maintain the marginal gap between 25 and 40 μm, according to American Dental Association (ADA) specification No.8, this is difficult to achieve. McLean studied the marginal fit of about 1000 PFM crown restorations and suggested that a marginal discrepancy of about 120 μm is the clinical threshold [25]. This finding indicates that the marginal discrepancy generated after a series of porcelain firing procedures in this study is within the clinically acceptable scope. The marginal discrepancy values generated during the porcelain firing procedures of this study tend to be higher than those of previous studies, which executed porcelain firing repeatedly without the porcelain build-up [38]. It appears that as the temperature drops during the porcelain firing thermal treatment procedure, additional distortion in marginal fit occurs due to the differences in the coefficient of thermal expansion between the metal coping and the porcelain superstructure.

Previous studies have reported that the marginal discrepancy occurs due to the release of residual stress generated during the porcelain thermal treatment, which is the first stage of porcelain firing [30,39]. According to Papazoglou et al., statistically meaningful distortion occurs during the first step of the porcelain firing thermal treatment, whereas Li Zeng et al. reported that the increase in marginal discrepancy was not statistically significant [38,40]. In this study, although the mean marginal discrepancy during the first porcelain firing heat treatment did not show a statistically meaningful difference, a clear discrepancy was observed when the marginal discrepancy was compared before and after the firing. The Noritake Super Porcelain EX-3 used in this study is high-fusing dental porcelain that requires an increase in temperature as high as 1000 °C during the firing cycle. The decline in the alloy under this high temperature can induce distortion in the metal coping, which subsequently, under repeated thermal treatments at high temperatures, results in marginal distortion [41,42]. In conclusion, even though there was no significant marginal discrepancy generated during the first stage of porcelain firing, it can be assumed that the marginal distortion of the actual metal coping accumulates.

The casting method showed higher mean marginal discrepancy compared with the milling and SLM groups in all stages of the porcelain firing procedures, which showed statistically meaningful differences ($p < 0.05$). Even though milling showed a lower mean marginal discrepancy in all stages of porcelain firing, it was not statistically meaningful ($p > 0.05$). Örtorp et al. reported that when the marginal fit of Co-Cr alloy coping manufactured with casting, milling, and SLM techniques was compared, the marginal gap increased in the order of laser sintering, casting, and milling techniques [43]. Additionally, Quante et al. reported on the marginal fit when using a silicone replica, wherein the marginal discrepancy increased in the order of casting, milling, and laser sintering techniques [44]. The result of this study was inconsistent with those of previous studies, which can be attributed to the difference in the accuracy of the CAM unit during milling, a lack of heat generation in the manufacturing environment, differences in the technical expertise of the laboratory technician when using the complex procedures of casting using the lost-wax technique, and differences in the laser beam conditions and layering systems of SLM. However, it is difficult to create a detailed internal shape using the milling technique, and this entails complex restoration manufacturing procedures. In addition, the milling technique decreases precision due to the fast wear of the cutting instrument during milling and the significant waste of materials during the manufacturing procedures.

The SLM technique showed a higher internal gap compared with the other groups prepared using different techniques, which was consistent with previous studies [45,46]. It can be inferred that the difference in the accuracy of the equipment used and the distortion due to the heat-generating environment during manufacturing contributed to this result. Comparing the coping manufactured using the SLM technique and the casting technique, the internal fit of the SLM did not show statistically meaningful differences in terms of casting axially, but the occlusal portion showed a statistically meaningful difference ($p < 0.05$), which is contrary to the previous result that shows a better marginal fit of SLM. When evaluating the SLM technique specimens compared with the casting technique specimens, the internal gap decreases from the occlusal to the marginal part due to the accompanying cervical contraction of the metal alloy manufactured using the SLM technique in previous studies [38]. This result may be attributed to repeated high heat and rapid cooling treatments during the preparation of the specimen for the SLM technique, during which the metal coping with a supporter designed in the occlusal direction increases distortion due to the heat in the marginal portion, which is relatively far from the supporter. As a result, contraction in the marginal part is more prominent than in the occlusal part, and when measuring the internal fit, the marginal contraction restricts the applied silicone impression to the occlusal part, resulting in a high occlusal gap. For the milling group, higher occlusal gaps than axial ones were explained by the reduced scanner accuracy in occlusal areas, which have limited access for the milling burs [47]. Increased gaps are not desirable and result in increased cement thickness; these may also interfere with the accurate fit of a restoration. However, the internal gap value in this study exhibited a range of 167.86 μm to 189.26 μm. According to the previous study, the acceptable clinical occlusal gap range is 100 to 200 μm [46]. Thus, it is within a clinically acceptable internal gap range.

In this study, in all groups of casting, milling, and SLM technique specimens, the mechanical traits and marginal fit showed different results, and the null hypothesis was rejected. In the case of the SLM technique, it is possible to make restorations with complex forms, generate results without voids, and create Co-Cr alloy copings with better mechanical traits than those made using conventional techniques. Thus, a Co-Cr alloy coping manufactured with the SLM technique can be used to maintain long-term marginal soundness and is deemed appropriate for clinical application. Accordingly, the fitness of restoration and mechanical traits were excellent when fabricating PFM restorations via the SLM technique. Even though the fabrication of restorations using additive manufacturing techniques is limited by the exorbitant price of the equipment, the results of this study suggest the significant potential for dental applications in the future.

4. Materials and Methods

4.1. Mechanical Properties

All of the specimen compositions for SLM, milling, and casting techniques follow the ASTM F-75 standard so as to minimize the errors due to compositional differences in the materials (Table 4) [45]. The specimen preparation for the SLM technique was conducted using a 3D printer (ProX® DMP 100, 3D systems Inc., Rock Hill, SC, USA), and SLM was used under 50 μm of Co-Cr metal power with the following parameters: P = 200 W, λ = 1070 nm, layer thickness = 20 μm, and speed = 40 mm/s. The specimen preparation in the milling technique was carried out using a milling machine (Arden 5X-WM, TPS Korea LTD, Gwang-Ju, Republic of Korea) and a Co-Cr Bar measuring 75 mm and weighing 4 kg apiece (R516045, Remelt Sources Inc., Cleveland, OH, USA), which was used for subtractive manufacturing. The specimen preparation for the casting technique was conducted by creating resin patterns with a 3D resin printer (Meg printer, Megagen, Dae-gu, Republic of Korea) using the Digital Light Processing (DLP) method, followed by burnout and casting procedures.

Table 4. Chemical compositions of Co-Cr dental alloys.

Group	Co	Cr	Mo	Ma	Si	Fe	C	Product Name
Casting	64	28	6	<1	<1	<1	<1	Dental Alloy Products
Milling	Balance	27.9	5.79	0.35	0.83	0.30	0.24	Super 6 Remelt Sources
SLM	Balance	29	5.5	<1	<1	<1	<1	R516045
ASTM F75-18 (Max)	Balance	30	7	1	1	0.75	0.35	SinT-Tech ST2724G
ASTM F75-18 (Min)	Balance	27	5	-	-	-	-	

The Co-Cr alloy specimens used for the tensile strength test were formatted with a dumbbell shape in accordance with the ISO 22674;2016 standard and designed using a CAD program (AutoCAD®, Autodesk, CA, USA), which was then converted to an STL file format. Each of the 5 specimens was prepared using SLM, milling, and casting techniques [48]. The prepared dumbbell-shaped specimens were subjected to a tensile strength of 10 mm/min crosshead speed on the universal testing machine (AG-Xplus 50 kN, Shimadzu, Kyoto, Japan), and each specimen was recorded at the time of fracture.

The Co-Cr alloy specimens used for the Vickers microhardness test (disc-shaped specimen measuring 10 mm in diameter and 4 mm in height) were designed using CAD (AutoCAD®, Autodesk, CA, USA). The surface hardness of the specimens in all groups was measured with a micro-Vickers durometer (TUKON-1202, Wilson, IL, USA) and the hardness measurements were made below 1 kgf for a duration of 10 s. After removing the load, the diagonal indentation was observed with a microscope attached to a hardness tester, which was used to calculate the Vickers microhardness (n = 12).

All of the Co-Cr alloy specimens used in the shear bond strength experiment were polished on the surface to which the porcelain was attached using water spray. Next, the specimens were sandblasted with Al_2O_3 particles measuring 125 μm on the surface. Porcelain was stacked up to 8 mm in diameter and 4 mm in height. Noritake Super Porcelain EX-3 (Kuraray Noritake Dental Inc., Tokyo, Japan) was used for the porcelain build-up, and the porcelain furnace (Austromat 3001, DEKEMA, Freilassing, Germany) was used in accordance with the manufacturer's instructions. The shear bond strength was measured using a universal testing machine (Model 1125, Instron, Canton, MA, USA) under the given load of testing speed at 1.0 mm/min until each specimen was fractured. To quantify the area fraction of adherence porcelain (AFAP), the specimen surface was studied in the center to determine the content of Si using SEM (SEM; scanning electron microscope, SNE-4500M plus, Dae-duk image, Dae-jeon, Republic of Korea) attached with EDS (EDS: energy-dispersive X-ray spectrometer, ESPRIT Compact, Brukernano GmbH, Berlin, Germany) at 200 x magnification. The atomic percentage of silicone (Si) was tested after sandblasting with 125 μm Al_2O_3, thermal oxidation (Sim), the application of opaque porcelain (Sio), and the elimination of porcelain from the specimens (Sif). The AFAP was estimated as AFAP = (Sif − Sim)/(Sio − Sim) (n = 5) [49].

The Co-Cr alloy specimens used for the surface roughness test measured 10 mm in diameter and 4 mm in height (n = 10). The surface roughness was measured under the following 4 conditions: (1) raw, (2) raw + sandblasting treatment, (3) after polishing, and (4) polishing + sandblasting treatment. In the surface roughness test of the specimens under the raw and raw + sandblasting treatment condition, the outer surfaces of the raw specimens made with the SLM technique were not treated entirely compared with the specimens obtained from the casting and milling techniques that were finished using a bur. Therefore, the surface roughness was measured in the following two groups: (1) specimens whose outer surfaces were depleted of metal pearl using stone bur (SLM-bur) and (2) specimens whose outer surfaces were not removed (SLM). All of the specimens were cleaned ultrasonically for 10 min using acetone, ethyl alcohol, and distilled water prior to the surface roughness test. The surface roughness of the specimens in all groups was measured using a 2D contact stylus profilometer (DIAVITE DH-8, DIAVITE AG, Feldstrasse, Switzerland).

To observe the metal structure, the specimen polished with No. 2000 SiC abrasive paper under water spray was photographed with SEM (SEM: scanning electron microscope, SNE-4500M plus, Dae-duk image, Dae-jeon, Republic of Korea). The specimens from all groups were observed at ×100 and ×500 magnification, and in the case of the Co-Cr alloy specimen made using the SLM technique, the surface observation was made at a ×2000 magnification for clearer observation of the surface structure.

4.2. Dental Restoration Fitness Test

Right mandibular first molars were selected from the standard resin tooth model (Dental model, NISSIN, Kyoto, Japan), which were then prepared by reducing by 1.5 mm on the labial, axial, and occlusal surfaces with a deep chamfer margin and 4~10 degrees of axial inclination using a diamond rotary bur. For the 3-unit bridge, a right mandibular first molar and premolar were prepared. The impression was obtained using a polysiloxane silicone impression material (Honigum-Light, DMG, Hamburg, Germany), which was cast with Type IV hard stone (Fujirock EP, GC Corp., Tokyo, Japan) to obtain the master cast.

The prepared abutment tooth working model was scanned using a 3D model scanner (Freedom HD, DOF, Seoul, Republic of Korea), which was then converted into an STL file format. The scanned 3D model was processed with a dental CAD program (Exocad, exocad GmbH, Darmstadt, Germany) to decide the margin and was designed using the conventional PFM restoration coping, with a cement gap of 30 µm.

The designed data were proposed to facilitate SLM Co-Cr alloy coping with a 3D printer (ProX® DMP 100, 3D systems Inc., Rock Hill, SC, USA) using the SLM technique. The milling Co-Cr alloy coping was facilitated with the milling machine (Arden 5X-WM, TPS Korea LTD, Gwang-Ju, Republic of Korea). The Co-Cr alloy coping using the casting technique was achieved using the lost-wax technique. All of the procedures were undertaken by a single experienced laboratory technician in a uniform fashion. Ten Co-Cr alloy copings (single crown and a 3-unit bridge) were made for each of the 3 techniques, using the right mandibular first molar as an abutment ($n = 10$).

Porcelain was fused to the metal restoration specimens of all groups according to the manufacturer's instructions. Noritake Super Porcelain EX-3 (Kuraray Noritake Dental Inc., Tokyo, Japan) was used for the porcelain build-up, and a porcelain furnace (Austromat 3001, Dekema, Freilassing, Germany) was used in accordance with the manufacturer's instructions. (Figure 8) The porcelain build-up was held in 5 stages of 'oxidizing–opaque–dentin–glazing', with each stage including the first state, which was the 'initial' stage (Figure 8).

The marginal fit of the Co-Cr alloy coping determined with 3 different techniques was measured using metallurgic microscopy (EGVM-452M, Easytech, An-yang, Republic of Korea) at ×300 magnification following the marginal fit defined by Holmes [25]. The measurement was made at the center of the marginal buccal, lingual, mesial, and distal areas. The margin fit was measured at every 5 stages of porcelain firing, and each measurement was made at the same magnification and location and using the same method (Figure 9). Every group was measured with 10 specimens per measurement area, and the mean value of these was determined as the marginal fit of the respective area.

The internal fit was measured in all groups of specimens finished until the glazing stage of the porcelain firing procedures to compare the internal fit based on the different manufacturing techniques. The silicone replica technique, which is widely used in many studies and proven to be credible and accurate, was used to determine the internal space measurement [50,51]. The internal surface of each specimen was filled with polysiloxane silicone impression material (Honigum-Light, DMG, Hamburg, Germany) to determine the internal space. Each specimen filled with soft silicone material was mounted on the corresponding abutment models, which were then pressured in the direction of the tooth's longitudinal axis. After the silicone was set, the specimens were removed, and a low-flowable polysiloxane silicone impression material (Honigum-Mono, DMG, Hamburg, Germany) was applied. After the final setting, the silicone was cut in a mesio–distal and linguo–buccal direction with a scalpel. The cross-section of the cut piece was observed at ×100 magnification using metallurgical microscopy (EGVM-452M, Easytech, An-yang, Republic of Korea), and the thickness of the silicone was measured to determine the internal gap (Figure 10). The internal gap of all specimens was measured in the mesio–distal and bucco–lingual part of the axial wall, cusp, and occlusal areas.

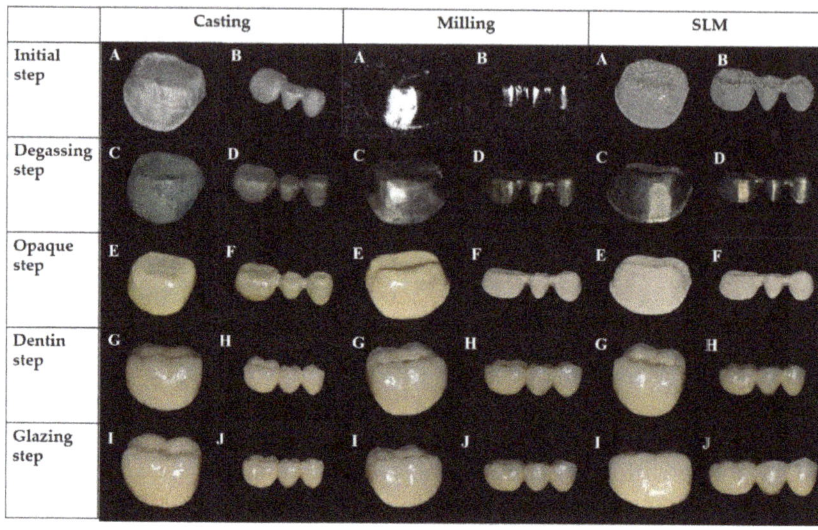

Figure 8. PFM dental restoration using casting, milling, and SLM methods. (**A**): single metal coping, (**B**): 3-unit bridge metal coping, (**C**): single coping after oxidizing process, (**D**): 3-unit bridge coping after the oxidizing process. (**E**): single coping after opaque porcelain firing procedure, (**F**): 3-unit bridge coping after opaque porcelain firing procedure, (**G**): single coping after dentin porcelain firing procedure, (**H**): 3-unit bridge coping after dentin porcelain firing procedure, (**I**): single coping after glazing procedure, (**J**): 3-unit bridge coping after glazing procedure.

Figure 9. Marginal fit of PFM dental restoration after final glazing procedure at ×300 magnification (**A**) casting, (**B**) milling, (**C**) SLM.

4.3. Statistical Analysis

In the case of the tensile strength test, the shear bond strength, and AFAP, a Kruskal–Wallis test was conducted to compare the mean values using the manufacturing techniques, followed by Mann–Whitney U test for post hoc analysis. In the case of the tests for Vickers microhardness, fitness, and surface roughness, A Shapiro–Wilk test was conducted to determine the normal distribution of the results. Following the test of normality, a one-way analysis of variance (ANOVA) was performed, followed by Tukey's test for post hoc analysis. All of the statistical analyses were carried out using SPSS ver. 25.0 for Windows (SPSS Inc., Chicago, IL, USA).

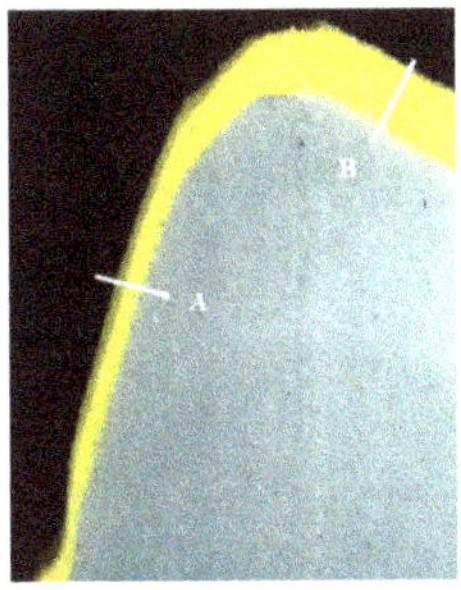

Figure 10. Internal fit of PFM dental single restoration at ×100 magnification. A: axial internal fit, B: occlusal internal fit.

5. Conclusions

In this study, a new method of manufacturing dental prosthetics using the SLM method was studied.

In the tensile strength test, the SLM technique exhibited the highest ultimate tensile strength and a 0.2% yield strength. This means that when the Co-Cr alloy specimen manufactured using the SLM technique is applied intra-orally, it has the lowest possibility of encountering permanent distortion due to masticatory force. Since the Co-Cr alloy using the SLM technique did not show a statistically significant difference in metal–porcelain bond strength compared to the conventional casting technique specimens, it is appropriate for clinical applications. The SLM technique exhibited better marginal fit compared to the conventional casting technique in all stages of porcelain firing procedures. This indicates that the Co-Cr alloy coping manufactured using the SLM technique can have more outstanding clinical performance. Co-Cr alloy coping manufactured using the SLM technique showed the highest internal gap.

Comprehensively compared to traditional casting methods, the dental prosthesis manufactured using the SLM method showed improved physical properties as well as improved marginal fit results, suggesting the possibility of clinical use.

Author Contributions: Conceptualization, K.Y.; methodology, H.K.; software, Y.Y.; validation, S.P.; formal analysis, H.K.; investigation, Y.Y.; resources, H.K.; data curation, E.-C.K.; writing—original draft preparation, Y.Y.; writing—review and editing, Y.-S.L.; visualization, K.Y.; supervision, Y.-S.L.; project administration, K.Y.; funding acquisition, K.Y. All authors have read and agreed to the published version of the manuscript.

Funding: This research received no external funding.

Institutional Review Board Statement: Not applicable.

Informed Consent Statement: Not applicable.

Data Availability Statement: Not applicable.

Conflicts of Interest: The authors declare no conflict of interest.

References

1. Krug, K.P.; Knauber, A.W.; Nothdurft, F.P. Fracture behavior of metal-ceramic fixed dental prostheses with frameworks from cast or a newly developed sintered cobalt-chromium alloy. *Clin. Oral Investig.* **2015**, *19*, 401–411. [CrossRef] [PubMed]
2. Viennot, S.; Dalard, F.; Lissac, M.; Grosgogeat, B. Corrosion resistance of cobalt-chromium and palladium-silver alloys used in fixed prosthetic restorations. *Eur. J. Oral Sci.* **2005**, *113*, 90–95. [CrossRef] [PubMed]
3. Hildebrand, H.F.; Veron, C.; Martin, P. Nickel, chromium, cobalt dental alloys and allergic reactions: An overview. *Biomaterials* **1989**, *10*, 545–548. [CrossRef] [PubMed]
4. van Noort, R. The future of dental devices is digital. *Dent. Mater.* **2012**, *28*, 3–12. [CrossRef]
5. Miyazaki, T.; Hotta, Y. CAD/CAM systems available for the fabrication of crown and bridge restorations. *Aust. Dent. J.* **2011**, *56*, 97–106. [CrossRef] [PubMed]

6. Witkowski, S.; Komine, F.; Gerds, T. Marginal accuracy of titanium copings fabricated by casting and CAD/CAM techniques. *J. Prosthet. Dent.* **2006**, *96*, 47–52. [CrossRef]
7. Stawarczyk, B.; Eichberger, M.; Hoffmann, R.; Noack, F.; Schweiger, J.; Edelhoff, D.; Beuer, F. A novel CAD/CAM base metal compared to conventional CoCrMo alloys: An in-vitro study of the long-term metal-ceramic bond strength. *Oral Health Dent. Manag.* **2014**, *13*, 446–452.
8. Al Jabbari, Y.S.; Koutsoukis, T.; Barmpagadaki, X.; Zinelis, S. Metallurgical and interfacial characterization of PFM Co–Cr dental alloys fabricated via casting, milling or selective laser melting. *Dent. Mater.* **2014**, *30*, e79–e88. [CrossRef]
9. Petzold, R.; Zeilhofer, H.F.; Kalender, W.A. Rapid prototyping technology in medicine—Basics and applications. *Comput. Med. Imaging Graph.* **1999**, *23*, 277–284. [CrossRef]
10. Dikova, T.S.; Dzhendov, D.; Simov, M. Microstructure and hardness of fixed dental prostheses manufactured by additive technologies. *J. Achiev. Mater. Manuf. Eng.* **2015**, *71*, 60–69.
11. Koutsoukis, T.; Zinelis, S.; Eliades, G.; Al-Wazzan, K.; Rifaiy, M.A.; Al Jabbari, Y.S. Selective laser melting technique of Co-Cr dental alloys: A review of structure and properties and comparative analysis with other available techniques. *J. Prosthodont.* **2015**, *24*, 303–312. [CrossRef] [PubMed]
12. Zhang, M.; Gan, N.; Qian, H.; Jiao, T. Retentive force and fitness accuracy of cobalt-chrome alloy clasps for removable partial denture fabricated with SLM technique. *J. Prosthodont. Res.* **2022**, *66*, 459–465. [CrossRef] [PubMed]
13. Yager, S.; Ma, J.; Ozcan, H.; Kilinc, H.I.; Elwany, A.H.; Karaman, I. Mechanical properties and microstructure of removable partial denture clasps manufactured using selective laser melting. *Addit. Manuf.* **2015**, *8*, 117–123. [CrossRef]
14. Ren, X.W.; Zeng, L.; Wei, Z.M.; Xin, X.Z.; Wei, B. Effects of multiple firings on metal-ceramic bond strength of Co-Cr alloy fabricated by selective laser melting. *J. Prosthet. Dent.* **2016**, *115*, 109–114. [CrossRef] [PubMed]
15. Torabi, K.; Farjood, E.; Hamedani, S. Rapid Prototyping Technologies and their Applications in Prosthodontics, a Review of Literature. *J. Dent.* **2015**, *16*, 1–9.
16. Özcan, M. Fracture reasons in ceramic-fused-to-metal restorations. *J. Oral Rehabil.* **2003**, *30*, 265–269. [CrossRef]
17. Joias, R.M.; Tango, R.N.; Junho de Araujo, J.E.; Junho de Araujo, M.A.; Ferreira Anzaloni Saavedra Gde, S.; Paes-Junior, T.J.; Kimpara, E.T. Shear bond strength of a ceramic to Co-Cr alloys. *J. Prosthet. Dent.* **2008**, *99*, 54–59. [CrossRef]
18. Alexandrino, D.; Antunes, L.H.; Munhoz, A.L.; Filho, A.P.; Silva, W.J. Mechanical and surface properties of Co-Cr alloy produced by additive manufacturing for removable partial denture frameworks. *J. Prosthet. Dent.* **2022**, *in press*. [CrossRef]
19. Pereira, A.L.; Medeiros, A.K.; Santos, K.; Almeida, É.; Barbosa, G.A.; Carreiro, A. Accuracy of CAD-CAM systems for removable partial denture framework fabrication: A systematic review. *J. Prosthet. Dent.* **2021**, *125*, 241–248. [CrossRef]
20. Yan, X.; Lin, H.; Wu, Y.; Bai, W. Effect of two heat treatments on mechanical properties of selective-laser-melted Co-Cr metal-ceramic alloys for application in thin removable partial dentures. *J. Prosthet. Dent.* **2018**, *119*, 1028.e1–1028.e6. [CrossRef]
21. Peng, P.W.; Hsu, C.Y.; Huang, H.Y.; Chao, J.C.; Lee, W.F. Trueness of removable partial denture frameworks additively manufactured with selective laser melting. *J. Prosthet. Dent.* **2022**, *127*, 122–127. [CrossRef] [PubMed]
22. NaBadalung, D.P.; Powers, J.M. Effectiveness of adhesive systems for a Co-Cr removable partial denture alloy. *J. Prosthodont.* **1998**, *7*, 17–25. [CrossRef] [PubMed]
23. Salim, S.; Salleh, N.M.; Abidin, Z.Z.; Yunus, N.; Rahmat, R.; Ibrahim, N. Physicomechanical properties of cobalt-chromium removable partial denture palatal major connectors fabricated by selective laser melting. *J. Prosthet. Dent.* **2022**, *128*, 530.e1–530.e7. [CrossRef] [PubMed]
24. Daou, E.E.; Özcan, M. Evaluation of ceramic adherence to cobalt-chromium alloys fabricated by different manufacturing techniques. *J. Prosthet. Dent.* **2022**, *128*, 1364.e1–1364.e8. [CrossRef]
25. Holmes, J.R.; Bayne, S.C.; Holland, G.A.; Sulik, W.D. Considerations in measurement of marginal fit. *J. Prosthet. Dent.* **1989**, *62*, 405–408. [CrossRef]
26. McLean, J.W.; von Fraunhofer, J.A. The estimation of cement film thickness by an in vivo technique. *Br. Dent. J.* **1971**, *131*, 107–111. [CrossRef]
27. Hung, S.H.; Hung, K.S.; Eick, J.D.; Chappell, R.P. Marginal fit of porcelain-fused-to-metal and two types of ceramic crown. *J. Prosthet. Dent.* **1990**, *63*, 26–31. [CrossRef]
28. Ural, C.; Burgaz, Y.; Saraç, D. In vitro evaluation of marginal adaptation in five ceramic restoration fabricating techniques. *Quintessence Int.* **2010**, *41*, 585–590.
29. Hamaguchi, H.; Cacciatore, A.; TuellerHamaguchi, V.M. Marginal distortion of the porcelain-bonded-to-metal complete crown: An SEM study. *J. Prosthet. Dent.* **1982**, *47*, 146–153. [CrossRef]
30. Anusavice, K.J.; Carroll, J.E. Effect of incompatibility stress on the fit of metal-ceramic crowns. *J. Dent. Res.* **1987**, *66*, 1341–1345. [CrossRef]
31. Alqahtani, A.S.; AlFadda, A.M.; Eldesouky, M.; Alnuwaiser, M.K.; Al-Saleh, S.; Alresayes, S.; Alshahrani, A.; Vohra, F.; Abduljabbar, T. Comparison of Marginal Integrity and Surface Roughness of Selective Laser Melting, CAD-CAM and Digital Light Processing Manufactured Co-Cr Alloy Copings. *Appl. Sci.* **2021**, *11*, 8328. [CrossRef]
32. Takaichi, A.; Suyalatu; Nakamoto, T.; Joko, N.; Nomura, N.; Tsutsumi, Y.; Migita, S.; Doi, H.; Kurosu, S.; Chiba, A.; et al. Microstructures and mechanical properties of Co–29Cr–6Mo alloy fabricated by selective laser melting process for dental applications. *J. Mech. Behav. Biomed. Mater.* **2013**, *21*, 67–76. [CrossRef] [PubMed]

33. Kim, H.R.; Jang, S.H.; Kim, Y.K.; Son, J.S.; Min, B.K.; Kim, K.H.; Kwon, T.Y. Microstructures and mechanical properties of Co-Cr dental alloys fabricated by three CAD/CAM-based processing techniques. *Materials* **2016**, *9*, 596. [CrossRef] [PubMed]
34. Zhou, Y.; Li, N.; Yan, J.; Zeng, Q. Comparative analysis of the microstructures and mechanical properties of Co-Cr dental alloys fabricated by different methods. *J. Prosthet. Dent.* **2018**, *120*, 617–623. [CrossRef] [PubMed]
35. Serra-Prat, J.; Cano-Batalla, J.; Cabratosa-Termes, J.; Figueras-Àlvarez, O. Adhesion of dental porcelain to cast, milled, and laser-sintered cobalt-chromium alloys: Shear bond strength and sensitivity to thermocycling. *J. Prosthet. Dent.* **2014**, *112*, 600–605. [CrossRef]
36. Sced, I.R.; McLean, J.W. The strength of metal-ceramic bonds with base metals containing chromium. A preliminary report. *Br. Dent. J.* **1972**, *132*, 232–234. [CrossRef]
37. Schweitzer, D.M.; Goldstein, G.R.; Ricci, J.L.; Silva, N.R.; Hittelman, E.L. Comparison of bond strength of a pressed ceramic fused to metal versus feldspathic porcelain fused to metal. *J. Prosthodont.* **2005**, *14*, 239–247. [CrossRef]
38. Zeng, L.; Zhang, Y.; Liu, Z.; Wei, B. Effects of repeated firing on the marginal accuracy of Co-Cr copings fabricated by selective laser melting. *J. Prosthet. Dent.* **2015**, *113*, 135–139. [CrossRef]
39. Fonseca, J.C.; Henriques, G.E.; Sobrinho, L.C.; de Góes, M.F. Stress-relieving and porcelain firing cycle influence on marginal fit of commercially pure titanium and titanium–aluminum–vanadium copings. *Dent. Mater.* **2003**, *19*, 686–691. [CrossRef]
40. Papazoglou, E.; Brantley, W.A.; Johnston, W.M. Evaluation of high-temperature distortion of high-palladium metal-ceramic crowns. *J. Prosthet. Dent.* **2001**, *85*, 133–140. [CrossRef]
41. Bertolotti, R.L.; Moffa, J.P. Creep rate of porcelain-bonding alloys as a function of temperature. *J. Dent. Res.* **1980**, *59*, 2062–2065. [CrossRef] [PubMed]
42. Gemalmaz, D.; Alkumru, H.N. Marginal fit changes during porcelain firing cycles. *J. Prosthet. Dent.* **1995**, *73*, 49–54. [PubMed]
43. Örtorp, A.; Jönsson, D.; Mouhsen, A.; Vult von Steyern, P. The fit of cobalt–chromium three-unit fixed dental prostheses fabricated with four different techniques: A comparative in vitro study. *Dent. Mater.* **2011**, *27*, 356–363. [CrossRef] [PubMed]
44. Quante, K.; Ludwig, K.; Kern, M. Marginal and internal fit of metal-ceramic crowns fabricated with a new laser melting technology. *Dent. Mater.* **2008**, *24*, 1311–1315. [CrossRef] [PubMed]
45. Huang, Z.; Zhang, L.; Zhu, J.; Zhang, X. Clinical marginal and internal fit of metal ceramic crowns fabricated with a selective laser melting technology. *J. Prosthet. Dent.* **2015**, *113*, 623–627. [CrossRef]
46. Nesse, H.; Ulstein, D.M.; Vaage, M.M.; Øilo, M. Internal and marginal fit of cobalt-chromium fixed dental prostheses fabricated with 3 different techniques. *J. Prosthet. Dent.* **2015**, *114*, 686–692. [CrossRef]
47. American Society for Testing and Materials. *Standard Specification for Cobalt-28 Chromium-6 Molybdenum Alloy Castings and Casting Alloy for Surgical Implants (UNS R30075)*; ASTM International: West Conshohocken, PA, USA, 2018.
48. *BS ISO 22674:2006*; Dentistry–Metallic Materials for Fixed and Removable Restorations and Appliances. British Standards Institute: London, UK, 2006.
49. Lee, K.M.; Cai, Z.; Griggs, J.A.; Guiatas, L.; Lee, D.J.; Okabe, T. SEM/EDS evaluation of porcelain adherence to gold-coated cast titanium. *J. Biomed. Mater. Res. B Appl. Biomater.* **2004**, *68*, 165–173. [CrossRef]
50. Park, J.K.; Lee, W.S.; Kim, H.Y.; Kim, W.C.; Kim, J.H. Accuracy evaluation of metal copings fabricated by computer-aided milling and direct metal laser sintering systems. *J. Adv. Prosthodont.* **2015**, *7*, 122–128. [CrossRef]
51. Kim, M.J.; Choi, Y.J.; Kim, S.K.; Heo, S.J.; Koak, J.Y. Marginal Accuracy and Internal Fit of 3-D Printing Laser-Sintered Co-Cr Alloy Copings. *Materials* **2017**, *10*, 93. [CrossRef]

Disclaimer/Publisher's Note: The statements, opinions and data contained in all publications are solely those of the individual author(s) and contributor(s) and not of MDPI and/or the editor(s). MDPI and/or the editor(s) disclaim responsibility for any injury to people or property resulting from any ideas, methods, instructions or products referred to in the content.

International Journal of *Molecular Sciences*

MDPI

Review

Functional Surface Coatings on Orthodontic Appliances: Reviews of Friction Reduction, Antibacterial Properties, and Corrosion Resistance

Ruichu Zhang [1,2], Bing Han [1,2,*] and Xiaomo Liu [1,2,*]

[1] Department of Orthodontics, School and Hospital of Stomatology, Peking University, Beijing 100081, China
[2] National Center of Stomatology & National Clinical Research Center for Oral Diseases & National Engineering Laboratory for Digital and Material Technology of Stomatology & Beijing Key Laboratory for Digital Stomatology & Research Center of Engineering and Technology for Computerized Dentistry Ministry of Health & NMPA Key Laboratory for Dental Materials, Beijing 100081, China
* Correspondence: kqbinghan@bjmu.edu.cn (B.H.); liuxiaomo@pkuss.bjmu.edu.cn (X.L.)

Abstract: Surface coating technology is an important way to improve the properties of orthodontic appliances, allowing for reduced friction, antibacterial properties, and enhanced corrosion resistance. It improves treatment efficiency, reduces side effects, and increases the safety and durability of orthodontic appliances. Existing functional coatings are prepared with suitable additional layers on the surface of the substrate to achieve the abovementioned modifications, and commonly used materials mainly include metal and metallic compound materials, carbon-based materials, polymers, and bioactive materials. In addition to single-use materials, metal-metal or metal-nonmetal materials can be combined. Methods of coating preparation include, but are not limited to, physical vapor deposition (PVD), chemical deposition, sol-gel dip coating, etc., with a variety of different conditions for preparing the coatings. In the reviewed studies, a wide variety of surface coatings were found to be effective. However, the present coating materials have not yet achieved a perfect combination of these three functions, and their safety and durability need further verification. This paper reviews and summarizes the effectiveness, advantages and disadvantages, and clinical perspectives of different coating materials for orthodontic appliances in terms of friction reduction, antibacterial properties, and enhanced corrosion resistance, and discusses more possibilities for follow-up studies as well as for clinical applications in detail.

Keywords: coated materials; corrosion; friction; microorganisms; orthodontic brackets; orthodontic wires; orthodontics

check for updates

Citation: Zhang, R.; Han, B.; Liu, X. Functional Surface Coatings on Orthodontic Appliances: Reviews of Friction Reduction, Antibacterial Properties, and Corrosion Resistance. *Int. J. Mol. Sci.* **2023**, *24*, 6919. https://doi.org/10.3390/ijms24086919

Academic Editor: Mary Anne Melo

Received: 29 January 2023
Revised: 19 March 2023
Accepted: 31 March 2023
Published: 7 April 2023

Copyright: © 2023 by the authors. Licensee MDPI, Basel, Switzerland. This article is an open access article distributed under the terms and conditions of the Creative Commons Attribution (CC BY) license (https://creativecommons.org/licenses/by/4.0/).

1. Introduction

Surface coating technology relies on various metals, polymers, and composite materials by preparing a suitable additional layer on the surface of the substrate and applying different coatings to the surface of the brackets and archwires to modify their surface morphology, mechanical properties, and antibacterial properties [1]. This is currently an important way to improve the performance of orthodontic appliances, enabling clinical effects such as friction reduction, antibacterial effects, and corrosion resistance.

It is necessary to reduce the friction between the archwire and brackets, improve the antibacterial properties of orthodontic appliances, and increase their corrosion and wear resistance using surface coatings, in order to improve treatment efficiency, reduce the consumption of anchorage, reduce the incidence of caries and periodontal disease caused by orthodontics, and increase the safety and durability of orthodontic attachments. Some studies have shown that friction can offset 12–60% of the orthodontic force during orthodontic treatment [2], reducing the efficiency of tooth movement and increasing the consumption of anchorage, which is detrimental to the treatment [2,3]. Moreover, a white spot lesion

(WSL) is often observed on the enamel surface around orthodontic attachments [4–6], and plaque-induced gingivitis [7] appears occasionally. These are due to difficulties in the mechanical removal of plaque and increased plaque retention [8]. Furthermore, metal orthodontic appliances are susceptible to corrosion in the intraoral environment [9–11]. In addition to electrochemical corrosion, chewing, brushing teeth, and archwires sliding along the brackets, the mechanical forces on the surfaces also accelerate the process of corrosion [10,12], causing the release of various metal ions—especially nickel (Ni) ions [13–15]. These metal ions act as allergens, increasing the likelihood of metal allergy for orthodontic patients [16,17]. Allergic reaction of the oral mucosa to nickel can cause distress and pain to patients; this is significantly more common in females than in males, especially between the ages of 16 and 35 years. The discomfort to patients includes burning sensation of the oral mucosa, cheilitis, parageusia, periodontitis, and even erythema multiforme and popular perioral rash [17]. These conditions can reduce patients' quality of life and oral health, which can adversely affect treatment compliance and therapeutic effects. Thus, surface coatings are important to be put into clinical use.

However, the existing coating materials and methods for orthodontic fixed appliances are numerous and complicated, while the purpose and focus point of the coatings differ. Based on the above considerations, this article mainly reviews and summarizes the effects, advantages, disadvantages, and clinical prospects of different coating materials with respect to friction reduction, antibacterial effects, and corrosion resistance, in order to guide the selection and preparation of suitable surface coatings for clinical applications. Coatings on brackets and archwires can be seen in Figure 1.

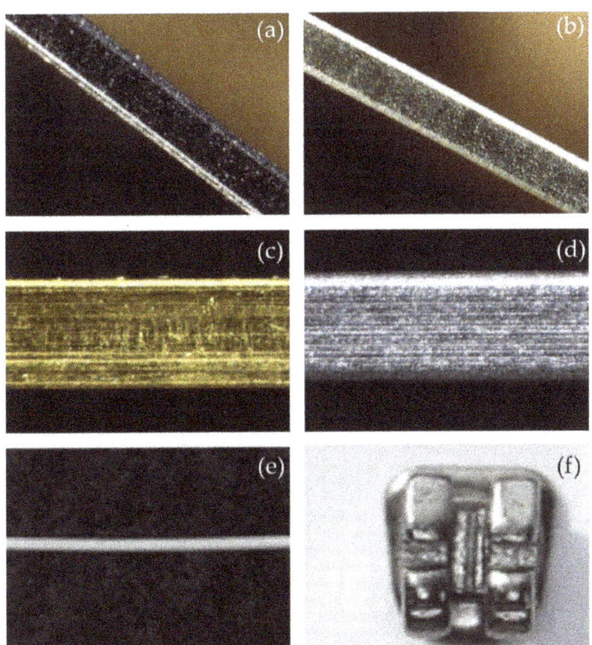

Figure 1. Functional coatings that have been applicated on brackets and archwires. (**a**) nickel (Ni) + molybdenum disulfide (MoS_2)-coated stainless steel (SS) archwire; (**b**) Ni + tungsten disulfide (WS_2)-coated SS archwire; (**c**) titanium nitride (TiN)-coated SS archwire; (**d**) TiN-coated nickel-titanium (NiTi) archwire; (**e**) polymer-coated wires (microcoated stainless steel wire®, G&H Wire Company, Franklin, IN, USA); (**f**) polydopamine and blue fluorescent hollow carbon dots (PDA-HCD) brackets. Adapted with permission from [18–21].

2. Friction Reduction Coatings

Existing orthodontic friction reduction coatings can be divided into two major categories—metallic and non-metallic—and each major category can be divided into various single coatings and composite coatings. Among metal and metallic compound friction reduction coatings, tungsten disulfide (WS_2) was the first to be applied [22]. Since then, silver (Ag) coatings and metallic compound coatings, such as zinc oxide (ZnO) [23–27], titanium nitride (TiN) [28–33], and aluminum oxide (Al_2O_3) [30], have also received good attention. Nonmetals, including carbon-based materials [34], polymers [35–39], and bioactive materials [40], are relatively new coating materials. In addition, preparing nanoparticles and depositing them on orthodontic attachment surfaces via magnetron sputtering [30,33], evaporation [41], and immersion are also common techniques. These materials play an antifriction role mainly by filling the grooves on the surface of the appliances and forming a lubricating layer, which effectively reduces the dynamic and static friction between the archwire and the brackets [42]. The main friction reduction coating materials discussed in this section are shown in Figure 2.

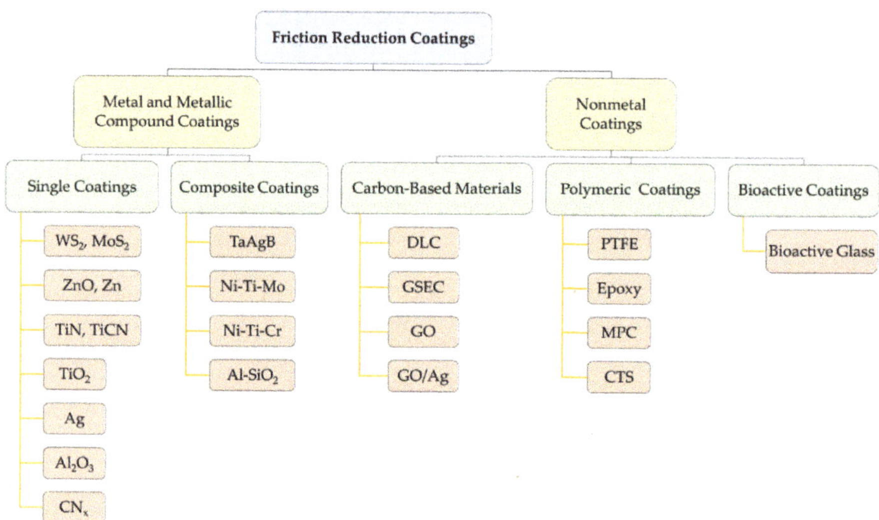

Figure 2. Summary of friction reduction coating materials. WS_2: tungsten disulfide; MoS_2: molybdenum disulfide; ZnO: zinc oxide; Zn: zinc; TiN: titanium nitride; TiCN: titanium carbonitride; TiO_2: titanium dioxide; Ag: silver; Al_2O_3: aluminum oxide; CN_x: carbon nitride; TaAgB: silver-doped tantalum boride; Ni-Ti-Mo: nickel-titanium-molybdenum; Ni-Ti-Cr: nickel-titanium-chromium; Al-SiO_2: aluminum-silicon dioxide; DLC: diamond-like carbon; GSEC: graphene-sheet-embedded carbon; GO: graphene oxide; GO/Ag: graphene oxide/silver; PTFE: polytetrafluoroethylene; MPC: 2-methacryloyloxyethyl phosphorylcholine; CTS: chitosan.

2.1. Inorganic Fullerene-like Nanoparticles of Tungsten Disulfide Coatings

A. Katz et al. used nickel-phosphorus (Ni-P) electroless films impregnated with inorganic fullerene-like nanoparticles of tungsten disulfide (IF-WS_2) to coat stainless steel (SS) archwires and showed that the IF-WS_2 coating significantly reduced archwire friction and mitigated adverse complications [22]. Similarly, the results of M. Redlich et al. showed a 54% reduction in the friction of the coated archwire compared to the uncoated SS archwires [42]. However, despite the biocompatibility of IF-WS_2, its cytotoxicity has not been studied in detail [42], which may have a negative impact on the organization. In 2019, a study applied molybdenum disulfide (MoS_2) and tungsten disulfide (WS_2) coatings to orthodontic SS archwires via electrodeposition and found that both materials reduced the friction between the archwires and the brackets under dry conditions [18]. However, when fixed appliances are placed inside the oral cavity, the environment to which they are exposed is wet. The wettability and durability of WS_2 and MoS_2 has been discussed in other fields [43–45], which needs to be taken into account when considering their application in the dental field. Therefore, the ability of MoS_2 and WS_2 coatings to reduce friction on archwires and brackets in wet environments—for instance, when immersed in artificial saliva or applied in animals' oral cavities—needs to be investigated further.

WS_2 nanoparticles can form a homogeneous and thickness-appropriate coating on the surface of the archwires and brackets; they can also have a similar effect in combination with Ni and Ni-P [22]. The film of nanoparticles peels off when exposed to friction, forming a solid lubricant film between the interfaces, thereby reducing the friction—especially when the pressure is high [42]. Therefore, WS_2 coatings are more suitable apply during the alignment process where the contact angle is large, i.e., the tipping and uprighting type of tooth movement [22]. Still, it is worth noting that large-angle bending can damage the coating and, therefore, WS_2 is not suitable for large-angle bending in clinical applications [42,46]. In addition, it has demonstrated good biocompatibility in preliminary studies, but its cytotoxicity is still unproven, and there are no experiments simulating the intraoral environment, so subsequent studies need to focus on these aspects.

2.2. Zinc and Zinc Compound Coatings

ZnO is of interest because of its low toxicity and good biocompatibility. A human safety review showed that nanostructured ZnO is safe for humans [47], and its suitability for biomedical applications has led to an increasing number of studies on ZnO nanocoatings for reducing friction on fixed appliances. Mojghan Kachoei et al. conducted an in vitro study and found that coating SS archwires with ZnO nanoparticles resulted in a 39–51% [23] (using deposition-precipitation) or 64% [24] (using the sol-gel technique) reduction in friction force on the coated archwires compared to uncoated archwires, and this positive effect could be retained under variations in the angle between the slot and the archwire (0–10°) [23]. Another experiment showed an even better result: when the angle increased, its ability to reduce friction force was enhanced [25]. However, a similar study conducted by Ahmad Behroozian et al. [26] showed that there was no significant reduction in friction force after coating ZnO nanoparticles on SS archwires, but coating ceramic brackets with ZnO nanoparticles could reduce friction force. More interestingly, having both SS archwires and ceramic brackets coated with ZnO nanoparticles did not contribute to the reduction in friction force [26]. This phenomenon hints that when using SS archwires and ceramic brackets as a pair, there is no need to apply ZnO coatings on both the archwires and the brackets, as this would increase the cost and decrease the friction-reducing effect. In addition to SS, nickel titanium (NiTi) archwires also need reduced friction due to their wide usage. Coating ZnO nanoparticles on NiTi archwires using the chemical deposition method reduces surface friction by nearly 21% [25], while using electrochemical deposition reduces friction force by 34% [27].

ZnO NPs not only provide good biocompatibility and antibacterial properties, but also have better friction reduction than compact ZnO [23], so they should be prepared in the form of nanoparticles when the ZnO coating is performed. In addition, ZnO coatings prepared using the sol-gel technique can be applied chairside [24], making this one of the most feasible preparation methods in the clinic. Currently, many studies have demonstrated its ability to exert good friction reduction on 0.019×0.025 SS archwires [23,26]. In fact, it has worked well in various sizes and materials of archwires and brackets, including ceramic brackets, while its utility in self-ligating brackets has not yet been determined. However, pairing ZnO-coated SS archwires with ZnO-coated ceramic brackets is not recommended, as it is not effective in reducing friction [26].

The use of physical vapor deposition (PVD) for zinc (Zn) plating on the surface of SS archwires is thought to significantly reduce surface friction. Whether the angle between the archwire and the slot is 0° or 10°, the change in angle does not significantly affect this result [23]. In addition, tensile strength and three-point bending strength also increase significantly after Zn plating, indicating that Zn plating has a good effect on changing the mechanical properties of SS archwires and is one of the potentially available materials [48]. One of the advantages of Zn coating over ZnO coating is that fewer cracks are formed, and the mechanical properties are better maintained after bending, which indicates that it can be used in clinical situations where the archwire is to be bent at large angles.

2.3. Zirconia Compound Coatings

In clinical medicine, zirconium oxide (ZrO_2) nanoparticle coatings are used on a variety of implant surfaces, such as artificial joints, skin implants, interventional catheters, and other surgical instruments, due to their high stability, corrosion resistance, and excellent biocompatibility [49–51]. Therefore, their use in orthodontic appliances has been investigated for performance improvements. However, among the NiTi, SS, and beta-titanium (TMA) archwires, ZrO_2 could only be applied to the TMA wire, and no coating was observed on the other two archwires using scanning electron microscopy (SEM) [52]. Moreover, ZrO_2 on the surface of TMA archwires did not have a significant effect on friction reduction, although it increased the surface smoothness of the archwires [52]. This differs from the outcomes of orthopedic implants studied in another field, which applied zirconia nanomaterials to titanium alloys and obtained lower friction [53]. This alloy has a different composition from TMA archwires, which could be the reason for the different results. Therefore, the composition of the substrate also needs to be taken into consideration when selecting the coating material. Based on the above conclusions, the clinical application of ZrO_2 coatings in orthodontic appliances is of low value.

2.4. Titanium and Titanium Compound Coatings

The use of titanium compound coatings in orthodontics is becoming more widespread. Ion-plated TiN coatings have the characteristics of high hardness, wear resistance, corrosion resistance, and surface lubrication. They can form a passivated titanium dioxide (TiO_2) layer on their surface, so TiN coatings are commonly used on various dental instruments and materials [54–56]. In an earlier study, the friction force of TiN-coated SS brackets coated via the hollow cathode discharge method and of uncoated SS archwires before and after corrosion in fluorinated solutions was investigated. It was found that the friction force of the coated brackets was not significantly lower than that of the uncoated brackets, whether before or after corrosion [28]. It has also been shown that TiN coated via ion beam-assisted deposition (IBAD) on SS materials can even increase the friction force [29]. Interestingly, a recent study by Arici N. et al. concluded that coating TiN using radio frequency magnetron sputtering on SS brackets can reduce the coefficient of friction (CoF) by 50% when applied in combination with uncoated SS archwires, while the combination of TiN-coated archwires and uncoated brackets increases the CoF whether the material of the archwires is NiTi or SS [30]. That is, when TiN is coated on SS brackets and archwires, there can be different influences on friction force. The different conclusions might be related to different coating methods or other conditions. Thus, more research is needed in order to determine and verify the reasons.

The effectiveness of TiN in reducing friction on the surface of SS archwires and brackets is still controversial. However, based on some of the studies where it was considered effective, it is easy to see that its clinical application requires pairing with specific brackets or archwires. For example, uncoated brackets are not recommended when TiN-coated archwires are used [30]. This suggests that attention should be paid to the application of TiN coatings when selecting brackets and archwires.

The current trend tends to combine TiN with other application materials. TiN-derived coating materials such as titanium carbonitride (TiCN) [31,32], multilayer titanium nitride/titanium (TiN/Ti) [33], and nickel-titanium-molybdenum (Ni-Ti-Mo) composite coatings [57] not only have a more exact friction reduction effect than TiN, but are also more beneficial to clinical application due to other advantages.

In a recent study, Suciu V et al. used direct-current (DC) reactive magnetron sputtering to cover the surface of SS archwires with TiN and TiCN films and varied the ambient nitrogen content during the film fabrication process to investigate the effects of different nitrogen contents on the properties of the produced films. They found that the higher the nitrogen and carbon contents, the lower the CoF. In particular, the TiCN coating could reduce the CoF to 0.30 and make the CoF more stable, while having better hardness and corrosion resistance than TiN. Therefore, overall, the TiCN film had better surface properties than TiN in the tested samples. More interestingly, the researchers also found that the colors of TiN and TiCN films have decorative aesthetic properties, so these two coatings may be among the future directions for the development of aesthetic friction reduction archwires [31]. Similarly, Jie Zhang et al. found that TiN and TiCN coatings on SS surfaces can reduce surface friction, especially in artificial saliva [32]. Liyuan Sheng et al. prepared different layers of TiN/Ti multilayer coatings on titanium-aluminum-vanadium alloy with 6% aluminum and 4% vanadium (Ti_6Al_4V) substrates with the same deposition time via high-power direct-current reactive magnetron sputtering, and they found that the CoF of 1- and 2-layer TiN/Ti multilayer coatings maintained an increasing trend, but the CoF of 4-, 8-, and 12-layer TiN/Ti multilayer coatings showed a decrease [33]. Thus, TiCN is superior to TiN in terms of friction reduction, hardness, corrosion resistance, and aesthetic properties, offering better clinical prospects. For multilayer coatings, researchers also found that in order to obtain the best clinical results, it is necessary to restrict the TiN/Ti coatings to no more than two layers.

The above studies are all about SS archwires and brackets, while there are relatively few studies about titanium compound coatings on orthodontic archwires made from other materials. A study by Haruki Sugisawa et al. showed that TiN coatings can reduce the friction of NiTi and SS archwires, increase the tensile strength and stiffness of SS archwires, and reduce the elasticity of NiTi archwires [58]. In addition, Vinod Krishnan et al. attempted to apply a titanium aluminum nitride (TiAlN) coating to β-titanium orthodontic archwires using PVD to investigate its effect on friction, but the results showed no significant reduction compared to uncoated archwires [59]. Not many further studies on the usage of TiAlN coatings in orthodontic appliances have been followed up. In addition, a recent study investigated TiO_2 and silicon dioxide (SiO_2) nanoparticles on the surface of chromium-nickel (Cr-Ni) archwires and found that TiO_2 did not show a significant ability to reduce friction [60]. Conversely, SiO_2 can be used to reduce friction between the brackets and archwires under both dry and wet conditions, while making the surface of the archwires smoother than uncoated ones [60]. A systematic review and meta-analysis suggested that nanoparticle-coated orthodontic archwires can be considered to significantly reduce frictional resistance [61]. In this regard, it can be presumed that TiO_2 nanoparticles are also theoretically able to reduce the friction force. In fact, the reason for the inability of TiO_2 nanocoatings to significantly reduce friction may be that the coating is thicker than other coatings [60]. Therefore, the pressure on the inner wall of the slot is greater, resulting in greater friction force. In addition, the deformation and peeling of the surface morphology are also among the reasons [62]. Similarly, the application of TiO_2 coatings on SS brackets using PVD does not reduce friction [63].

Therefore, the TiO_2 coating may not exhibit the desired ability to reduce friction, because of its thickness [60]. However, is still worth developing as a very versatile material and, therefore, has a certain research value. In future research, other coating methods to reduce the thickness could be considered. Based on the consideration of film thickness, the preparation of a thinner film of TiO_2 nanocoatings could be attempted in future studies. Recent studies have proven that low-pressure metal-organic chemical vapor deposition (LPMOCVD) can control the film thickness by adjusting the deposition time [64]. In fact, this method of controlling the thickness of the coating can be applied to many deposition-based coating methods; thus, future research could take it into consideration when finding the most suitable coating conditions.

In addition to titanium compounds, composite coatings of titanium with other metals are also being investigated. In the study of Yılmaz H et al. (2021), nickel ions and TiO_2 nanoparticles were used as standard coating materials on the surface of NiTi archwires, while the addition of 0.0015 g/dm^3 sodium molybdate (Na_2MoO_4) on this basis led to the formation of a Ni-Ti-Mo coating on the surface of the archwires; thus, a reduction in the CoF from 0.288 to 0.252 was observed. However, although the addition of a certain amount of molybdenum ion (Mo^{6+}) to the bath solution can reduce the surface friction of the archwires after the formation of the coating, the reduction in surface friction gradually becomes less effective as the concentration of Mo^{6+} increases. The standard NiTi coating does not reduce friction, and the addition of different concentrations of chromium (Cr) to the standard coating material to form a nickel-titanium-chromium (Ni-Ti-Cr) coating also does not reduce friction. [57]

In summary, the trends of Ti compound coatings are nitriding [30], incorporation of other metal ions [57], multi-layered materials [33], and low-thickness materials [64]. Most of these are still some distance away from being applied in clinical practice, but attempts can be made to find the most suitable preparation conditions and methods to obtain coating materials with satisfactory properties in all respects. The biggest disadvantage of this type of coating, however, may be its high price [65]. If the price of orthodontic supplies is increased by the application of the coating, then there is no advantage compared to other coatings with similar properties, and patients are less likely to choose to use these materials. Therefore, although the materials are excellent in various respects, in order to put them into use, there is still a need to find ways to reduce the cost of their production.

2.5. Silver and Silver Compound Coatings

Nanosilver coatings can reduce the surface roughness of SS brackets [63] or NiTi archwires [66], but the change in friction between brackets and archwires is not obvious [63]. Moreover, silver (Ag) plating on 0.019 × 0.025-inch coated SS archwires significantly reduces the friction between the brackets and the archwires, while 0.017 × 0.025-inch SS archwires do not significantly reduce the friction [67]. The 0.019 × 0.025-inch SS archwires, whether coated or not, have higher friction than the 0.017 × 0.025-inch SS archwires. The play between 0.019 × 0.025-inch SS archwires and the slots is smaller, so the pressure is higher when the slots are of the same size. After applying the coating, the friction coefficient of the surface should be the same regardless of the size of the archwires, so the alteration of the friction force of the archwires subjected to more pressure may be relatively more obvious when the other conditions are the same [67]. However, this explanation has not yet been clearly confirmed. Therefore, more research is still needed in order to confirm its capabilities. Still, considering the antibacterial properties of silver [63], nanosilver coatings are still a popular research topic for surface coatings of orthodontic appliances, even without clear friction reduction properties.

Currently, there are only a few types of materials related to silver used in the surface coating of orthodontic appliances, which are generally silver nanoparticles (AgNPs). Interestingly, Jia Wang et al. [68] successfully prepared silver-doped tantalum boride (TaAgB) solid solution coatings on SS blocks using magnetron sputtering technology with Ag doping. They verified the effectiveness of the coating in reducing the CoF of the SS material to 0.08, as shown in Figure 3, and suggested that it could be used for orthodontic archwires and brackets because of its good hardness, strength, and biocompatibility. Therefore, as a surface coating material with low friction, good antibacterial properties, high wear resistance, and high corrosion resistance, it has strong potential value for application on fixed appliances.

As a precious metal, silver coatings are also a relatively high-value material. However, nanosilver coatings have multifunctional properties that may be accepted by the general public even if their price is high [69]. Therefore, it is necessary to develop more friction reduction, antibacterial, and other properties of nanosilver coatings.

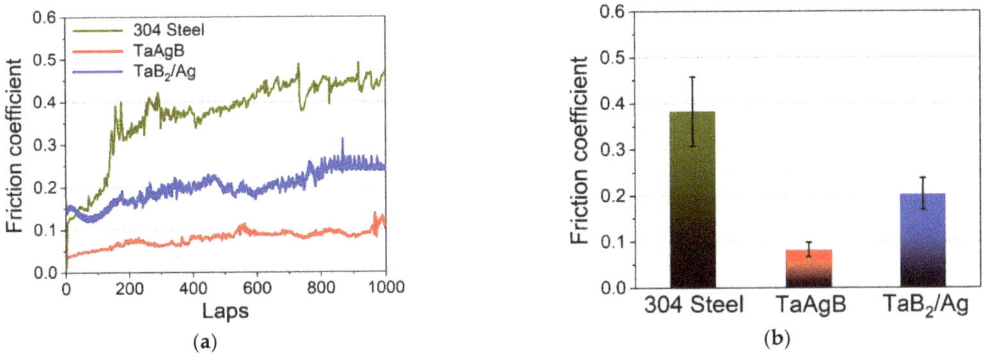

Figure 3. (**a**) Dynamic friction curves for TaAgB coating, TaB_2/Ag coating, and uncoated 304 steel. (**b**) Mean friction coefficient for TaAgB coating, TaB_2/Ag coating, and uncoated 304 steel. TaB_2: tantalum boride. Adapted with permission from [68]. Copyright 2022, Elsevier.

2.6. Aluminum and Aluminum Compound Coatings

Al_2O_3 coatings are oxide coatings that are common in various fields. Research by Nursel Arici et al. [30] in 2021 demonstrated the friction reduction effect of Al_2O_3 coatings in detail. When an Al_2O_3 coating is applied on the surface of SS brackets, whether the material of the combined archwire is made of SS or NiTi, and regardless of whether the archwires have an Al_2O_3 coating or not, the CoF can be reduced significantly. Among all combinations, the combination of Al_2O_3-coated SS brackets with the same coated archwire was able to obtain the lowest CoF. That is, the combination of coated SS brackets-coated SS archwire is better than other combinations. It is also worth mentioning that in almost all cases involving coated archwire-uncoated bracket combinations, an increase in CoF was observed for both NiTi and SS archwires, being much higher than for both uncoated combinations. This phenomenon may be related to the peeling of the coating [30]. Therefore, when testing friction, the pressure of the coating environment, time, immersion conditions, and peeling should also be taken into account to obtain more accurate conclusions.

Others have combined aluminum (Al) and SiO_2 to form Al-SiO_2 composite coatings. The excellent friction reduction ability of SiO_2 has been mentioned above [60]. In contrast, the friction reduction properties of Al have not been adequately studied. A reduction in the CoF is observed in Al-SiO_2-coated NiTi and SS archwires compared to the uncoated ones [70]. Therefore, Al-SiO_2 composite coatings may also be used to reduce the friction of orthodontic archwires in the future.

Al compound coatings are effective in reducing friction, retain high aesthetic value, and are relatively inexpensive [30]. However, studies about them are still scarce, especially since their performance changes after long-term use are not confirmed. In addition, Al_2O_3 needs to be considered in clinical applications for the combination of coated and uncoated archwires and brackets. Combining coated archwires with uncoated brackets is not recommended, but it is possible to use coated brackets only [30]. Ideally, both the brackets and the archwire should be coated, which will result in the best friction reduction.

2.7. Other Metal and Metallic Compound Coatings

Nursel Arici et al. [30] noted that the CrN coating was ineffective in reducing friction between the SS brackets and the SS or NiTi archwires, whether both were coated or only one was coated. SEM images after friction tests showed that the CrN coating peels off from the surface. Therefore, residual coating material at the contact area between the surfaces may be responsible for the increase in CoF. This phenomenon is not found in the application of hard chrome carbide plating (HCCP) coatings. HCCP coatings significantly reduce the friction of SS archwires, and mechanical properties such as flexural strength and flexural modulus remain unchanged [21]. HCCP is a promising coating material that can potentially be used clinically to reduce friction. However, its shortcomings include aesthetics that need to be improved and biocompatibility that has not been explicitly tested.

Another common metal coating is rhodium. As a coating based on aesthetic considerations, it is generally applied to archwires. Some recent studies have begun to explore its usage beyond aesthetics. A study by Tahereh Hosseinzadeh Nik et al. found that the frictional force generated during sliding between ceramic brackets and rhodium-plated SS archwires was significantly less than that of ceramic brackets and SS archwires. Even after immersion in 0.05% sodium fluoride (NaF) mouthwash, the increase in friction was much less in the former than in the latter [71]. This indicates that rhodium plating can be used not only as a method to increase the aesthetic effect of the archwires, but also to reduce the friction between the archwires and the brackets, constituting a method to increase the efficiency of the orthodontic treatment and maintain good performance even when the patient uses NaF mouthwash. In addition, they demonstrated that rhodium-plated SS archwires showed a smaller increase in roughness than uncoated SS archwires after treatment in mouthwash [71]. Of course, the correlation between surface roughness and friction remains controversial. Some think that the two can be directly correlated [72,73], while others do not [35,74], so it is not yet possible to conclude that the smaller increase in friction of rhodium-plated archwires is related to the smaller increase in surface roughness. Friction-reducing coatings made of other metallic monomers are also gradually being investigated. Recent studies have attempted to prepare niobium (Nb), tantalum (Ta), and vanadium (V) coatings on 316 SS substrates using the plasma sputtering method and found that V coatings have an effect on the roughness of the SS surface [75]. However, other experiments on the tribology of these new materials have not yet been conducted.

2.8. Carbon-Based Coatings

Carbon-based coating materials mainly contain diamond-like carbon (DLC) and graphene.

DLC is a diamond-like carbon film that exhibits extremely high hardness, a low coefficient of friction, chemical inertness, and high corrosion resistance [34]. Research has demonstrated the biomedical applications of DLC films. For example, implants with protective films can extend the life of the implant by reducing corrosion and wear, which can benefit patients [76].

DLC-coated SS brackets exhibit lower static [77] and dynamic friction [78], and their effect is better than that of TiN-coated brackets [79]. However, some studies have also found no reduction in friction on the surface of DLC-coated SS brackets [80]. In addition to brackets, DLC-coated SS archwires [78,80–82] or NiTi archwires [83,84] can be applied to reduce friction. Hao Zhang et al. concluded that the use of the plasma-enhanced chemical vapor deposition (PECVD) method to cover the surface of SS archwires with a DLC coating can reduce the coefficient of dynamic friction by 40.71% and significantly improve the surface hardness of the archwire, as well as reducing friction and exhibiting good biocompatibility [82]. It has also been shown that the reduction in friction is effective whether the angle between the archwire and the bracket is 0° or 10° [81].

Furthermore, the application of DLC via different coating methods—such as plasma-enhanced chemical vapor deposition (PECVD) [79,82,85], plasma-based ion implantation/deposition (PBIID) [81,84,86], and mirror confinement electron cyclotron resonance (MCECR) plasma sputtering [80,83]—has shown similar friction-reducing effects. It has also been shown that for DLC-coated SS brackets, the size and material of the archwire, the contact angle, and the dryness or humidity of the experimental conditions have no influence on the friction reduction effect of DLC coatings [77]. However, although changes in the above conditions have no significant influence on the effect of DLC coatings in reducing friction, the effect of the thickness of the coating is still not negligible. Muguruma et al. [78] investigated the mechanical properties of DLC coatings of different thicknesses. They found that a thin DLC coating reduces surface friction, which is the same result as previously described, yet the relatively thick DLC layer tends to break from the DLC-steel interface.

In addition, thicker layers have a negative effect on friction force. DLC coatings also have the advantage of being resistant to high-fluorine environments. Even at high fluoride ion concentrations, the friction force remains relatively stable, i.e., there is no significant increase [83].

DLC coatings with different properties were prepared under different conditions as DLC-1 and DLC-2. DLC-1 uses a 10 kV target voltage, an acetylene-toluene gas environment, and deposits for 3 min to form a structure with a diamond-rich outer surface and a graphite-rich inner surface; DLC-2 uses a 7 kV target voltage, a toluene gas environment, and deposits for 4 min to form a structure with a graphite-rich outer surface and a diamond-rich inner surface. The results showed that DLC-2 produces significantly lower friction than the uncoated case when wet and at an angle of 10°, which was not observed for DLC-1. The reasons for this were that partial cracks in the DLC-1 coating led to increased friction, while no cracks or breakages of the coating were observed in DLC-2. Therefore, we can conclude that the DLC-2 coating is more valuable than DLC-1 in practical applications because of its good flexibility and adhesion [87].

In addition to pure DLC, fluorine-doped DLC (F-DLC) and silicon-doped DLC (Si-DLC) coatings can also reduce friction on the surface of SS brackets. The F-DLC-coated brackets exhibit low static friction between the brackets and the archwire under moist conditions, which is lower than that under dry conditions. However, doping DLC with fluorine or silicon results in a significant reduction in surface hardness. Even though F-DLC is most effective in reducing the friction between the archwire and the brackets in humid conditions, this disadvantage needs to be taken into account in practical applications [85]. Titanium-doped DLC (Ti-DLC) coatings are also ideal friction-reducing films for orthodontic brackets, as they exhibit a lower CoF than TiCN and TiN in most cases, under both artificial saliva and dry conditions [32].

Recently, research on graphene coatings has gradually emerged. Commonly used graphene materials include graphene oxide and graphene-sheet-embedded carbon (GSEC). Graphene oxide (GO) coatings can reduce the CoF of NiTi alloy [88,89], while the addition of AgNPs and the formation of graphene oxide/silver nanoparticle (GO/Ag) coatings can further reduce CoF [88]. In artificial saliva, the GSEC surface coating of SS archwires coated with MCECR reduced the CoF to 0.12 [90]. Pengfei Wang et al. later discovered ways to make the friction on the surface of GSEC coatings even lower. In this study, they found that the thickness of the coating increased as the GSEC film deposition time increased. After 80 min of deposition, both stable and low friction coefficients could be obtained, even reaching a level of less than 0.10 when running against three-row microgroove-textured SS brackets in an artificial saliva environment [46]. The possible mechanisms are shown in Figure 4. Although immersion causes a rebound in CoF, after 30 days of immersion, the CoF remains below 0.30, consistently exhibiting very good friction reduction [90]. This illustrates the effectiveness of the graphene coating in reducing the friction on the surface of the archwire.

In summary, carbon-based friction-reducing coatings have good overall durability and biocompatibility [82], and DLC coatings are resistant to corrosion by fluoride-containing mouthwashes [83]. The coating method and conditions are less restrictive compared to metal coatings, but there are strict requirements for the thickness of the coating layers, requiring a thickness that is both thin and effective [87]. GSEC coatings can be used for up to one month, and the optimal coating deposition length has been determined and may be ready for clinical application soon.

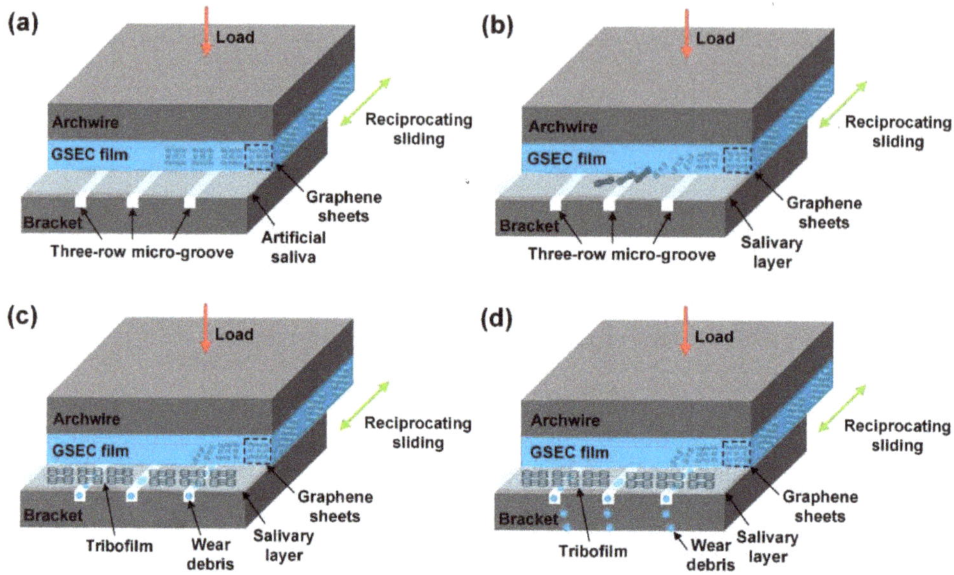

Figure 4. The low-friction and low-wear mechanisms of the GSEC-film-coated stainless steel archwires running against three-row microgroove-textured stainless steel brackets in an artificial saliva environment. (**a**) Initial state of the contact combination, (**b**) stable low friction with the formation of graphene-rich tribofilm and salivary adsorbed layer, (**c**) accumulation of wear debris detached from the GSEC film with microgroove, and (**d**) flow out of wear debris with artificial saliva from the microgroove. Adapted with permission from [46].

2.9. Polymeric and Bioactive Coatings

Teflon (polytetrafluoroethylene (PTFE)), an aesthetic white coating for dental instruments, has been shown to provide a smoother surface and reduce frictional resistance and frictional damage [35–39] on NiTi [91] or SS archwires [92]—especially when paired with metal slots in ceramic brackets [38]—in both dry and wet conditions. However, when combined with ceramic brackets, PTFE- and epoxy-coated NiTi archwires will increase the friction between the archwires and the brackets [93]. This is probably because ceramics have a rougher texture and a more porous surface than stainless steel. Another research shows the contrast, which whether combined with ceramic or metal brackets, PTFE-, or epoxy-coated SS archwires always show higher friction resistance [94]. However, the latest study by K. Ranjan and R. Bhat et al. contradicts this finding; the study concludes that whether paired with SS brackets, ceramic brackets, or ceramic brackets with metal slots, the PTFE-coated archwire produces lower frictional resistance compared to uncoated SS archwires [92]. In addition, research has investigated suitable preparation conditions. PTFE coatings prepared at 200 °C offer lower friction, higher wear resistance, lower coloration, and higher resistance to microbial adhesion than those prepared at 380 °C [95].

In addition to PTFE, other polymer surface coatings have been used to explore the ability to reduce friction on fixed appliances, such as chitosan (CTS) [24], 2-methacryloyloxyethyl phosphorylcholine (MPC), parylene, and epoxy.

Elhelbawy N. et al. investigated CTS as a newer and potentially useful coating material for reducing friction on fixed appliance surfaces in comparison to ZnO. This comparison revealed no significant difference in the degree of friction reduction between CTS and ZnO (64 and 53%, respectively). Interestingly, when CTS and ZnO were applied to the brackets or archwires alone, and the matching archwires or brackets were not coated, the degree of friction reduction was similar to that of having both brackets and archwires coated with CTS or ZnO, suggesting that coating both the brackets and the archwires did not significantly improve this aspect of friction reduction [24].

In a recent study, Ryo Kunimatsu et al. found a promising application of MPC in fixed appliances. They covered SS archwires with an MPC coating and performed tensile tests using a bracket-archwire combination to demonstrate the coating's ability to reduce friction. Furthermore, they directly measured the efficiency of tooth movement using an in vitro experimental tooth movement model at 50 g and 100 g traction force, and finally confirmed that the SS archwire coating with MPC improves the efficiency of tooth movement by reducing the occurrence of friction [96]. This experimental design has provided ideas for subsequent studies. Simply confirming that a certain surface coating can reduce friction is not enough to fully prove its clinical value. Its positive effect on improving tooth movement efficiency needs to be further determined through model experiments, animal experiments, and even clinical trials.

Chin-Yu Lin et al. [97] coated SS archwires with parylene, epoxy, or PTFE and tested the friction between the archwires and ceramic brackets after immersion in water for different lengths of time. Compared to the other groups, the epoxy-coated archwires presented higher or no significant difference in friction resistance and were the only coated archwires whose maximal resistance to sliding at a 3° contact angle (MRS3) was consistently lower than that of the uncoated control group. Parylene, on the other hand, had relatively high friction at all water immersion times (0 to 4 weeks) and angles (0 to 3°). Therefore, epoxy-coated archwires may even be a better choice than PTFE.

In addition to polymers, bioactive glass (BG) coatings have recently been prepared on the surface of SS archwires using electrophoretic deposition. Although BG has higher hardness, elastic modulus, and surface smoothness, it failed to reduce the friction in the experiment [40]. Overall, studies on the application of bioactive coatings on fixed appliances are extremely rare and could perhaps be explored more in the future.

Carbon-based materials [82], natural polymeric materials [24], and bioactive materials [40] are safer than metal and metallic compound materials and can be used to replace other coating materials that are potentially cytotoxic [42]. CTS coatings can replace ZnO and are relatively more convenient and cost-effective, as they do not require coating on both the archwire and the bracket surface [24]. Aesthetic coatings such as PTFE are suitable for both SS and NiTi archwires, but when used with ceramic brackets, it is better to choose ceramic brackets with metal slots [92] and other commonly used polymeric aesthetic archwires. Among the other common aesthetic polymer coatings on the market, only epoxy shows a stronger friction reduction effect than PTFE [97]. However, the popularity of its use and the difficulty of its preparation have not been conclusively proven, so the most suitable polymer aesthetic coating for clinical friction reduction is still PTFE.

Recent studies on friction reduction coatings for orthodontic appliances are summarized in Table 1.

Table 1. Recent studies on friction reduction coatings for orthodontic appliances.

Ref.	Coating Materials	Coating Technique	Substrate	Study Type	Roughness	Friction Reduction Effectiveness	Wear Mechanism	Wear Resistance	Other Effectiveness
[67]	Ag	Direct current sputtering	SS archwires	In vitro (dry condition)	NA	Compared to uncoated archwires, the friction force of 0.019 × 0.025-inch SS archwires reduced 1 N after coating ($p = 0.032$), but no significant difference between coated and uncoated 0.017 × 0.025-inch SS archwires ($p = 0.854$).	NA	NA	NA
[30]	Al_2O_3	Radio frequency magnetron sputtering	SS brackets, NiTi archwires, SS archwires	In vitro (dry condition)	Surface roughness decreased with the coating process. R_a = 331.53 nm, R_q = 426.17 nm (coated brackets); R_a = 361.64 nm, R_q = 466.01 nm (coated NiTi archwires); R_a = 95.86 nm, R_q = 128.01 nm (coated SS archwires)	NiTi archwires: When both brackets and archwires were coated and only brackets were coated, CoF reduced from 0.316 to 0.238 and 0.251, respectively. CoF increased from 0.316 to 0.400 when only archwires were coated. SS archwires: When both brackets and archwires were coated, only brackets were coated, and only archwires were coated, CoF reduced from 0.552 to 0.227, 0.2,35, and 0.445, respectively.	No peeling off	The coatings did not peel off after friction, thermal, and brushing tests.	NA
[70]	Al-SiO_2	Magnetron sputtering	NiTi and SS archwires	In vitro (in artificial saliva)	NA	CoF of NiTi archwires and SS archwires reduced from 0.68 to 0.46 and from 0.58 to 0.45, respectively.	NA	Corrosion-resistant	Corrosion-resistant effectiveness Biocompatibil-ity
[40]	BG	Electrophoretic deposition	SS archwires	In vitro (dry condition)	Sa increased from 0 to 0.46–0.79.	Friction forces showed no significant reduction.	NA	None of the coatings were damaged after three-point bending and they sustained good interfacial adhesion.	Aesthetic effect
[84]	CF_4	Plasma-based ion implantation and deposition	NiTi archwires	In vitro (dry condition)	NA	Friction forces reduced from 129.48 to 104.97 gf, which showed no significant reduction.	NA	NA	NA
[84]	CH_4	Plasma-based ion implantation and deposition	NiTi archwires	In vitro (dry condition)	NA	Friction forces reduced from 129.48 to 87.30 gf, which showed no significant reduction.	NA	NA	NA

Table 1. *Cont.*

Ref.	Coating Materials	Coating Technique	Substrate	Study Type	Roughness	Friction Reduction Effectiveness	Wear Mechanism	Wear Resistance	Other Effectiveness
[29]	CN_x	Ion beam-assisted deposition	304L SS disks	In vitro (friction test: NA; antibacterial test: in bacterial suspension)	The films slightly reduce the surface roughness parameter. R_a reduced from 0.181 to 0.140 µm	CoF reduced from 0.431 to 0.188 ($p < 0.05$).	NA	NA	Antibacterial Effectiveness Biocompatibility
[30]	CrN	Radio frequency magnetron sputtering	SS brackets, NiTi archwires, SS archwires	In vitro (dry condition)	Surface roughness decreased with the coating process. R_a = 276.85 nm, R_q = 360.24 nm (coated brackets); R_a = 354.35 nm, R_q = 454.66 nm (coated NiTi archwires); R_a = 190.28 nm, R_q = 229.26 nm (coated SS archwires)	NiTi archwires: When both brackets and archwires were coated, only brackets were coated, and only archwires were coated, CoF increased from 0.316 to 0.443, 0.324, and 0.505, respectively. SS archwires: When both brackets and archwires were coated and only archwires were coated, CoF increased from 0.552 to 0.598 and 0.586, respectively. CoF reduced from 0.552 to 0.410 when only brackets were coated.	Peeling off	Large areas of peeling could be seen after friction, thermal, and brushing tests.	NA
[24]	CTS (NPs)	Sol-gel dip coating	SS archwires and SS brackets	In vitro	NA	Friction force decreased by ~53%.	NA	NA	NA
[84]	DLC	Plasma-based ion implantation and deposition	NiTi archwires	In vitro (dry condition)	NA	The friction force reduced from 129.48 to 86.13 gf ($p = 0.039$).	NA	NA	NA
[87]	DLC	Plasma-based ion implantation and deposition	SS archwires	In vitro (in artificial saliva and dry condition)	NA	When coated with DLC-2, the static friction reduced from 2.39 to 2.09 N in artificial saliva and from 2.49 to 2.25 N under dry conditions, and the kinetic friction reduced from 2.37 to 1.99 N in artificial saliva and from 2.55 to 2.21 N under dry conditions, which showed a significantly lower frictional force than the uncoated archwires, while DCL-1 showed no significant difference compared with uncoated samples.	Rupture	No rupture was observed for the DLC-2 condition after the drawing-friction testing.	NA
[97]	Epoxy	NA (commercial)	SS archwires	In vitro (in distilled-deionized water)	NA	The average resistance under 0° bracket-wire angle reduced from 1.63 to 1.13 N immediately after being coated. The average resistance under a 3° bracket-wire angle reduced from 5.12 to 4.27 N immediately after being coated.	NA	NA	Durability (>4 weeks)

Table 1. *Cont.*

Ref.	Coating Materials	Coating Technique	Substrate	Study Type	Roughness	Friction Reduction Effectiveness	Wear Mechanism	Wear Resistance	Other Effectiveness
[94]	Epoxy	NA (commercial)	SS archwires	In vitro (in artificial saliva and dry condition)	NA	The friction force increased from 3.00–9.00 N to 16.00–20.50 N in both wet and dry conditions.	NA	NA	NA
[93]	Epoxy	NA (commercia)	NiTi archwires	In vitro (dry condition)	NA	Friction forces increased from 49.287 to 53.316 gf.	NA	NA	NA
[89]	GO	Silane coupling	NiTi archwires	In vitro (friction test: NA; antibacterial test: in bacterial suspension; corrosion test: artificial saliva)	The surface of the samples coated with 2 mg/mL GO concentrations was smooth with a uniformly coated area.	CoF reduced from ~0.9 to 0.2–0.4.	Grooves in the same direction as the gliding	The samples coated with 0.5 mg/mL GO concentrations had fewer grooves, but a small amount of wear debris was present.	Antibacterial effectiveness Corrosion-resistant effectiveness Biocompatibility
[88]	GO/Ag (NPs)	Electrophoretic deposition	NiTi alloy	In vitro (dry condition)	R_a ranged from 50.72–69.93 nm.	CoF reduced from 0.060 to 0.006, and increased coating time led to lower CoF.	NA	NA	NA
[46,90]	GSEC	Electron cyclotron resonance plasma sputtering	SS archwires	In vitro (in artificial saliva)	NA	CoF reduced to ~0.10 and remained under 0.30 after 30 days.	Peeling off	The corresponding wear rate was strongly decreased from 4.84×10^{-6} to 0.11×10^{-6} mm^3/Nm.	NA
[21]	HCCP	Electroplating	SS archwires	In vitro (in PBS solution and dry condition)	There were very small protrusions on the surfaces of the coated archwires, while the surfaces of uncoated archwires were smooth.	The friction force reduced from 147.15 to 124.61 gf ($p = 0.0076$) and from 143.55 to 121.41 gf ($p = 0.04$) under dry and wet conditions, respectively.	Scratches	After the friction test, scratches were seen on the surfaces on the coated surfaces.	Aesthetic effect
[18]	Ni+MoS_2 (NPs)	Electrochemical co-deposition	SS archwires	In vitro (in artificial saliva and dry condition)	NA	Dry conditions: CoF reduced from 0.58–1.43 to 0.50–1.19. In artificial saliva: CoF reduced from 0.95–2.52 to 0.94–2.35.	NA	NA	NA

Table 1. *Cont.*

Ref.	Coating Materials	Coating Technique	Substrate	Study Type	Roughness	Friction Reduction Effectiveness	Wear Mechanism	Wear Resistance	Other Effectiveness
[18]	Ni+WS_2 (NPs)	Electrochemical co-deposition	SS archwires	In vitro (in artificial saliva and dry condition)	NA	Dry conditions: CoF reduced from 0.58–1.43 to 0.42–1.06. In artificial saliva: CoF reduced from 0.95–2.52 to 0.66–1.46.	NA	NA	NA
[57]	Ni-Ti-Cr	Chronopotentiometry	NiTi archwires	In vitro (in artificial saliva)	NA	CoF showed no significant reduction.	NA	NA	Corrosion-resistant effectiveness Durability
[57]	Ni-Ti-Mo	Chronopotentiometry	NiTi archwires	In vitro (in artificial saliva)	NA	CoF reduced from 0.288 to 0.252–0.265.	NA	NA	Corrosion-resistant effectiveness Durability (>60 days)
[97]	Parylene	NA (commercial)	SS archwires	In vitro (in distilled-deionized water)	NA	The average resistance under 0° bracket-wire angle increased from 1.63 to 5.39 N immediately after being coated. The average resistance under the 3° bracket-wire angle increased from 5.12 to 11.38 N immediately after being coated.	NA	NA	Durability
[97]	PTFE	NA (commercial)	SS archwires	In vitro (in distilled-deionized water)	NA	The average resistance under 0° bracket-wire angle reduced from 1.63 to 1.15 N immediately after being coated. The average resistance under a 3° bracket-wire angle showed no significant change immediately after being coated.	NA	NA	Durability
[95]	PTFE	Spraying	SS, Ni-Ti, and β-titanium archwires	In vitro	R_a of coated archwires increased from 0.02–0.21 to 0.53–0.58 μm.	The friction force of coated archwires reduced from 123.94–152.61 to 102.98–124.40 gf compared to uncoated archwires and 200 °C coating resulted in less friction against brackets than did the conventional 380 °C coating.	Scratches	PTFE-coating at 200 °C resulted in good microbial adhesion and tolerance of wear.	Durability (>3 months) Aesthetic effect

Table 1. *Cont.*

Ref.	Coating Materials	Coating Technique	Substrate	Study Type	Roughness	Friction Reduction Effectiveness	Wear Mechanism	Wear Resistance	Other Effectiveness
[94]	PTFE	NA (commercial)	SS archwires	In vitro (in artificial saliva and dry condition)	NA	The friction force was higher than uncoated archwires but lower than epoxy and rhodium-coated archwires in wet and dry conditions. Dry conditions: The friction force increased from 3.80–9.00 to 5.50–12.80 N. In artificial saliva: The friction force increased from 3.00–8.20 to 3.80–11.60 N.	NA	NA	NA
[93]	PTFE	NA (commercial)	NiTi archwires	In vitro (dry condition)	NA	Friction forces increased from 49.287 to 61.427 gf.	NA	NA	NA
[92]	PTFE	NA (commercial)	SS archwires	In vitro	The surface roughness increased after the friction test.	When compared with ceramic with a metal slot, the friction force reduced from 1.61 to 1.03 N after the archwires were coated with PTFE.	NA	NA	NA
[21]	Rhodium	NA (commercial)	SS archwires	In vitro (in PBS and dry condition)	There were protrusions on the surfaces of the coated archwires, while the surfaces of uncoated archwires were smooth.	The friction force increased from 147.15 to 216.29 gf ($p < 0.001$) and from 143.55 to 210.21 gf ($p < 0.001$) under dry and wet conditions, respectively.	Scratches	After the friction test, scratches were seen on the surfaces on the coated surfaces.	Aesthetic effect
[94]	Rhodium	NA (commercial)	SS archwires	In vitro (in artificial saliva and dry condition)	NA	The friction force increased from 3.00–9.00 N to 17.30–21.70 N in both wet and dry conditions.	NA	NA	NA
[71]	Rhodium	NA (commercial)	SS archwires	In vitro (in artificial saliva and 0.05% NaF mouthwash)	Surface roughness showed no significant reduction with the coating process.	The friction force was reduced from 2.22 to 1.49 N in artificial saliva and reduced from 2.72 to 2.17 N in 0.05% sodium fluoride mouthwash.	NA	NA	NA
[60]	SiO_2 (NPs)	NA (commercial)	Cr-Ni archwires	In vitro (in artificial saliva and dry condition)	Surface roughness decreased with the coating process. R_a reduced from 0.673 to 0.040 µm.	Friction force reduced from 0.51 to 0.38 N and from 0.56 to 0.44 N under wet and dry conditions, respectively.	NA	NA	NA

Table 1. *Cont.*

Ref.	Coating Materials	Coating Technique	Substrate	Study Type	Roughness	Friction Reduction Effectiveness	Wear Mechanism	Wear Resistance	Other Effectiveness
[68]	TaAgB	Magnetron sputtering with Ag-doping	304 SS sheets	In vitro (in artificial saliva)	NA	CoF reduced from 0.38 to 0.08.	No serious wear phenomenon	The wear rate was 6.51 × 10^{-15} m^3/Nm.	Antibacterial effectiveness Biocompatibility
[32]	TiCN	PVD	316L SS plates	In vitro (in artificial saliva and dry condition)	R_a = 29.86 nm; root mean square roughness: R_q = 39.09 nm	CoF reduced from 0.20 to less than 0.06 under dry conditions with loads of 5 N.	NA	NA	NA
[31]	TiCN	Direct current reactive magnetron sputtering	316L SS and (100)-oriented Si substrates	In vitro (in artificial saliva)	NA	CoF reduced to less than 0.30, and the addition of N_2 and C in the Ti matrix further lowered the CoF.	NA	The lowest wear rate was 5.6 × 10^{-6} mm^3/Nm	Corrosion-resistant effectiveness Aesthetic effect
[32]	Ti-DLC	PVD	316L SS plates	In vitro (in artificial saliva and dry condition)	The surface of the Ti-DLC film has many nanocrystal clusters that cause substantial pitting and layer defects. R_a = 10.40 nm; R_q = 13.43 nm	CoF remained less than 0.04 under both dry and artificial saliva conditions.	NA	NA	NA
[29]	TiN	Ion beam-assisted deposition	304L SS disks	In vitro (friction test: NA; antibacterial test: in bacterial suspension)	The films slightly reduce the surface roughness parameter. R_a reduced from 0.181 to 0.162 μm.	CoF increased from 0.431 to 0.469 ($p < 0.05$).	NA	NA	Antibacterial Effectiveness Biocompatibility
[58]	TiN	Ion plating	SS and NiTi archwires	In vitro (friction test: NA; corrosion test: in 0.9% NaCl solution)	R_a of TiN-coated SS archwire increased from 0.023 to 0.046 μm. R_a of TiN-coated NiTi archwire remained at 0.001 μm.	At angles of 0 degrees, the friction forces of the TiN-coated NiTi archwires reduced, but SS archwires showed no significant difference. At angles of 10 degrees, the friction forces of the TiN-coated SS and NiTi archwires were both reduced.	NA	NA	Corrosion-resistant effectiveness
[32]	TiN	PVD	316L SS plates	In vitro (in artificial saliva and dry condition)	The surface of the TiN film has some particles and pinholes. R_a = 26.59 nm; R_q = 32.62 nm	CoF reduced from 0.08 to less than 0.03 under dry conditions with loads of 5 N.	NA	NA	NA

Table 1. *Cont.*

Ref.	Coating Materials	Coating Technique	Substrate	Study Type	Roughness	Friction Reduction Effectiveness	Wear Mechanism	Wear Resistance	Other Effectiveness
[30]	TiN	Radio frequency magnetron sputtering	SS brackets, NiTi, and SS archwires	In vitro (dry condition)	R_a = 354.12 nm, R_q = 458.74 nm (coated brackets); R_a = 391.99 nm, R_q = 534.36 nm (coated NiTi archwires); R_a = 105.15 nm, R_q = 132.67 nm (coated SS archwires)	NiTi archwires: When both brackets and archwires were coated, only brackets were coated, and only archwires were coated, CoF increased from 0.316 to 0.399, 0.331, and 0.446, respectively. SS archwires: When both brackets and archwires were coated and only brackets were coated, CoF reduced from 0.552 to 0.372 and 0.237, respectively. CoF increased from 0.552 to 0.818 when only archwires were coated.	Peeling off	After friction, thermal, and brushing tests, the coatings peeled off in some small areas.	NA
[31]	TiN	Direct current reactive magnetron sputtering	316 L SS and (100)-oriented Si substrates	In vitro (in artificial saliva)	NA	CoF was reduced to less than 0.30, and the addition of N_2 in the Ti matrix further lowered the CoF.	NA	The lowest wear rate was 7.8 × 10^{-6} mm^3/Nm	Corrosion-resistant effectiveness
[29]	TiO_2	Ion beam-assisted deposition	304L SS disks	In vitro (friction test: NA; antibacterial test: in bacterial suspension)	NA	NA	NA	The TiO_2 film wore out in a few seconds.	Antibacterial effectiveness Biocompatibility
[60]	TiO_2 (NPs)	NA (commercial)	Cr-Ni archwires	In vitro (in artificial saliva and dry condition)	Surface roughness decreased with the coating process. R_a reduced from 0.673 to 0.042 µm.	Friction force showed no significant reduction or even increased from 0.56 to 0.61 N compared to uncoated archwires.	NA	NA	NA
[48]	Zn	PVD	SS archwires	In vitro	NA	The friction force reduced from 2.98 to 2.03 N (p = 0.001) and from 3.51 to 1.72 N (p < 0.0001) at 0° and 10° angles, respectively.	Cracks and scratches	The scratches and cracks could be seen clearly.	NA
[46]	ZnO (NPs)	Sol-gel dip coating	SS archwires and SS brackets	In vitro	NA	Friction force decreased by ~64%.	NA	NA	NA
[27]	ZnO (NPs)	Electrochemical deposition	NiTi archwires	In vitro (friction test: NA; antibacterial test: on nutrient agar plates)	NA	Friction force decreased by 34%.	NA	NA	Antibacterial effectiveness

Table 1. *Cont.*

Ref.	Coating Materials	Coating Technique	Substrate	Study Type	Roughness	Friction Reduction Effectiveness	Wear Mechanism	Wear Resistance	Other Effectiveness
[75]	Niobium (NPs)	Plasma sputtering	316L SS	In vitro	R_a reduced from 157.8 to 133.3–156.1 nm after being coated ($p > 0.05$).	Potential friction reduction effectiveness	NA	NA	NA
[75]	Tantalum (NPs)	Plasma sputtering	316L SS	In vitro	R_a reduced from 157.8 to 110–130.6 nm after being coated ($p > 0.05$).	Potential friction reduction effectiveness	NA	NA	NA
[75]	Vanadium (NPs)	Plasma sputtering	316L SS	In vitro	R_a reduced from 157.8 to 83.4–96.8 nm after being coated ($p = 0.002$).	Potential friction reduction effectiveness	NA	NA	NA

Ag: silver; NA: not applicable; SS: stainless steel; NiTi: nickel titanium; R_a: average roughness; R_q: root mean square deviation; CoF: coefficient of friction; Al-SiO_2: aluminum-silicon dioxide; BG: bioactive glass; S_a: mean surface roughness; CF_4: tetrafluoromethane; CN_x: carbon nitride; NPs: nanoparticles; CTS: chitosan; DLC: diamond-like carbon; DLC-1: a DLC structure with a diamond-rich outer surface and a graphite-rich inner surface; DLC-2: a DLC structure with a graphite-rich outer surface and a diamond-rich inner surface; GO/Ag: graphene oxide/silver; GSEC: graphene-sheet-embedded carbon; HCCP: hard chrome carbide plating; PBS: phosphate-buffered solution; Ni: nickel; MoS_2: molybdenum disulfide; WS_2: tungsten disulfide; Ni-Ti-Mo: nickel-titanium-molybdenum; Ni-Ti-Cr: nickel-titanium-chromium; PTFE: polytetrafluoroethylene; SEM: scanning electron microscopy; SiO_2: silicon dioxide; Cr-Ni: chromium-nickel; TaAgB: silver-doped tantalum boride; TiCN: titanium carbonitride; PVD: physical vapored deposition; Si: silicon; N_2: nitrogen; C: carbon; Ti: titanium; Ti-DLC: titanium-doped DLC; TiN: titanium nitride; TiO_2: titanium dioxide; ZnO: zinc oxide; Zn: zinc; NaF: sodium fluoride.

3. Antibacterial Coatings

Among orthodontic antimicrobial coatings, the most valued are Ag and ZnO coatings, as they have been shown to exert good antimicrobial activity and are safe for humans in other areas of medicine [98,99]. Recently, TiO_2 photosensitive antimicrobial coatings, polymeric coatings originally used for aesthetic purposes, and bioactive lysozyme coatings have also gradually entered the trend and have been explored for their potential as antibacterial coatings. As for the research on the antimicrobial properties of carbon-based material coatings, only Dai et al. have investigated this topic, and the material used was GO [89]. This study found that when the GO concentration was low, the GO coating could not completely cover the NiTi substrate, and the tribological and anticorrosion properties were barely improved, while the antibacterial properties—although statistically significantly different—only reduced the survival of *Streptococcus mutans* by 20%. Increasing the GO concentration enhanced the antimicrobial properties of the GO coatings, but the biocompatibility of the GO-coated NiTi substrate decreased [89]. Based on the above results, the GO concentration should be controlled within a suitable range when making GO coatings, and the recommended value of this concentration has not been determined at present, while the biocompatibility of GO coatings to the human body has also not been clearly confirmed thus far. In contrast, Ag and ZnO coatings not only have reliable antimicrobial properties, but also have been applied in various medical disciplines with a higher guarantee of safety. Therefore, the earliest Ag and ZnO may still be the most suitable materials for antibacterial coatings. The main antibacterial coating materials discussed in this section are shown in Figure 5.

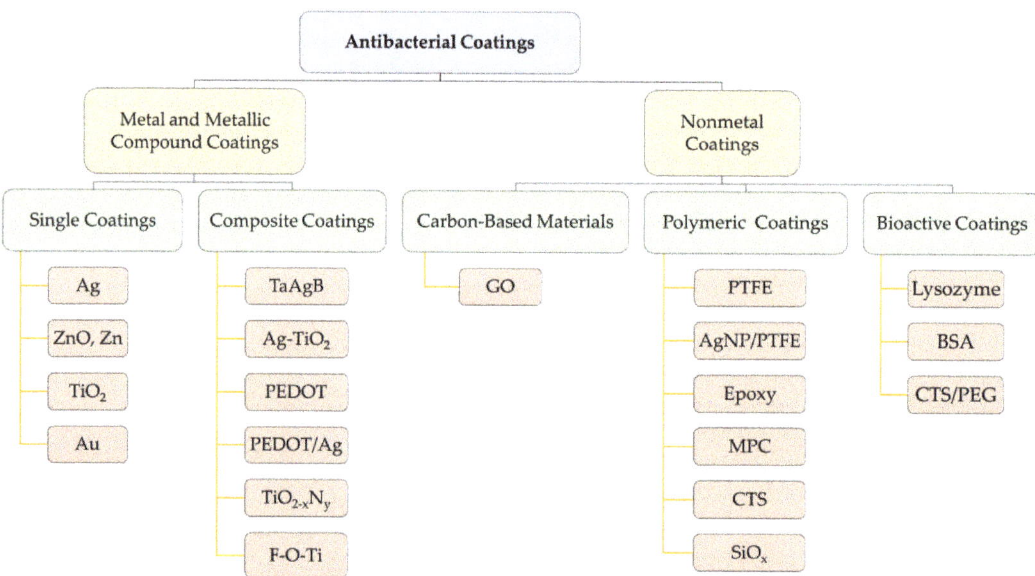

Figure 5. Summary of antibacterial coating materials. Au: gold; Ag-TiO_2: silver-titanium dioxide; PEDOT: poly(3,4-ethylenedioxythiophene); PEDOT/Ag: poly(3,4-ethylenedioxythiophene)/silver; $TiO_{2-x}N_y$: nitrogen-doped titanium dioxide; F-O-Ti: fluorine and oxygen double-deposited titanium; AgNP/PTFE: nanosilver/polytetrafluoroethylene; SiO_x: silicon oxides; BSA: bovine serum albumin; CTS/PEG: chitosan/polyethylene glycol.

3.1. Silver and Silver Compound Coatings

Silver has a wide range of antibacterial properties, protecting the surface of the material from microbial adhesion through different mechanisms, and has a certain antibacterial effect [100,101]. AgNPs can continuously release silver ions, which cause damage to bacterial cell walls and cell membranes. They can also penetrate cells and interfere with the synthesis of proteins and deoxyribonucleic acid (DNA). Inactivation of respiratory enzymes and the production of reactive oxygen species in bacterial cells are other mechanisms that can kill the bacteria. In addition, silver nanoparticles themselves can adhere to the cell wall and the membrane surface, directly causing cell membrane denaturation and perforation, and disrupting the signal transduction of bacterial cells [99,102]. In addition to the direct killing and inhibition of bacteria, AgNPs also regulate inflammatory responses, further inhibiting the reproductive survival of microorganisms [103]. Therefore, it can be assumed that by applying a silver coating to orthodontic appliances, bacterial colonization of the dental surfaces can be reduced, reducing plaque and caries formation during orthodontic treatment.

In a study by Mhaske et al., the authors found that silver plating applied by thermal evaporation on the surface of SS and NiTi archwires reduced the adhesion of *Lactobacillus acidophilus* on the surface of archwires and showed antibacterial activity against *Lactobacillus acidophilus* [41]. AgNPs coated on SS archwire via hydrothermal synthesis also showed significant inhibition of *Staphylococcus aureus* and *Streptococcus mutans* [104].

Nanosilver-coated SS brackets also have significant antibacterial activity. Valiollah Arash et al. prepared silver particles via an electroplating method and verified the antibacterial activity of silver plating against *Streptococcus mutans* using direct contact tests and disk diffusion tests, concluding that its antibacterial activity could last for up to 30 days after application [105]. Tania Ghasemi et al. demonstrated direct inhibition of *Streptococcus mutans* reproduction by preparing nanosilver coatings on the surface of brackets using PVD [63]. In addition to its antiadhesive and antibacterial properties against *Streptococcus mutans* and *Streptococcus distortus* [106], the nanosilver coating was also effective against *Staphylococcus aureus* and *Escherichia coli* [107]. Moreover, this effect was not limited to SS brackets, but also applied to ceramic and cobalt-chromium (Co-Cr) alloy brackets [107].

In addition to in vitro tests, there have also been animal studies demonstrating the antibacterial activity of silver-coated brackets. Gamze Metin-Gürsoy et al. found that nanosilver-coated orthodontic brackets can inhibit *Streptococcus mutans* in dental plaque and produce low levels of nanosilver ion release in saliva, as well as reducing smooth-surface caries. However, they did not have any significant effect on the incidence of occlusal caries [108]. This is the only study to date in which surface-coated fixed appliances have been applied in animals' oral cavities; thus, we need more animal experiments in further studies.

Different surface coating methods for silver have different effects on the improvement of antibacterial properties. In a clinical trial, Viktoria Meyer-Kobbe et al. first investigated the use of plasma-immersion ion implantation and deposition (PIIID) in SS brackets. They concluded that the antibacterial effect of PIIID silver-modified surfaces is as significant as that of electroplated silver layers and PVD silver coatings, reducing biofilm volume and surface coverage, and producing an even stronger bactericidal performance [109]. Coating of SS brackets with silver using PIIID has proven good antibacterial properties in clinical applications, and direct evidence is next needed to show its ability to reduce the incidence of white spots and plaque gingivitis caused by orthodontic appliances. In addition to this, TaAgB also reduces the attachment and growth of *Streptococcus mutans* on the SS surface. This is evidenced by the results observed via SEM; not only was there a reduction in *Streptococcus mutans* on the TaAgB surface, but the morphology of the bacteria was also altered, further confirming its antibacterial activity [68].

Silver compound coatings have been studied for many years. To increase the corrosion resistance of silver coatings, a hard coating of silver-platinum (Ag-Pt) alloy on stainless steel surfaces has been developed and demonstrated significant antibacterial efficacy against *Streptococcus mutans* and *Aggregatibacter actinomycetemcomitans,* as well as good biocompatibility [110]. However, this study did not use a bracket as the substrate, but rather an SS block. This design may not fully simulate the bracket when it is performing its function in the oral cavity, as the morphology and ligation of the brackets and archwires are not taken into account.

To further increase the antibacterial properties of the silver coating, some studies have applied Ag together with TiO_2 to SS brackets. They demonstrated better antiadhesive and antibacterial properties against *Streptococcus mutans* and *Porphyromonas gingivalis* than Ag alone, helping to prevent dental caries and plaque accumulation, with satisfactory biocompatibility [111].

Both Ag and ZnO nanoparticle-coated SS brackets exhibit antibacterial effects against *Streptococcus mutans* and *Lactobacillus acidophilus*, and the antibacterial effects are maintained for at least 3 months. In addition, silver/zinc oxide nanoparticle composite-coated brackets had a stronger antibacterial effect on *Streptococcus mutans* and *Lactobacillus acidophilus* compared to silver nanoparticles and zinc oxide nanoparticles alone [112].

Other silver particle composite coatings are still being explored. A 2021 study concluded that a silver-polymer-coated archwire was not antibacterially effective. Its colony counts did not differ significantly from those of uncoated archwires in 0% sucrose and 3% sucrose environments [113]. Of course, this study only added a single species of bacteria (*Streptococcus mutans*) as the test bacteria, whereas in reality, the environment and strain composition in the oral cavity are more complex. To verify the accuracy of this result, further experiments under conditions that more closely simulate the real conditions of the oral cavity are needed.

A composite nanocoating was synthesized by Bor-Shiunn Lee et al. [114]. They applied a layer-by-layer deposition method with materials consisting of polydopamine, functionalized poly(3,4-ethylenedioxythiophene) (PEDOT), and AgNPs, and demonstrated the antifouling and antibacterial properties of this coating when applied to the surface of stainless steel materials. Its release of Ag ions does not have a harmful effect on humans, as it is biocompatible. This may be used in the future as a coating for orthodontic brackets and archwires.

Silver has been used as an antibacterial material for a long time and has also been shown to be biocompatible on numerous occasions [115]. Silver coatings are antibacterial against a number of species, including cariogenic and periodontal pathogenic bacteria, and provide a definite resistance to smooth-surface caries [108]. The long-lasting antibacterial effect allows for the use of brackets, which are an infrequently replaced component [112]. However, the shortcomings of silver coatings are their poor corrosion resistance, low hardness, and high price. The upside is that silver composite coatings with other materials have similar antibacterial effects [112], so silver composite coatings can be developed to discover potential improvements in other properties.

3.2. Titanium and Titanium Compound Coatings

TiO_2, as a photosensitive material, is capable of generating hydroxyl (OH) radicals and reactive oxygen species (ROS) when irradiated by ultraviolet (UV) light, which are highly reactive when exposed to organic compounds [116,117]. Based on this principle, the antibacterial properties of TiO_2 have been receiving attention.

Several studies have confirmed that SS and NiTi archwires coated with TiO_2 show antiadhesive effects against *Streptococcus mutans* [118–121], as well as bactericidal effects on *Streptococcus mutans* [118–121], *Porphyromonas gingivalis* [119], *Candida albicans*, and *Enterococcus faecalis* [120]. For brackets, TiO_2 coatings also show antiadhesive and antibacterial properties against *Streptococcus mutans* [122,123], *Candida albicans* [122], and *Lactobacillus acidophilus* [124].

Rutile, anatase, and brookite are crystal structures in which TiO_2 exists. Anatase is formed by anodic oxidation (AO), while rutile is formed by thermal oxidation (TO). Studies have shown that anatase films applied to titanium (Ti) and titanium silver (TiAg) plates are more effective than rutile films against *Streptococcus mutans* [125]. In contrast, Roshen Daniel Baby et al. applied TiO_2 coatings with different crystal structures to SS brackets but came to the opposite conclusion. Their results showed that both structures of TiO_2 have antibacterial effects against *Streptococcus mutans*, but the rutile phase has a stronger bactericidal effect and more significant cytotoxicity than the anatase phase [126]. All of these considerations mean that rutile should be avoided regardless of which crystal type has the better antibacterial effect, as its cytotoxicity reduces the biocompatibility of the material.

Recently, there have been some clinical studies of TiO_2 coatings. Keerthi Venkatesan et al. showed that TiO_2 nanoparticle coatings on NiTi archwires have good antiadhesive and antibacterial effects against *Streptococcus mutans*. However, by the end of 1 month, 60% of the TiO_2 coating was lost, and the roughness of the archwire was similar to that of the uncoated one [127]. Nevertheless, the adhesion of *Streptococcus mutans* to TiO_2-coated archwires was still lower compared to uncoated archwires, which can be attributed to the antibacterial properties of the nanoparticles [127]. In addition, the TiO_2 nanoparticle coating on the SS archwire also showed an effect against *Streptococcus mutans* in the first and third weeks [128]. Whether on NiTi or SS archwires, the TiO_2 coating was effective in reducing the initial bacterial adhesion [127,128]. However, although the TiO_2 coating was shown to have a direct antiadhesive and bactericidal effect, its effect on the prevention of enamel demineralization was not significant [127]. Therefore, more experiments are needed to prove its preventive effects against enamel demineralization and periodontal disease in practice.

In addition to the TiO_2 coating applied directly to the surface of the material, the surface of Ti or titanium alloys also oxidizes to form a TiO_2 layer, which is responsible for the coatings of titanium compound materials having certain antibacterial properties [125]. As mentioned already, Ag is also an effective antibacterial agent, so it has also been investigated to be added to Ti to prepare TiAg alloys, which it is hoped will have a stronger antibacterial effect than a Ti coating alone. For *Streptococcus mutans*, TiO_2 coatings on TiAg alloy plates have a significantly stronger and faster antibacterial effect than TiO_2 coatings on Ti plates, suggesting that the addition of Ag to Ti can result in a synergistic enhancement of the effect against *Streptococcus mutans* [125]. However, for *Lactobacillus acidophilus*, although both showed antibacterial activity, the TiAg-coated samples showed no difference in resistance to *Lactobacillus acidophilus* compared to Ti [129], which means that the TiAg coating did not have stronger activity against *Lactobacillus acidophilus* than the Ti coating. Nevertheless, irrespective of their antibacterial effects, neither of them exhibited cytotoxicity [129]. Therefore, both materials have the potential for application on fixed appliances, although further research is needed to determine whether the addition of Ag will give the desired antibacterial effect. Ti coatings and TiAg coatings have not yet been applied to archwires or brackets, which is a gap for the time being. In addition, apart from the clinical studies, the above studies have not taken into account the mechanical forces to which the archwires are subjected in practice. Thus, the ability of the coatings to withstand mechanical forces has not been tested.

The shortcoming of TiO_2 is that it only produces hydroxyl radicals under UV light irradiation. When UV light is filtered and only visible light is used, the photocatalytic activity of TiO_2 is poor [130]. On the other hand, the nitrogen-doped (N-doped) TiO_2 thin film substrate, as a visible-light-sensitive photocatalyst, has a significant bactericidal rate under visible-light irradiation [121]. Nitrogen-doped titanium dioxide ($TiO_{2-x}N_y$) films coated on the surface of SS brackets using RF magnetron sputtering produced 95.19%, 91.00%, 69.44%, and 98.86% inhibition against *Streptococcus mutans*, *Lactobacillus acidophilus*, *Actinomyces viscous*, and *Candida albicans*, respectively [131]. In addition, the coating exhibited strong antibacterial properties against *Streptococcus mutans* over a 90-day timeframe. In other words, surface coating of SS brackets with N-doped TiO_2 films inhibits *Streptococcus mutans* for up to 3 months [132]. To obtain coatings with the highest visible-light photocatalytic activity and antibacterial activity against *Lactobacillus acidophilus* and *Candida albicans*, a sputtering temperature of 300 °C for 180 min, an aigon (Ar):nitrogen (N) ratio of 30:1, and an annealing temperature set at 450 °C can be performed [133]. The preparation of TiO_2 nanofilms using PVD has also been examined and has shown similar results on the antibacterial effect of TiO_2, regardless of whether the coating thickness was 60 μm or 100 μm [63]. In some studies, however, the visible-light exposure time was set to 24 h [121,131]. This condition is impossible to achieve in practical applications. Therefore, simulation experiments that are closer to the clinical situation, animal experiments, or clinical trials are needed to further demonstrate the effectiveness of N-doped TiO_2 coatings in improving the antibacterial properties of fixed appliances. As shown in a randomized controlled clinical trial by Avula Monica et al. [134], N-doped TiO_2-coated SS brackets are effective in reducing the extent of the increase in *Streptococcus mutans* concentration when exposed to natural visible light and dental surgical light. This effect holds for up to 60 days and is better at 30 days after the placement of orthodontic appliances than at 60 days. However, even this latest related experiment did not directly detect and document enamel demineralization around the brackets or gingival inflammation. This will need to be followed up in future studies.

When TiO_2 is applied in combination with other materials, it is found that Ag applied with TiO_2 on SS brackets has better antiadhesive and antibacterial properties against *Streptococcus mutans* and *Porphyromonas gingivalis* [111].

Titanium nitride is also considered to have an antibacterial effect, although TiN is not as antibacterial as TiO_2 and carbon nitride (CN_x) [29]. However, the study by Licia Pacheco Teixeira et al. concluded that neither TiN nor titanium nitride doped with calcium phosphate (TNCP) films interfere with the attachment of *Streptococcus mutans* to the surface of the bracket, meaning that this coating has no antiadhesive effect and does not reduce the growth of *Streptococcus mutans* [135]. Moreover, the use of TiN as the surface coating may increase the surface friction [29]. In summary, the utility of TiN as a surface coating for orthodontic materials needs to be further verified.

Mian Chen et al. [136] used titanium treated with plasma-enhanced fluorine (F) and oxygen (O) mono/double chemical vapor deposition to obtain nanofunctional coatings with improved antibacterial properties and biocompatibility. The results showed that the fluorine-deposited samples can effectively exterminate *Staphylococcus aureus* with sufficient antibacterial properties, as shown in Figure 6. More importantly, the F and O double-deposited (F-O-Ti) coatings exhibited better sustained antibacterial performance than the F single-deposited coatings after 7 days of immersion in 0.9% sodium chloride (NaCl) solution. This result tells us that this type of material may be seen as an effective antibacterial coating for clinical applications.

Titanium compound coatings rely mainly on the photoreactivity of TiO_2 for their antibacterial action [130]. The advantages include their ability to resist caries and periodontal pathogenic bacteria [111,131,136], as well as their biocompatibility [125]. TiO_2 coatings are probably best used for short-term applications only, as they reduce initial bacterial adhesion [134] and undergo rapid peeling [127], making them unsuitable for long-term use. In addition, the increased friction of such coatings is detrimental to therapeutic efficiency, and the need for light causes inconvenience in their use [130]. Titanium has composite applications with silver [125], nitrogen [121], fluorine, and oxygen [136], which enhance its antibacterial properties, but the durability and high friction have not been addressed thus far. Currently, difficulties remain in clinical application.

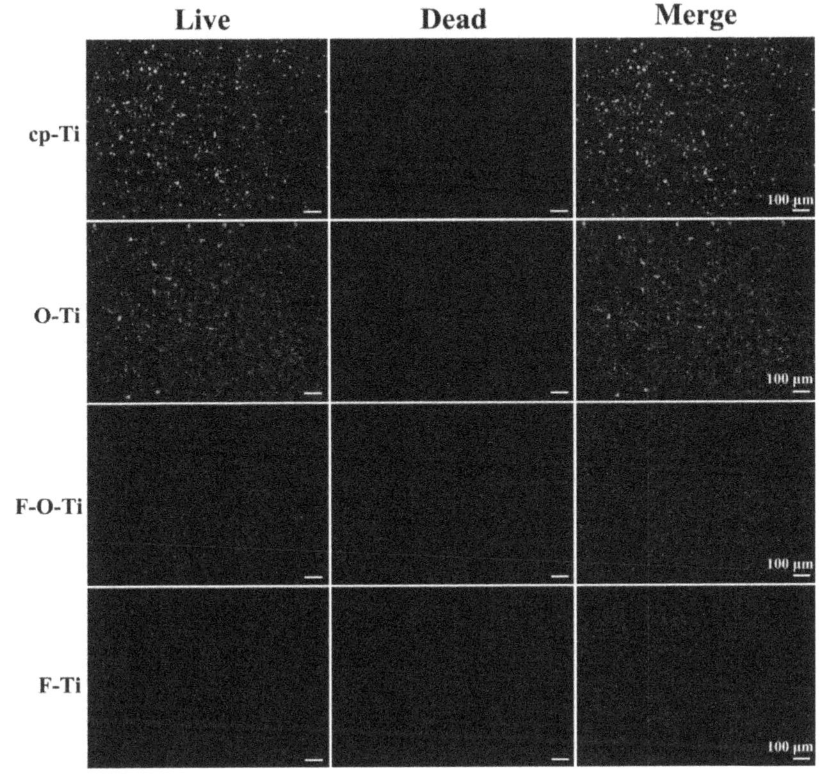

Figure 6. Live (green) and dead (red) staining of *Staphylococcus aureus* on different samples after culturing for 1 day. Adapted with permission from [136]. Copyright 2022, Elsevier.

3.3. Zinc and Zinc Compound Coatings

ZnO nanoparticles reduce biofilm formation, with low toxicity as well as good biocompatibility [137], making them suitable for biomedical applications. In a 2015 study, Baratali Ramazanzadeh et al. found that ZnO-nanoparticle-coated brackets prepared via spray pyrolysis had antibacterial effects, but they were not as effective as copper oxide (CuO) nanoparticle coatings and CuO-ZnO composite coatings, both of which reduced the number of *Streptococcus mutans* to 0 after 2 h [138]. However, the coatings were not tested for biocompatibility. If the addition of CuO leads to elevated cytotoxicity beyond the acceptable threshold for biocompatibility, it will be difficult to apply it directly in clinical applications, even if the antibacterial properties are good enough. Unlike CuO, ZnO has proven to be biocompatible [25]. Therefore, it is safe to use for a wide range of biomedical materials.

It has been shown that ZnO nanoparticle coatings on the surface of NiTi archwires have antibacterial effects against *Streptococcus mutans* [25,139], *Staphylococcus aureus*, *Streptococcus pyogenes*, and *Escherichia coli* [27]. Different coating methods can be used to obtain ZnO particles with different physicochemical properties, such as chemical precipitation, chemical vapor deposition (CVD), electrostatic spinning, polymer composite coating, and the sol-gel method. Comparisons also reveal that the smaller the particles, the larger the specific surface area of the particles, and the more the antibacterial performance is improved [139,140]. In addition, the color of the ZnO coating is white or light gold and does not tarnish in the short term, resulting in a higher aesthetic quality [25,138].

However, a study in 2021 concluded that the antibacterial effect of ZnO coatings was not significant [123]. In this study, the investigators prepared a photocatalytic ZnO coating on the surface of SS brackets via magnetron sputtering, and the results showed that the antibacterial effect of ZnO was not satisfactory. Based on these contradictory findings, more experiments are needed in order to determine the feasibility of the clinical application of ZnO nanoparticle coatings on orthodontic archwires or brackets.

In addition to the composite application of ZnO with CuO, it can also be used to form a coating with Ag nanoparticles. Noha K. Zeidan et al. [112] made a composite coating of ZnO with silver nanoparticles and applied it to SS brackets, finding that the antibacterial effect against *Streptococcus mutans* and *Lactobacillus acidophilus* exceeded that of the application of ZnO alone. Considering the antibacterial effectiveness of silver and the mechanical properties of ZnO, the combination of the two can bring the advantages of the coating to a more powerful level.

In summary, ZnO can be used as a multifunctional coating in clinical applications. It is antibacterial, friction reducing [26], and biocompatible [25], and it is safe and aesthetically pleasing to use [25,138]. On the downside, its corrosion resistance and durability for long-term application are not yet confirmed, and the antibacterial effect is still controversial. However, its application in combination with Ag nanoparticles can provide similar antibacterial performance and is more effective than ZnO alone [112]. In addition, future studies could focus on how to obtain finer ZnO nanoparticles in the preparation of coatings to enhance the antibacterial properties.

3.4. Other Metal and Metallic Compound Coatings

For antibacterial surface coatings prepared from other metallic materials, such as rhodium and gold, attempts have also been made on fixed appliances, although many research gaps remain.

There is no consensus on the effect of rhodium plating on the surface of archwire in terms of antibacterial properties. Some believe that the rhodium-coated NiTi archwires have increased antibacterial properties [141], while others consider that the risk of biofilm retention on rhodium-coated NiTi archwires is as high as that of uncoated NiTi archwires [113].

In 2014, a gold-plated aesthetic archwire was shown to reduce the surface roughness of SS archwires, but it caused no reduction in the adhesion of *Streptococcus mutans*. Moreover, surface free energy (SFE) may have a positive correlation with bacterial adhesion. If the SFE of the coating is the same as that of the archwire, the antiadhesive effect may not be significant, just as the SFE of gold-plated SS archwires is similar to that of SS archwires [141]. However, another study on other coatings suggested that surface roughness is positively correlated with biofilm adhesion [142]. This contradiction may be due to different experimental designs, or to the varying nature of different coatings and their antibacterial mechanisms. According to the "attachment point theory", biofoulers (in the context of this review, bioorganisms) have more attachment points on rough surfaces [143]. This theory may partially explain why some believe that bacteria are more likely to be attached on rough surfaces. However, smooth surfaces are not necessarily more resistant to adhesion than surfaces with undulations. Some researchers, inspired by bionics, have tried to create nanofolds on the surface of the substrate. These folds are like ripples with a certain wavelength and amplitude. When these two parameters are appropriate, the adhesion of biofoulers to the surface can be reduced. That is, although the surface is not completely smooth, its antiadhesion properties are significantly better than even that of a smooth surface when the gap between the folds is slightly smaller than the size of the bacteria [144]. This suggests that preparing nanofolds of suitable wavelengths and amplitudes on the surfaces of archwires and brackets might be a possible way to resist bacterial adhesion.

In addition, other gold materials—such as 4,6-diamino-2-pyrimidinethiol-modified gold nanoparticles (AuDAPT) [145] and quaternary ammonium (QA)-modified gold nanoclusters (QA-GNCs) [146]—can be used as antibacterial coatings for invisible orthodontic appliances. AuDAPT and QA-GNC have resistance to *Porphyromonas gingivalis* [145] and

Streptococcus mutans [146], respectively, and they have shown excellent biocompatibility, but they have not yet been investigated in fixed appliances.

3.5. Polymeric and Bioactive Materials

Epoxy has been clinically demonstrated to have antibacterial properties as an aesthetic coating on the surface of NiTi archwires [147]; however, this study took the plaque directly from the surface of the archwire during orthodontic treatment and counted the number of bacterial colonies in the culture, without distinguishing different strains. Two other in vitro experiments focused on *Streptococcus mutans* and showed that epoxy coating on the surface of SS or NiTi archwires reduces the adhesion of *Streptococcus mutans* and *Streptococcus distans* to the archwires in the short term [141,148]. Regarding the mechanism of this antiadhesive effect, the authors suggested that it was due to a reduction in the SFE of the archwire, rather than a change in surface roughness [141]. Conversely, Deise C. Oliveira et al. found that the amount of biofilm adhesion was significantly higher on epoxy-coated than on uncoated NiTi archwires. In addition, there was no significant difference in the number of bacterial colonies between coated and uncoated archwires [113]. This indicates that the epoxy coating does not increase the antibacterial properties of the NiTi archwire surface, and it may even decrease the antiadhesive properties. Based on these contradictory findings, multiple replicate experiments are needed in the future to determine the accuracy of the results. Efforts are also needed to make the in vitro experimental conditions more closely resemble the intraoral physiological environment, or to conduct animal and clinical experiments.

Another polymeric aesthetic coating material that has been used for a long time is polytetrafluoroethylene, also known as PTFE. It has been shown in clinical trials to reduce the amount of plaque biofilm adhesion on NiTi archwires [147] and SS brackets [149], and it is effective on every surface of the SS bracket [149]. Additionally, PTFE coatings prepared at low temperatures have better resistance to microbial adhesion than those prepared at high temperatures [95]. Still, these antibacterial and antiadhesive effects cannot be attributed to a reduction in initial bacterial adhesion [150]. However, similar to epoxy, it has been suggested that PTFE does not reduce the amount of biofilm adhesion on the surface of NiTi archwires. Interestingly, both results came from the same research [113], and only this single study reached the opposite conclusion from the previous ones. More similar studies may be needed in the future to verify the effects of PTFE and epoxy coatings on the antibacterial properties of NiTi archwire surfaces. Furthermore, the combined usage of AgNPs and PTFE through a similar method has been investigated, showing better antibacterial properties compared to applying PTFE coatings alone.

Regarding other polymeric coatings, Adauê S Oliveira et al. [151] prepared silicon oxides (SiO_x) hydrophobic/superhydrophobic coatings on SS and ceramic bracket surfaces via a sol-gel process. Hydrophobic coatings were made using cetyltrimethoxysilane diluted in ethanol, termed HS1; superhydrophobic coatings were made using 1H,1H,2H,2H-perfluorodecyltriethoxysilane diluted in dimethyl sulfoxide, termed HS2. The results showed that both HS1 and HS2 reduced the biofilm accumulation on the surface of the bracket in 24 h, and the superhydrophobic HS2 was more effective than the hydrophobic HS1. Based on the results of this study, when applying SiO_x coatings to increase the antibacterial properties of the brackets, HS2 superhydrophobic coatings can be prepared to obtain the best results [151]. The oral environment is moist, so the liquid–air interface formed between the hydrophobic surface and saliva prevents bacteria from adhering and forming biofilms [152], which explains the excellent antimicrobial properties of hydrophobic and superhydrophobic coatings. The abovementioned study was published in 2015, and there have been no subsequent studies on the surface coating of orthodontic brackets or archwires that present prospects worth exploring.

MPC polymer coatings inhibit the adhesion of *Streptococcus mutans* to SS archwires [96]. CTS has been used as an antibacterial material in the resin for bonding orthodontic attachments, with the purpose of controlling enamel demineralization, but it has not been tested for its antibacterial properties as a surface coating for orthodontic attachments, which is a direction that could be developed in the future [24]. As another polymer, polyethylene glycol (PEG)-coated SS archwires exhibited the best antiadhesive properties at a molecular mass of 5000, maintaining very low bacterial adhesion even after 10 h [153].

Bioactive materials such as lysozyme are also increasingly being used in orthodontics. New lysozyme-coated orthodontic composite archwires (CAWs) exhibit antibacterial properties in artificial saliva—mainly against *Staphylococcus aureus* [154]. In addition, the corrosion resistance and biocompatibility of the lysozyme coating are relatively considerable [154]. However, other strains of bacteria in the oral cavity—such as *Streptococcus mutans*, which often causes caries, and *Porphyromonas gingivalis*, which causes periodontal disease—have not been studied. Nevertheless, lysozyme as a bioactive material provides inspiration for follow-up studies. More bioactive materials can be explored for surface modification in the surface coating of fixed appliances. However, although lysozyme coatings have been shown to have an antibacterial effect against *Staphylococcus aureus*, their effect on Gram-negative bacteria is weaker than on Gram-positive bacteria, so there is a risk of disrupting the balance of the intraoral flora when applied to fixed orthoses. Further animal experiments are needed in order to confirm their clinical value.

There are also bioactive antimicrobial coatings that are based on antiadhesion. The protein molecule bovine serum albumin (BSA) significantly reduced bacterial adhesion on the surfaces of SS, ceramic, and resin brackets and SS archwires, with a maximum decrease of more than 95% on the surfaces of brackets, probably due to the BSA-mediated reduction in surface free energy [155]. In addition, in 2020, Peng et al. developed an antimicrobial hydrogel using a combination of PEG and chitosan (CTS/PEG) [156]. Archwires coated with these bioactive materials showed a significant increase in both antiadhesive and antimicrobial properties; thus, their possible uses in dental applications are considerable.

Recent studies have found that coatings of polydopamine (PDA) and blue fluorescent hollow carbon dots (HCDs) can maintain more than 50% antimicrobial performance against *Escherichia coli* and *Streptococcus mutans* for 14 days. The fluorescent nature of HCDs may also provide clinical visualization, allowing for timely replacement of defective coatings. In addition, the combination of low drug resistance, low toxicity, and high biosafety will allow this bioactive material to go far in the future [20].

Recent studies on antibacterial coatings for orthodontic appliances are summarized in Table 2.

Table 2. Recent studies on antibacterial coatings for orthodontic appliances.

Ref.	Coating Materials	Coating Technique	Substrate	Study Type	Roughness	Antibacterial Effectiveness	Other Effectiveness
[109]	Ag	Galvanic, PVD, PIIID, and deposition	SS brackets	Clinical trial	R_a for the untreated bracket material was 0.04 μm, R_a for the galvanic coating was 0.12 μm, R_a for the PVD coating was 0.08 μm, and R_a for the PIIID procedure coating was 0.06 μm.	The biofilm volume per test specimen for the control was 7.24×10^8 μm^3. For the galvanically applied silver coating, the biofilm volume decreased to 2.62×10^7 μm^3, for the PVD coating to 4.44×10^7 μm^3, and the PIIID procedure to 3.82×10^7 μm^3. The reduction of the biofilm volume compared to the control was statistically significant for all surface modifications. The percentage surface coverage per test specimen was 64.40% for the unmodified control and decreased to 16.97% for the galvanic silver surface, 23.81% for the PVD coating, and 23.63% for the PIIID-modified surface.	NA
[112]	Ag (NPs)	PVD	SS brackets	In vitro (in artificial saliva)	NA	The inhibition percent of *Streptococcus mutans* and *Lactobacillus acidophilus* were 27.60% and 62.02%, respectively.	Durability (>3 months)
[106]	Ag (NPs)	NA (commercial)	NA	In vitro (in suspensions of microorganisms)	NA	The silver coating decreased the adhesion of both *Streptococcus mutans* and *Streptococcus sanguinis* to the orthodontic brackets	NA
[104]	Ag (NPs)	Hydrothermal synthesis	SS archwires	In vitro (in suspensions of *Streptococcus mutans* and *Staphylococcus aureus*)	It was not possible to observe changes in roughness after coating.	Microbial adhesion and biofilm formation of *Staphylococcus aureus* and *Streptococcus mutans* reduced.	Aesthetic effect
[107]	Ag (NPs)	Synthesized in situ	SS and ceramic brackets	In vitro (on Mueller–Hinton agar plates)	NA	The inhibitory halos obtained by the in vitro evaluation of the antibacterial effect, in terms of brackets with silver nanoparticles with *Staphylococcus aureus* and *Escherichia coli*, showed an excellent inhibition of microbial growth compared to the bracket control, with a diameter between 9 and 10 mm.	NA
[140]	Ag (NPs)	Aqueous reduction	NiTi, CuNiTi, SS archwires, and SS brackets	In vitro (in suspensions of *Streptococcus mutans*)	Roughness values were increased in SS wires ($7.094 \times 10^3 + 1$ nm), followed by NiTi wires ($6.234 \times 10^3 + 1$ nm), and the lowest roughness value for CuNiTi wires ($3.116 \times 10^3 + 1$ nm).	Smaller Ag nanoparticles (16.7 μg/mL) had consistently better antimicrobial inhibition effects against the *Streptococcus mutans* strain compared to larger Ag nanoparticles (66.8 μg/mL), showing significant differences between them. The coated brackets had significantly better antiadherence activity (4.3 CFU/mL for smaller particles and 5 CFU/mL for larger particles) than the uncoated brackets (356 CFU/mL). Both sizes of Ag nanoparticles had statistically good adherence inhibition of the *Streptococcus mutans* strain for all types of orthodontic wires (SS = 26.1–52.6 CFU/mL, NiTi = 15.1–49.6 CFU/mL, and CuNiTi = 89.1–287.8 CFU/mL) compared with the control groups (SS = 346.7 CFU/mL, NiTi = 342.3 CFU/mL, and CuNiT = 376.2 CFU/mL).	NA

Table 2. *Cont.*

Ref.	Coating Materials	Coating Technique	Substrate	Study Type	Roughness	Antibacterial Effectiveness	Other Effectiveness
[112]	Ag/ZnO (NPs)	PVD	SS brackets	In vitro (in artificial saliva)	NA	The inhibition percent of *Streptococcus mutans* and *Lactobacillus acidophilus* were 45.32% and 80.29%, respectively.	Durability (>3 months)
[113]	Ag + polymer	NA (commercial)	NiTi archwires	In vitro (in suspensions of microorganisms + sucrose)	NA	No significant reduction in bacterial adhesion (0% sucrose) or biofilm accumulation (3% sucrose) was found when assessed by colony counting of *Streptococcus mutans*.	NA
[157]	AgNP/PTFE	Layer-by-layer deposition	316L SS plates	In vitro (in suspensions of *Escherichia coli*)	The surface roughness increased from 59.4 ± 6.1 nm to 158.1 ± 2.7 nm (deposition time of 6 h) and 177.3 ± 5.1 nm (deposition time of 12 h), respectively.	Coatings with 6 and 12 h deposition time could inhibit by ~75% and ~90% bacterial growth over the initial 3 days, respectively. After 7 days, coatings with 6 and 12 h deposition time still exhibited significant antibacterial activity, reducing by ~40% and ~50% of bacterial growth, respectively.	Corrosion-resistant effectiveness Biocompatibility
[158]	Al_2O_3 (NPs)	Atomic layer deposition	316L SS	In vitro (in simulation body fluid solution)	NA	The mean colonies forming units per milliliter were reduced by 20%; the diffusion zone was ~4 mm.	Corrosion-resistant effectiveness Biocompatibility
[158]	Al_2O_3/TiO_2 Multilayer	Atomic layer deposition	316L SS	In vitro (in simulation body fluid solution)	NA	The mean colonies forming units per milliliter were reduced by 40%; the diffusion zone was >6 mm.	Corrosion-resistant effectiveness
[155]	BSA	Chemical deposition	SS, ceramic, and resin brackets; SS archwires	In vitro (in suspensions of *Streptococcus mutans*)	The adsorbed BSA molecule was not uniform on the surface, thereby leading to slight surface roughness.	After integrating BSA molecules, the three kinds of brackets all showed more than 95.0% reduction in *Streptococcus mutans* adhesion (i.e., 98.3% for SS, 96.3% for ceramic, and 95.2% for resin). Compared with bare archwires, only a few bacteria (~7.5%) could be found on the BSA-coated archwires' surface, even after incubation in bacterial suspension for 300 min; and the optimal BSA concentration was 10 mg/mL.	NA
[156]	CTS/PEG hydrogel	Combining silane chemistry and subsequent copolymerization	SS archwires	In vitro (in suspensions of *Streptococcus mutans*)	R_q of CTS/PEG-coated SS archwires increased from 0.26 to 1.57 nm.	This biointerface showed superior activity in early-stage adhesion inhibition (98.8%, 5 h) and displayed remarkably long-lasting colony-suppression activity (93.3%, 7 d).	Durability (>7 days) Wear resistance Biocompatibility
[29]	CN_x	Ion beam-assisted deposition	304L SS disks	In vitro (friction test: NA; antibacterial test: in bacterial suspension)	The films slightly reduce the surface roughness parameter. R_a reduced from 0.181 to 0.140 μm	Bacteria density reduced from 13.002×10^3 /mm^2 to 4.030×10^3 /mm^2 ($p < 0.05$).	Friction reduction effectiveness Biocompatibility
[113]	Epoxy	NA (commercial)	NiTi archwires	In vitro (in suspensions of microorganisms + sucrose)	NA	No significant reduction in bacterial adhesion (0% sucrose) or biofilm accumulation (3% sucrose) was found when assessed by colony counting of *Streptococcus mutans*.	NA

Table 2. *Cont.*

Ref.	Coating Materials	Coating Technique	Substrate	Study Type	Roughness	Antibacterial Effectiveness	Other Effectiveness
[142]	Epoxy	NA (commercial)	NiTi archwires	In vitro (in suspensions of *Streptococcus mutans* and *Streptococcus sobrinus*)	R_a = 1.29 μm (higher than uncoated archwires)	Epoxy-coated wires demonstrated an increased adhesion of *Streptococcus mutans* (5.55 CFU/cm^2) and *Streptococcus sobrinus* (4.64 CFU/cm^2).	NA
[136]	F-O-Ti	Plasma-enhanced fluorine and oxygen mono/dual CVD	Commercially available pure titanium with 99.9% purity	In vitro (in suspensions of *Staphylococcus aureus* and artificial saliva)	A large amount of convex texture with 100–200 nm size distributes uniformly all over the surface.	The antibacterial rates were higher than 90%.	Corrosion-resistant effectiveness Durability (~7 days) Biocompatibility
[89]	GO	Silane coupling	NiTi archwires	In vitro (friction test: NA; antibacterial test: in bacterial suspension; corrosion test: artificial saliva)	The surface of the samples coated with 2 mg/mL GO concentrations was smooth with a uniformly coated area.	The bacterial CFU values for *Streptococcus mutans* were 0.77 (samples coated with 0.5 mg/mL GO concentrations), 0.40 (samples coated with 2 mg/mL GO concentrations), and 0.23 (samples coated with 5 mg/mL GO concentrations) relative to that on bare NiTi (1.00). The number of live bacteria decreased, and the number of dead bacteria increased, indicating that GO coating could effectively resist adherent bacteria. This effect had a positive correlation with GO concentration, indicating the concentration-dependent antibacterial ability of these GO-coated surfaces.	Friction reduction effectiveness Corrosion-resistant effectiveness Wear resistance Biocompatibility
[154]	Lysozyme	Liquid phase deposition	CAWs	In vitro (in suspensions of *Staphylococcus aureus* and artificial saliva)	The surface roughness increased after being coated according to two- and three-dimensional atomic force micrographs.	When coated with 20, 40, and 60 g/L lysozyme, Live/dead bacteria staining of *Staphylococcus aureus* reduced from 90% to 82%, 59%, and 61%, respectively.	Corrosion-resistant effectiveness Durability (~2 weeks) Wear resistance Biocompatibility
[134]	N-doped TiO_2	Radio frequency magnetron sputtering	SS brackets	Clinical trial	NA	Coated archwires (38.54 and 36.84) showed greater Ct values than uncoated wires (34.71 and 31.89) at 30 d and 60 d, respectively. Greater Ct values indicate lower *Streptococcus mutans* adhesion. Therefore, coated wires demonstrated lower *Streptococcus mutans* adhesion when compared with uncoated wires.	Durability
[132]	N-doped TiO_2	Radio frequency magnetron sputtering	SS brackets	In vitro (in suspensions of *Streptococcus mutans*)	NA	The coating decreased *Streptococcus mutans* colonies from 401.21 to 37.82 CFU/mL.	Durability (>90 days)

Table 2. *Cont.*

Ref.	Coating Materials	Coating Technique	Substrate	Study Type	Roughness	Antibacterial Effectiveness	Other Effectiveness
[114]	PEDOT/Ag	Layer-by-layer deposition	316L SS plates	In vitro (in suspensions of *Streptococcus mutans* and *Escherichia coli*)	NA	The antiadhesive and antibacterial activity against *Streptococcus mutans* and *Escherichia coli* significantly increased.	Biocompatibility
[20]	PDA and HCDs	Electrostatic adsorption	SS brackets	In vitro (in the soaking solution of the archwires)	NA	After soaking in artificial saliva for 7 and 14 days, the coatings formed by 50 µL and 100 µL of HCDs showed antibacterial rates against *Escherichia coli* and *Streptococcus mutans* that could still reach more than 50%, but the antibacterial properties of the coatings formed by 150 µL of HCDs were much weaker than those of the above two groups.	Durability (>7 days) Biocompatibility
[113]	PTFE	NA (commercial)	NiTi archwires	In vitro (in suspensions of microorganisms + sucrose)	NA	No significant reduction in bacterial adhesion (0% sucrose) or biofilm accumulation (3% sucrose) was found when assessed by colony counting of *Streptococcus mutans*.	NA
[142]	PTFE	NA (commercial)	NiTi archwires	In vitro (in suspensions of *Streptococcus mutans* and *Streptococcus sobrinus*)	R_a = 0.74 µm (higher than uncoated archwires)	PTFE-coated wires demonstrated an increased adhesion of *Streptococcus mutans* (4.76 CFU/cm^2) and *Streptococcus sobrinus* (3.73 CFU/cm^2).	NA
[157]	PTFE	NA	316L SS plates	In vitro (in suspensions of *Escherichia coli*)	The surface roughness increased from 59.4 ± 6.1 nm to 134.7 ± 3.9 nm after being coated.	The PTFE coating only demonstrated short-term antiadhesive activity, reducing by ~45% biomass adhesion on the first day as compared with 316L SS, while coatings with 6 h and 12 h deposition time inhibited by ~65% and ~80% of biomass formation.	Corrosion-resistant effectiveness Biocompatibility
[113]	Rhodium	NA (commercial)	NiTi archwires	In vitro (in suspensions of microorganisms + sucrose)	NA	No significant reduction in bacterial adhesion (0% sucrose) or biofilm accumulation (3% sucrose) was found when assessed by colony counting of *Streptococcus mutans*.	NA
[142]	Rhodium	NA (commercial)	NiTi archwires	In vitro (in suspensions of *Streptococcus mutans* and *Streptococcus sobrinus*)	R_a = 0.34 µm (higher than uncoated archwires)	Rhodium-coated wires demonstrated an increased adhesion of *Streptococcus mutans* (3.85 CFU/cm^2) and *Streptococcus sobrinus* (2.79 CFU/cm^2).	NA
[68]	TaAgB	Magnetron sputtering with Ag-doping	304 SS sheets	In vitro (in artificial saliva)	NA	The amount of *Streptococcus mutans* adhesion on the TaAgB surface was significantly reduced and the morphology changed according to the SEM image.	Friction reduction effectiveness Biocompatibility
[29]	TiN	Ion beam-assisted deposition	304L SS disks	In vitro (friction test: NA; antibacterial test: in bacterial suspension)	The films slightly reduce the surface roughness parameter. R_a reduced from 0.181 to 0.162 µm.	Bacteria density reduced from 13.002 × 10^3 /mm^2 to 3.888 × 10^3 /mm^2 ($p < 0.05$).	Friction reduction effectiveness Biocompatibility
[135]	TiN	Cathodic cage deposition	SS brackets	In vitro (in suspensions of *Streptococcus mutans*)	NA	The presence of coatings did not influence the formation of the *Streptococcus mutans* biofilm ($p = 0.06$).	NA

Table 2. *Cont.*

Ref.	Coating Materials	Coating Technique	Substrate	Study Type	Roughness	Antibacterial Effectiveness	Other Effectiveness
[29]	TiO_2	Ion beam-assisted deposition	304L SS disks	In vitro (friction test: NA; antibacterial test: in bacterial suspension)	NA	Bacteria density reduced from 13.002×10^3 /mm^2 to 1.368×10^3 /mm^2 ($p < 0.05$).	Wear resistance Biocompatibility
[123]	TiO_2	Radio frequency magnetron sputtering	SS brackets	In vitro (in suspensions of *Streptococcus mutans*)	NA	The "weight increase" reduced from 1.0100×10^3 mg to 0.6750×10^3 mg, which showed increased antiadhesive activity against *Streptococcus mutans*. The survival rate reduced from 3548.3350 to 2895.0000 CFU/mL, which showed increased antibacterial activity against *Streptococcus mutans*.	NA
[127]	TiO_2	Radio frequency magnetron sputtering	NiTi archwires	Clinical trial	R_a reduced from 1591.08 to 746.14 nm.	Coated archwires (30.97) showed greater Ct values than uncoated wires (37.00). Greater Ct values indicate lower *Streptococcus mutans* adhesion. Therefore, coated wires demonstrated lower *Streptococcus mutans* adhesion when compared with uncoated wires. This difference was statistically significant ($p = 0.0005$).	Durability (~1 month)
[128]	TiO_2	PVD	SS archwires	Clinical trial	NA	The coating decreased *Streptococcus mutans* colonies from 5122.0 to 1400.0 and from 1141.8 to 297.7 CFU/mL in the first and the third week, respectively.	Durability (>3 weeks) Biocompatibility
[158]	TiO_2 (NPs)	Atomic layer deposition	316L SS	In vitro (in simulation body fluid solution)	NA	The mean colonies forming units per milliliter were reduced by 15%; the diffusion zone was ~2 mm.	Corrosion-resistant effectiveness Biocompatibility
[135]	TNCP	Cathodic cage deposition	SS brackets	In vitro (in suspensions of *Streptococcus mutans*)	NA	The presence of coatings did not influence the formation of the *Streptococcus mutans* biofilm ($p = 0.06$).	NA
[123]	ZnO	Radio frequency magnetron sputtering	SS brackets	In vitro (in suspensions of *Streptococcus mutans*)	NA	The antiadhesive and antibacterial activity against *Streptococcus mutans* showed no significant increase.	NA
[112]	ZnO (NPs)	PVD	SS brackets	In vitro (in artificial saliva)	NA	The inhibition percent of *Streptococcus mutans* and *Lactobacillus acidophilus* were 17.54% and 28.85%, respectively.	Durability (>3 months)

Table 2. *Cont.*

Ref.	Coating Materials	Coating Technique	Substrate	Study Type	Roughness	Antibacterial Effectiveness	Other Effectiveness
[139]	ZnO (NPs)	CVD, chemical precipitation method, polymer composite coating, sol-gel synthesis, the electrospinning process	NiTi archwires	In vitro (in suspensions of *Streptococcus mutans*)	The sizes of the NPs were 59–61 nm, 30–150 nm, and 28 nm when coated during CVD, chemical precipitation method, and sol-gel synthesis process, respectively. Electrospinning gave a branch of fibers gathered together in a network, and polymer composite coating showed a nonuniform and excursive surface.	The highest *Streptococcus mutans* antibacterial effect with 98%, 96%, and 93% microbial cell reduction belonged to CVD, precipitation method, and sol-gel synthesis, respectively, and the lowest cell reduction was seen in the electrospinning method (72%)	NA
[27]	ZnO (NPs)	Electrochemical deposition	NiTi archwires	In vitro (friction test: NA; antibacterial test: on nutrient agar plates)	NA	The inhibition zone of *Staphylococcus aureus*, *Streptococcus pyogenes*, and *Escherichia coli* was formed around all ZnO nanoparticles coated archwires, with diameters of 4.25, 6.25, and 3.57 mm, respectively.	Friction reduction effectiveness
[159]	ZrO_2	EBPVD	316L SS	In vitro (in artificial saliva and artificial saliva containing NaF)	R_a reduced from 10 to 3 nm after being coated.	The bacterial adhesion reduced from 8 to 5.5×10^3 CFU/cm^2.	Corrosion-resistant effectiveness Wear resistance Biocompatibility

NA: not applicable; PIIID: plasma immersion ion implantation; CuNiTi: copper-nickel-titanium; CFU: colony-forming unit; Ag/ZnO: silver/zinc oxide; AgNP/PTFE: nanosilver/polytetrafluoroethylene; Al_2O_3: aluminum oxide; Al_2O_3/TiO_2: aluminum oxide/titanium dioxide; BSA: bovine serum albumin; CTS/PEG: chitosan/polyethylene glycol; F-O-Ti: fluorine and oxygen double-deposited titanium; GO: graphene oxide; CAWs: orthodontic composite arch wires; Ct: cycle threshold; N: nitrogen; PEDOT/Ag: poly(3,4-ethylenedioxythiophene)/silver; PDA: polydopamine; HCDs: blue fluorescent carbon dots; TNCP: titanium nitride doped with calcium phosphate; CVD: chemical vapor deposition; ZrO_2: zirconium oxide; EBPVD: electron beam-physical vapor deposition.

4. Corrosion-Resistant Coatings

Research on corrosion-resistant coatings has focused on nitride- and carbonized-titanium compound materials, as well as diamond-like material coatings. Although other carbon-based materials such as graphene are also available, their effectiveness is still not clearly recognized. In particular, a common phenomenon in corrosion-resistant coatings is that their durability has not been confirmed by long-term experiments. Other properties of aesthetic coated polymers such as epoxy and PTFE have recently been investigated in the literature, demonstrating their favorable corrosion resistance for orthodontic archwires. This versatility gives more space for the development of polymeric aesthetic coatings. The main corrosion-resistant coating materials discussed in this section are shown in Figure 7.

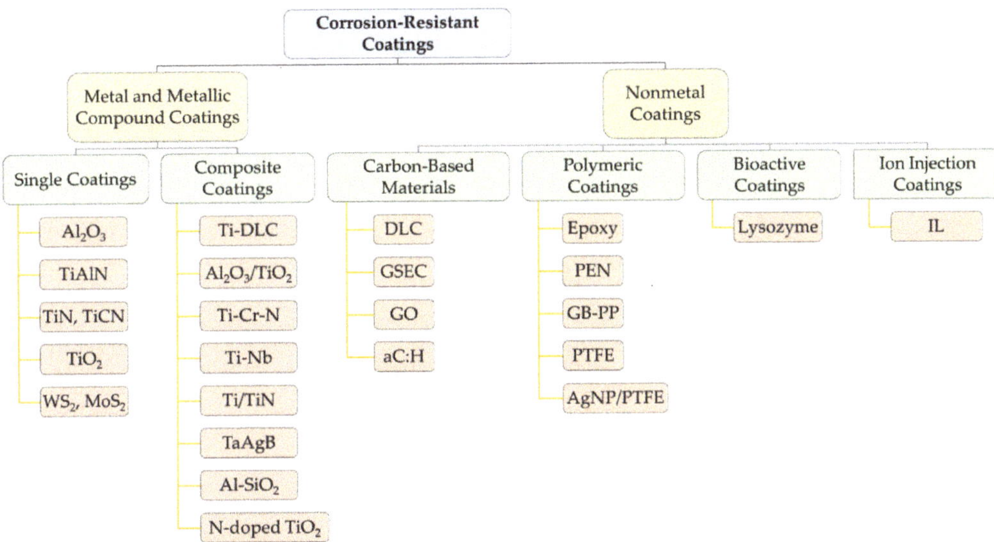

Figure 7. Summary of antibacterial coating materials. TiAlN: titanium aluminum nitride; Ti-DLC: titanium-diamond-like carbon; Al_2O_3/TiO_2: aluminum oxide/titanium dioxide; Ti-Cr-N: titanium-chromium-nitrogen; Ti-Nb: titanium-niobium; Ti/TiN: titanium/titanium nitride; N: nitrogen; aC:H: hydrogenated forms of diamond-like carbon; PEN: polyethylene naphthalate; GB-PP: grid-blasted plasma-polymerized; IL: ionic liquid.

4.1. Titanium and Titanium Compound Coatings

NiTi alloys have a tendency to corrode in chlorine- and fluorine-containing solutions [160], as does stainless steel. Some surface coatings can enhance their corrosion resistance, but coatings might wear and peel off after a period of application, resulting in increased roughness and reduced aesthetics [161]. Therefore, the surface coating of fixed appliances needs to be resistant to wear as well as corrosion in order to provide better durability.

As early as 2002, a study concluded that the corrosion resistance of ion-plated titanium nitride (TiN) SS brackets in artificial saliva was not significantly better than that of uncoated SS brackets [162]. Later, the opposite conclusion was reached, suggesting that TiN-plated SS brackets had increased corrosion resistance [28] and wear resistance [163] in artificial saliva. However, at that time, the corrosion resistance of TiN coatings in fluorine-containing solutions had not been confirmed in studies [28]. It was not until Xue-shun Yuan et al. prepared N-doped TiO_2 films on SS brackets via radio frequency magnetron sputtering that they were demonstrated to increase the corrosion resistance of SS brackets in both artificial saliva and 1.23% acidulated phosphate fluoride (APF) solution [164]. In addition to brackets, TiN-coated SS archwires have shown good corrosion resistance in both acidic saliva [19] and electrochemical corrosion experiments [58].

Recently, it has been found that TiN coatings have favorable wear resistance but are more likely to fracture due to bending [165]. However, for coatings on SS substrates, some new materials also have higher corrosion resistance than TiN. These coatings include titanium-diamond-like carbon (Ti-DLC) [166] and TiCN [31], and the corrosion resistance of TiCN increases with the nitrogen content and carbon content used in the preparation process [31]. In addition to SS, the use of TiCN coatings also reduced the amount of ions released from nickel-chromium-molybdenum (Ni-Cr-Mo) alloy substrates [167] and the mass loss of nickel-chromium (Ni-Cr) alloy substrates [168], indicating that the coatings improved the corrosion resistance of various metallic materials.

In addition to carbon, aluminum compounded with nitrogen and titanium also has a good effect. TiAlN coatings applied by cathodic arc physical vapor deposition (CA-PVD) are also capable of reducing the corrosive effect of fluoride on TMA wire and are more effective than tungsten carbide/carbon (WC/C) coatings [59].

For NiTi brackets, multilayer TiN/Ti coatings deposited by pulsed bias arc ion plating (PBAIP) have higher corrosion resistance than a single layer of TiN in artificial saliva [169]. For NiTi archwires, TiN coatings formed by nitriding [170,171] or electroplating [19] reduce the precipitation of metal ions, and ion-plated TiN coatings show a visible reduction in the area of corrosion spots on the surface of the archwire [58]. TiN/Ti coatings formed by PVD not only improve the corrosion resistance of the archwire, but also its resistance to friction loss [172]. Liu, Jia-Kuang et al. [172] further discovered that in TiN/Ti coatings, the TiN layer provides protection against mechanical damage, while the Ti layer improves the corrosion resistance. In multilayer TiN/Ti, the improvement of wear resistance is similar for a single layer, two layers, and four layers, and when the number of layers continues to increase, the wear resistance of the substrate is reduced instead [33]. Therefore, it is recommended to use multilayer TiN/Ti coatings with four or fewer layers to enhance wear resistance and corrosion resistance.

It has been shown that the corrosion resistance of composite archwires (CAWs) can be significantly improved by coating them with TiO_2 nanocrystalline films—especially by using N-doped TiO_2 nanocrystalline films [121]. However, a study by Kielan-Grabowska Z et al. [173] showed that TiO_2- and Ag-doped TiO_2-coated SS materials have reduced corrosion resistance. In addition, although Al_2O_3 coatings are rarely used alone for orthodontic brackets and archwires, they can be coated together with TiO_2. Muna Khethier Abbass et al. prepared nanocoated films of Al_2O_3, TiO_2, and multilayer aluminum oxide/titanium dioxide (Al_2O_3/TiO_2) on SS surfaces via atomic layer deposition (ALD) and found that all three films could effectively enhance corrosion resistance, and that multiple layers are better than a single layer, while Al_2O_3 is better than TiO_2 [158]. However, in a recent study, researchers formed nonporous, nickel-free TiO_2 coatings with a thickness of 50 nm via pulse anodization, as shown in Figure 8, which exhibited good hydrophilicity and corrosion protection [174]. To date, there has been no study comparing the differences in the performance of this nonporous nickel-free TiO_2 coating with compared with N-doped TiO_2 film and Al_2O_3/TiO_2 coating, which would be necessary for the consideration of practical applications.

In addition to titanium nitrides and oxides, metallic titanium and other composite coatings are also under research. Firstly, titanium sputter-coated NiTi archwire has good adhesion and corrosion resistance even after 30 days of exposure to artificial saliva [175]. Secondly, titanium-chromium-nitrogen (Ti-Cr-N) coatings obtained via radio frequency reactive sputtering deposition annealed at 400 and 700 °C can reduce the chromium released from SS substrates by ~67% in artificial saliva, and the effect is better at 400 °C than at 700 °C [176]. Another example is that titanium-niobium (Ti-Nb) coatings prepared via laser melting methods on NiTi substrates reduce the release of nickel ions [177]. These are surface coatings that can potentially be applied to orthodontic fixed appliances to prevent wear and metal sensitization of orthodontic attachments.

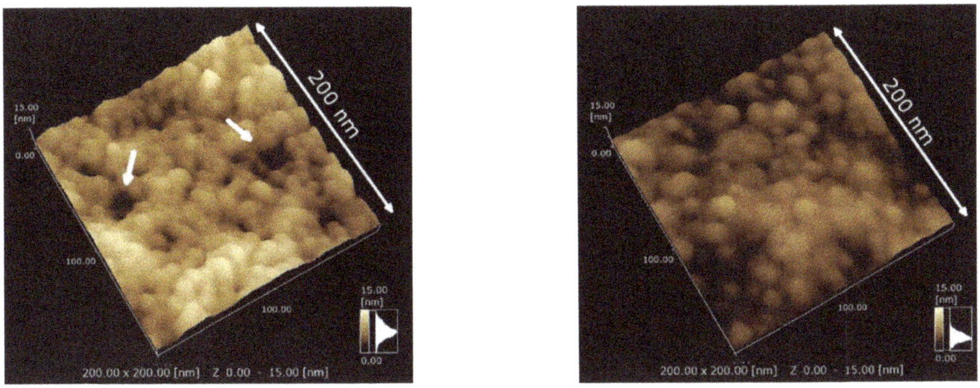

(**a**) (**b**)

Figure 8. Scanning probe microscope (SPM) images of (**a**) current-anodized and (**b**) pulse-anodized NiTi surfaces. The white arrows show the pores on the surfaces of the coating. Adapted with permission from [174]. Copyright 2022, Elsevier.

As TiO_2 is not the most optimal material for corrosion-resistant coatings, TiN is relatively more worthy of research [158,173]. TiN coatings have the advantage of being effective in enhancing the corrosion resistance of brackets and archwires in both chloride- and fluoride-containing solutions [164], but they also have disadvantages, such as the tendency to fracture when bending and the potential adverse effect on friction [165]. Based on these conditions, TiN can be used as a corrosion-resistant coating for fixed appliances in patients who normally use mouthwash. In addition, carbon or other metallic materials can be added to the material, or a double-layer titanium/titanium nitride (Ti/TiN) coating can be prepared to compensate for the deficiencies of TiN applied alone [33]. In terms of coating method, a coating method that can be applied chairside is preferable, so that the archwire can bend first before coating to prevent undesirable coating peeling and archwire corrosion.

4.2. Carbon-Based Coatings

Around 15 years ago, S. Kobayashi et al. [76] and Yasuharu Ohgoe et al. [178] used ion beam plating to prepare DLC films on the surface of NiTi archwires. They found that the DLC coating was able to reduce the release of nickel ions in physiological saline at 80–85 °C by as much as 80% after 5 days of immersion [178], while still remaining corrosion resistant after 14 days [76]. Additionally, the DLC-coated NiTi archwire still reduced the nickel ion concentration in the solution by 16.7% after 6 months of immersion in physiological saline at 37 °C [179]. DLC coating not only enhances the corrosion resistance of NiTi archwire in warm saline, but also alleviates the toxic effects of nickel ion release on cells, and shows good biocompatibility [178,179]. It also has excellent friction resistance and will not peel off the surface of the archwire easily [76]. Similarly, on the SS filament surface, DLC coatings can be deposited to provide comparable friction reduction and corrosion resistance [82].

The corrosion of metals by fluorine ions should not be neglected either. Therefore, a study investigated the performance of MCECR plasma sputtering coatings in high-fluoride-ion environments. The results indicated that the DLC coating of the NiTi archwire under a high-fluorine environment reduced the change in the surface roughness of the archwire due to corrosion by 91.3% [83], verifying the outstanding corrosion resistance of DLC coatings. DLC-coated NiTi archwire prepared by the same method has also been shown to have strong friction resistance, significantly reducing the micromotion wear of the archwire [80]. There is still room for improvement of this excellent performance of DLC coatings. For example, deep cryogenic treatment (DCT) at temperatures from −120 to −196 °C can improve adhesion strength from 30.08 N to up to 40.54 N [180]. Therefore, this post-treatment method may be a potential way to improve wear resistance.

Graphene is valued for its excellent mechanical properties in dental material applications. Both graphene oxide (GO) and GO/Ag coatings prepared by Viritpon Srimaneepong et al. using electrophoretic deposition (EPD) on the surface of NiTi alloys had lower corrosion rates than uncoated NiTi alloys, indicating that they can enhance the corrosion resistance of NiTi alloys, as shown in Figure 9 [181]. GO can also improve the wear resistance of NiTi archwires. Dai, Danni et al. found that many grooves caused by wear were produced on the surface of uncoated NiTi archwire, but the width of grooves on the surface of GO-coated archwire was less [89]. In addition, a GO coating concentration of 2 mg/mL resulted in less wear and a smoother surface after mechanical friction compared to GO coatings at concentrations of 0.5 or 5 mg/mL [89]. This indicates that coating with GO at a concentration of 2 mg/mL results in better wear resistance. Therefore, this is the reference value when GO coatings are applied to the surface of NiTi archwires.

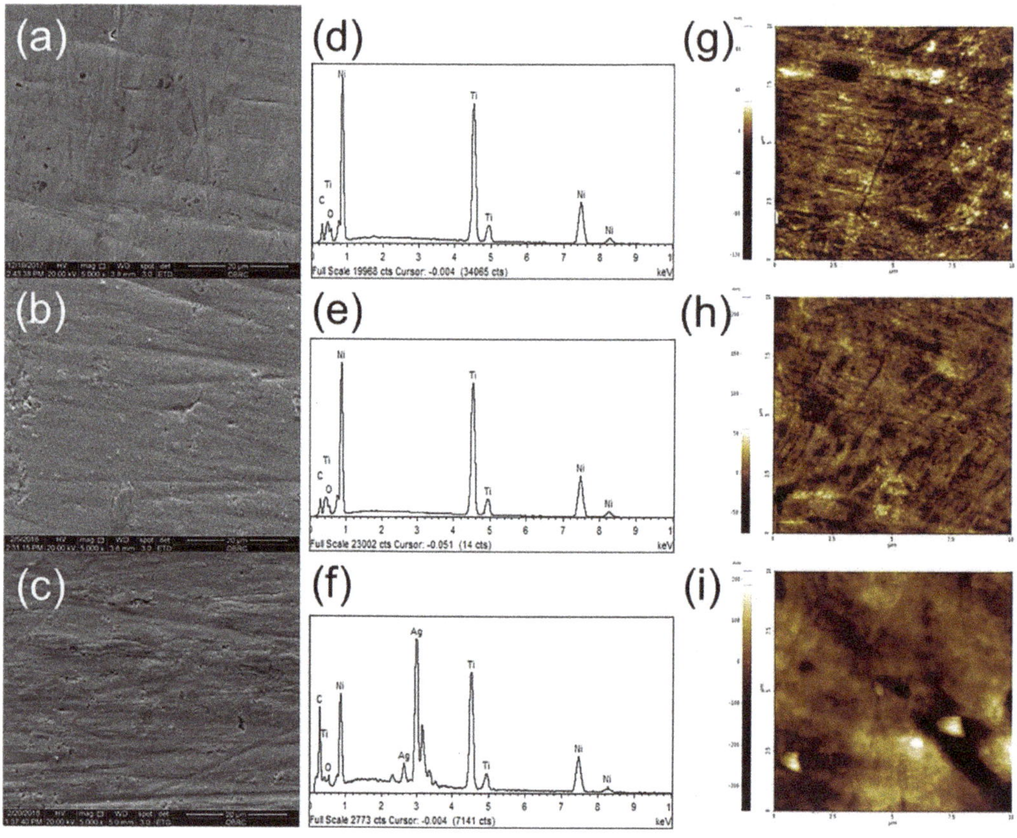

Figure 9. (**a**–**c**) Scanning electron microscope (SEM) images, (**d**–**f**) energy-dispersive spectroscopy (EDS) analysis, and (**g**–**i**) atomic force microscope (AFM) images of the bare NiTi, GO-coated NiTi, and GO/Ag-coated NiTi alloy, respectively. Adapted with permission from [181].

GSEC is another form of graphene as a surface coating, which also has good wear resistance. As observed by Pan et al. [90], GSEC-coated SS archwire produces significantly fewer wear marks than bare SS archwire after being subjected to friction [90], while the wear rate can reach a minimum of 0.11×10^{-6} mm^3/Nm [46].

Fróis, António et al. coated SS brackets and archwires with hydrogenated forms of diamond-like carbon (aC:H) sputter coating via the magnetron sputtering method. They demonstrated its chemical and mechanical inertness and ability to reduce the pitting corrosion that occurs on SS substrates. However, the leaching of metal ions in the aC:H coating was higher than that in the uncoated SS [182]. What should be considered in this case, if it is to be applied, is the need to reduce either the damage to the surface morphology or the metal allergy due to the leaching of metal ions. However, studies on this coating are still insufficient, and more research should be carried out in the future.

Among the carbon-based coatings, DLC, GO, and GSEC have similar corrosion resistance advantages. However, DLC shows the best clinical application prospects. The DLC coating is not only durable [179], but also allows fixed aligners to withstand certain thermal and fluoridated environments [83]. Specifically, it reduces the corrosion that occurs on the archwire and brackets under the influence of a hotter diet or fluoride mouthwash. DLC coatings, in turn, have the ability to reduce friction [80] and are more biocompatible [178,179]. Although the application of DLC coatings leads to a decrease in the surface hardness of the attachments, this can be enhanced and the corrosion resistance further increased by means of DCT [180].

4.3. Other Metal and Metallic Compound Coatings

Some other metal and metallic compound coatings have corrosion resistance, but are mainly used for friction reduction, such as WS_2 [43], ZnO [24–26], and TiN [29,31,32]. The potential of the remaining metallic materials as corrosion-resistant coatings is still not enough for clinical applications, so there is still space for corrosion-resistant coatings to be developed using metallic materials. Among all metal coatings, only TaAgB has good performance in friction reduction, antibacterial activity, and corrosion resistance, as a multifunctional coating with good application prospects [68].

IF-WS_2 has been applied for a long time to reduce the friction on the surface of archwires and has been proven to have good wear resistance. In a study by Redlich, Meir et al. [183], after 100 cycles under dry conditions, very little surface wear was observed on the Ni + IF-WS_2-coated SS archwire, and no peeling of the coating was observed—as opposed to the uncoated or Ni-coated archwires, which were severely worn after only 10 cycles [183]. This indicates that WS_2 is highly resistant to wear and tear. In recent years, Antonio Gracco et al. have come to similar conclusions and have further verified that WS_2 has good resistance to wear even under wet conditions [18]. Additionally, tungsten disulfide (MoS_2) has similar properties to WS_2 [18]. Both of them have good performance in reducing friction and wear. Therefore, the use of either MoS_2 or WS_2 as the surface coating of SS archwires can maintain the durability of their effects.

Solid TaAgB solution coating on SS blocks have high hardness, smoothness, and wear resistance, with a wear rate of only $6.51 \times 10^{-15} m^3/Nm$ [68]. Based on its low-friction and antibacterial properties described above, it could be applied in the future as a multifunctional coating for orthodontic fixed appliances.

Rhodium is commonly used as an aesthetic coating and, in addition to its aesthetic properties, also has the property of reducing friction on the surface of the archwire, as well as potential antibacterial properties. However, there is more distrust in the corrosion resistance of rhodium. Katić Višnja et al. [170] and Milena Carolina de Amorim et al. [184] both suggested that rhodium coating reduces the corrosion resistance and electrochemical resistance of NiTi archwires in artificial saliva, making them more susceptible to localized pitting corrosion on the surface. Lina M. Escobar et al. [185] described rhodium-plated NiTi archwire as having a significantly rougher surface and many scratches on its surface after testing. This indicates that the wear resistance of rhodium plating is also less than satisfactory.

Although copper-coated NiTi archwire has sufficient corrosion resistance [186], the biocompatibility of copper and copper oxide coatings is yet to be confirmed, due to the cytotoxicity of copper ions, among other characteristics. Until then, its use in clinical settings requires caution. Gold-plated SS archwire is resistant to friction damage, although it is not durable under bending [165]. The performance characteristics of these materials have not been completely evaluated, and more research is needed.

Zinc oxide [187], boron-doped hydroxyapatite (B-HAp) [188], and composites of both materials [189] have been well studied as surface coatings for titanium alloys. Their resistance to corrosion, wear and tear, and peeling has been demonstrated. Zirconium oxide and zirconium oxide-silica composite coatings have also been shown to have excellent corrosion resistance. The former was used to coat a stainless steel metal block surface using the electron-beam physical vapor deposition (EBPVD) technique [159], while the latter was deposited on a titanium metal block surface using the sol-gel method [190]. However, the above materials have not been verified in terms of corrosion and wear resistance in orthodontic archwires and brackets, but are more focused on friction reduction and antibacterial properties. More research should be conducted in the future on their performance in terms of wear resistance and corrosion resistance.

4.4. Polymeric and Bioactive Coatings

Epoxy resin has been proven to be effective in improving the corrosion resistance of NiTi archwires in artificial saliva [191], exhibiting a significant decrease in their electrochemical corrosion tendency [184,191] and nickel ion release [184,186]. Furthermore, its ability to reduce the release of nickel ions has been confirmed in a double-blind randomized clinical trial [81].

The corrosion resistance of PTFE is stronger than that of epoxy resin [191]. The corrosion rate of a PTFE-coated NiTi archwire was 10 times lower than that of uncoated NiTi substrates [192]. However, PTFE may exhibit slight cytotoxicity to fibroblasts—approximately 36% [185]. Therefore, careful consideration is needed when choosing this material. If a high level of biocompatibility is required, it is typically better to choose epoxy resins. Recently, Zhang, Shuai et al. immobilized PTFE nanoparticles in a sol-gel matrix and dip-coated them onto 316L SS via a mussel-inspired method, followed by AgNPs deposition. The nanosilver/polytetrafluoroethylene (AgNP/PTFE) coating obtained in this way can be used on metal implant surfaces because of its good antibacterial and anticorrosion properties [157]. This coating has not been applied in orthodontics thus far and could be subjected to trials.

Polyethylene naphthalate (PEN)-coated SS archwires exhibit virtually no chromium, manganese, iron, or nickel ion leaching in HCl compared to uncoated archwires, and they consistently provide significant corrosion resistance over a bending range of 0–120° [165]. It can be assumed that the PEN coating on the SS surface has good corrosion resistance and durability to withstand large-angle bending of the archwire.

The research applications of superhydrophobic coatings are also on the rise. Cheng-Wei Lin et al. prepared a double-layer grid-blasted plasma-polymerized (GB-PP) superhydrophobic coating on an SS surface. Even after a period of toothbrush cleaning action or food chewing, the water contact angle (WCA) still exceeded 90°, indicating that the hydrophobic properties are always maintained. Moreover, its surface morphology and microstructure were still similar to those at the beginning of the coating, indicating its good durability [193].

There are a few bioactive materials used in orthodontic appliances, such as lysozyme. Longwen He et al. [154] applied a lysozyme coating to a CAW surface via liquid-phase deposition and examined its multiple properties. After 2 weeks of immersion in artificial saliva at 37 °C, the substrate with 40 g/L lysozyme showed the least corrosion pitting, cracking, and roughness. Therefore, the best concentration to use as a corrosion- and wear-resistant coating for CAW surfaces may be 40 g/L. However, no similar studies have been conducted for common SS and NiTi archwires and brackets. In the future, we may try to widen the application range of such coatings.

Studies on corrosion- and friction-resistant polymers and bioactive surface coatings are relatively new and few in number. Epoxy coatings have better corrosion resistance and friction reduction than PTFE [191], but they have not been compared to other polymeric coatings. As no obvious defects have been found so far, epoxy is one of the most promising multifunctional coatings while the performance of other polymeric coatings is not conclusive in all respects. However, the various studied polymeric coatings have good friction and corrosion resistance, which may lead to them becoming a research trend in the future.

4.5. Ion Injection Coatings

In 2021, Rasha A. Ahmed et al. [194] formed an ionic liquid (IL) coating on the surface of a kind of nickel-titanium-cobalt shape memory alloy ($Ni_{47}Ti_{49}Co_4$) archwire and demonstrated the excellent corrosion resistance of the coating through electrochemical experiments and 0.1% sodium fluoride (NaF) immersion experiments. Moreover, the addition of 0.015–0.05% albumin to any IL concentration could further improve the corrosion resistance of the coated surface through the deposition of albumin on the surface of coated or uncoated archwires. However, this is the only study on the corrosion and wear resistance of IL coatings, and it is also a relatively recent development. This field is still in a nascent state and deserves more exploration.

Recent studies on corrosion-resistant coatings for orthodontic appliances are summarized in Table 3.

Table 3. Recent studies on corrosion-resistant coatings for orthodontic appliances.

Ref.	Coating Materials	Coating Technique	Substrate	Study Type	Roughness	Corrosion Resistant Effectiveness	Other Effectiveness
[182]	aC:H	Reactive magnetron sputtering	SS brackets, tubes, and bands	In vitro (in Fusayama-Meyer artificial saliva)	The coating presented a low R_a of ~7 nm.	Right after 7 days of immersion, the Cr release rate from coated samples was ~40% higher than that from uncoated samples, but was similar at day 30. Ni release from coatings was ~10% lower than uncoated samples after 7 days but ~55% higher after 30 days. Finally, Fe release was ~15% lower and similar to uncoated samples after 7 and 30 days, respectively. However, there were no segregation, metallic inclusion, delamination, or detachments on coated samples after 30 days of immersion.	Wear resistance Durability (>30 days)
[157]	AgNP/PTFE	Layer-by-layer deposition	316L SS plates	In vitro (in suspensions of *Escherichia coli*)	The surface roughness increased from 59.4 ± 6.1 nm to 158.1 ± 2.7 nm (deposition time of 6 h) and 177.3 ± 5.1 nm (deposition time of 12 h).	The I_{corr} values reduced from 2.01×10^{-6} to $1.58–2.51 \times 10^{-7}$ A/cm^2, indicating enhanced corrosion protection. However, the corrosion protection was still lower than the PTFE coatings.	Antibacterial effectiveness Durability (>7 days) Biocompatibility
[158]	Al_2O_3 (NPs)	Atomic layer deposition	316L SS	In vitro (in simulation body fluid solution)	NA	The corrosion rate calculated by current density was 1.913 mpy for an uncoated sample and reduced after coated to 0.203 mpy for titania film, 0.174 mpy for alumina film, and 0.164 mpy for multilayer. The corrosion resistance was effectively enhanced by thin films, multilayer proved to be more corrosion protection than single layers, and Al_2O_3 had better corrosion resistance than TiO_2.	Antibacterial effectiveness Biocompatibility
[158]	Al_2O_3/TiO_2 Multilayer	Atomic layer deposition	316L SS	In vitro (in simulation body fluid solution)	NA	The corrosion rate calculated by current density was 1.913 mpy for an uncoated sample and reduced after coated to 0.203 mpy for titania film, 0.174 mpy for alumina film, and 0.164 mpy for multilayer. The corrosion resistance was effectively enhanced by thin films, multilayer proved to be more corrosion protection than single layers, and Al_2O_3 had better corrosion resistance than TiO_2.	Antibacterial effectiveness
[70]	Al-SiO_2	Magnetron sputtering	NiTi and SS archwires	In vitro (in artificial saliva)	NA	The I_{corr} values decreased from 23.72 to 1.21 $\mu A/cm^2$ and from 0.22 to 0.06 $\mu A/cm^2$ after coating with Al-SiO_2 on the NiTi archwires and SS archwires, respectively.	Friction reduction effectiveness Wear resistance Biocompatibility
[188]	B-HAp	Electrophoretic deposition	NiTi alloy	In vitro (in simulated body fluid)	NA	The corrosion rate of 10 wt.% B-HAp and 15 wt.% B-Hap coatings reduced from 0.055 to 0.046 and 0.036 mpy, respectively, which showed better corrosion-resistant effectiveness than other concentrations and only-HAp coatings.	Adhesion strength (up to 20–30 Mpa)

Table 3. *Cont.*

Ref.	Coating Materials	Coating Technique	Substrate	Study Type	Roughness	Corrosion Resistant Effectiveness	Other Effectiveness
[184]	Epoxy	NA (commercial)	NiTi archwires	In vitro (in artificial saliva)	NA	The Ni release was reduced from 8.36 to 0.57 mg/L.	Durability (>30 days)
[136]	F-O-Ti	Plasma-enhanced fluorine and oxygen mono/dual chemical vapor deposition	Commercially available pure titanium with 99.9% purity	In vitro (in suspensions of *Staphylococcus aureus* and artificial saliva)	A large amount of convex texture with 100–200 nm size distributes uniformly all over the surface.	I_{corr} reduced from 0.22 to 0.09 μA/cm^2	Antibacterial effectiveness Durability (~7 days) Biocompatibility
[193]	GB-PP	Pulsed-direct current plasma-enhanced chemical vapor deposition	AISI 304 SS	In vitro (in modified Fusayama artificial saliva)	R_a increased from 0.062 to 10.64 μm after being coated.	The coating particles were retained on surfaces even after being worn 500 times by a toothbrush, peanut, and nougat, while the uncoated substrates showed some scratching and surface pitting or cavity after the wear tests, especially by peanut and nougat.	Wear resistance Durability
[89]	GO	Silane coupling	NiTi archwires	In vitro (friction test: NA; antibacterial test: in bacterial suspension; corrosion test: artificial saliva)	The surface of the samples coated with 2 mg/mL GO concentrations was smooth with a uniformly coated area.	When coated with 0.5, 2, or 5 mg/mL GO concentrations, nickel ion release decreased from 20.75 to 19.75, 18.00, and 17.75 μg/L cm^2, weight loss decreased from 0.31% to 0.29%, 0.25%, and 0.23%, I_{corr} decreased from 0.696 to 0.547, 0.381 and 0.504 μA/cm^2, upon application of 4 mm dislocation, respectively.	Friction reduction effectiveness Antibacterial effectiveness Wear resistance Biocompatibility
[181]	GO	Electrophoretic deposition	NiTi alloy	In vitro (in 3.5% NaCl solution)	R_a increased from 7.55 to 11.67 nm after being coated.	I_{corr} reduced from 0.158 to 0.017 μA/cm^2, and E_{corr} increased from −0.170 to 0.031 V vs. SCE, which exhibited good corrosion resistance.	Biocompatibility
[181]	GO/Ag	Electrophoretic deposition	NiTi alloy	In vitro (in 3.5% NaCl solution)	R_a increased from 7.55 to 18.43 nm after being coated.	I_{corr} reduced from 0.158 to 0.002 μA/cm^2, and E_{corr} increased from −0.170 to 0.008 V vs. SCE, which exhibited good corrosion resistance.	Biocompatibility
[154]	Lysozyme	Liquid phase deposition	CAWs	In vitro (in suspensions of Staphylococcus aureus and artificial saliva)	The surface roughness increased after being coated according to two- and three-dimensional atomic force micrographs.	When coated with 20, 40, and 60 g/L lysozyme, the copper ion release of the archwires reduced from 0.225 to 0.20, 0.1,5, and 0.125 μg, respectively.	Antibacterial effectiveness Wear resistance Durability (>2 weeks) Biocompatibility

Table 3. *Cont.*

Ref.	Coating Materials	Coating Technique	Substrate	Study Type	Roughness	Corrosion Resistant Effectiveness	Other Effectiveness
[174]	Ni-free oxide layer	Pulsed anodization	NiTi plate	In vitro (in PBS solution)	R_a of the anodized surfaces, calculated from the SPM image, was 1.78 nm for the current-anodized surface and 1.16 nm for the pulse-anodized surface. The current-anodized surface included tiny pores (~10 nm), where the depth was determined as ~10 nm; these pores did not appear on the pulse-anodized surface.	The pores on the surface reduced, and Ni release of a pulse-anodized surface at 168 h reduced from ~0.55 to ~0.2 μg·cm^{-2}; however, Ni release of a current-anodized surface at 168 h increased from ~0.55 to ~1.0 μg·cm^{-2}.	Durability (>168 h) Biocompatibility
[57]	Ni-Ti-Cr	Chronopotentiometry	NiTi archwires	In vitro (in artificial saliva)	NA	E_{corr} of electrodes of coated archwires (−427–328 mV) were higher than the uncoated archwires (−447 mV), which represented the corrosion-resistant effectiveness of the coatings.	Friction reduction effectiveness Durability (>60 days)
[57]	Ni-Ti-Mo	Chronopotentiometry	NiTi archwires	In vitro (in artificial saliva)	NA	E_{corr} of electrodes of coated archwires (−395−366 mV) were higher than the uncoated archwires (−447 mV), which represented the corrosion-resistant effectiveness of the coatings.	Friction reduction effectiveness Durability (>60 days)
[157]	PTFE	NA	316L SS plates	In vitro (in suspensions of *Escherichia coli*)	The surface roughness increased from 59.4 ± 6.1 nm to 134.7 ± 3.9 nm after being coated.	The PTFE coating exhibited the best substrate protection as the I_{corr} parameter was over one order of magnitude lower in value than the 316L SS substrate.	Antibacterial effectiveness Biocompatibility
[185]	PTFE	NA (commercial)	NiTi archwires	In vitro (using before-use and after-use samples)	NA	A significant reduction (17.2%) in the percentage of Ni was observed between before- and after-use archwires, which exhibited an unsatisfactory corrosion resistance.	Wear resistance Durability (>2 months) Biocompatibility
[185]	Rhodium	NA (commercial)	NiTi archwires	In vitro (using before-use and after-use samples)	NA	A significant reduction (9.6%) in the percentage of Ni was observed between before- and after-use archwires, which exhibited an unsatisfactory corrosion resistance.	Wear resistance Durability (>2 months) Biocompatibility
[184]	Rhodium	NA (commercial)	NiTi archwires	In vitro (in artificial saliva)	NA	The Ni release was reduced from 8.36 to 1.52 mg/L.	Durability (>30 days)
[31]	TiCN	Direct current reactive magnetron sputtering	316L SS and (100)-oriented Si substrates	In vitro (in artificial saliva)	NA	The corrosion resistance of TiCN increased with the nitrogen content and carbon content of the preparation process when the N_2 flux was larger than 2.5 sccm.	Friction reduction effectiveness Aesthetic effect
[166]	TiCN	Multi-arc ion plating	316L SS plates	In vitro (in artificial saliva)	NA	The TiCN film showed a corrosion current density of ~7 μA/cm^2, which showed better corrosion resistance than TiN, but not as good as Ti-DLC.	NA

Table 3. *Cont.*

Ref.	Coating Materials	Coating Technique	Substrate	Study Type	Roughness	Corrosion Resistant Effectiveness	Other Effectiveness
[167]	TiCN	Magnetron sputtering	Ni-Cr-Mo disks	In vitro (dry condition)	R_a increased from 0.33 to 0.38–0.45 μm after being coated.	There were no visible areas with an elevated content of nickel or chromium, which proved that the base had not been revealed.	Wear resistance
[168]	TiCN	Magnetron sputtering	Ni-Cr alloy	In vitro (in neutral salt spray and seawater acetic acid)	NA	The mass loss of the examined samples reduced from 0.37 to ~0.20 mg/mm^2 × 10^{-4}.	Durability (>30 days)
[176]	Ti-Cr-N	Radiofrequency reactive sputtering	304 SS	In vitro (in artificial saliva)	The RMS roughness increased from 5 to 105–20 nm after being coated.	Ti-Cr-N coatings reduced 304 SS's release of chromium species by ~67%, and annealed at 400 °C displayed higher corrosion resistance.	Durability (>90 days)
[166]	Ti-DLC	Multi-arc ion plating	316L SS plates	In vitro (in artificial saliva)	The surface roughness was 10.4 nm after being coated.	The Ti-DLC film showed the lowest corrosion current density (~4.577 μA/cm^2) and thickness reduction (~0.12 μm) in different electrolytes compared to TiN and TiCN.	Wear resistance
[31]	TiN	Direct current reactive magnetron sputtering	316 L SS and (100)-oriented Si substrates	In vitro (in artificial saliva)	NA	The corrosion resistance of TiN decreased with the nitrogen content and carbon content used in the preparation process.	Friction reduction effectiveness Wear resistance
[58]	TiN	Ion plating	SS and NiTi archwires	In vitro (friction test: NA; corrosion test: in 0.9% NaCl solution)	R_a of TiN-coated SS archwire increased from 0.023 to 0.046 μm. R_a of TiN-coated NiTi archwire remained at 0.001 μm.	Corrosion resistance increased in coated archwires. The pitting corrosion was reduced compared to uncoated archwires. The breakdown potential of the non-coated and the TiN-coated SS wire was 0.46 and 0.61 V ($p < 0.05$), and that of the non-coated and the TiN-coated Ni-Ti wire was 1.20 V and more than 2.0 V, respectively.	Friction reduction effectiveness
[166]	TiN	Multi-arc ion plating	316L SS plates	In vitro (in artificial saliva)	NA	The TiN film showed a corrosion current density of ~8.5 μA/cm^2, which showed lower corrosion resistance than TiCN and Ti-DLC.	NA
[19]	TiN	Hollow cathode discharge	SS and NiTi archwires	In vitro (in physiological saline, sterile water, 35% hydrochloric acid, and 88% lactic acid)	When SS archwires were coated, R_a increased from ~0.025 to ~0.04 μm; when NiTi archwires were coated, R_a increased from ~0.14 to ~0.15 μm.	The acid-mediated corrosion and the elution of Ni ions from the wire surface reduced from 2.38, 3.37, 215, 499.84, and 2.29 μg/L to 1.35, 4.16, 13536.28, and 1.44 μg/L after being immersed for 30 min in sterile water, physiological saline, 35% hydrochloric acid, and 88% lactic acid, respectively.	NA
[33]	TiN/Ti	High-power magnetron sputtering deposition	Ti_6Al_4V plate	In vitro	The average grain size was 9.1–11.6 μm after being coated.	The 1-layer, 2-layer, and 4-layer TiN/Ti multilayer coatings had similar wear rates which were less than 1/20 of the wear rate of the Ti6Al4V substrate, while the wear rates of the 8-layer and 12-layer TiN/Ti multilayer coatings were higher than that of the Ti6Al4V substrate.	Wear resistant (1-layer, 2-layer, and 4-layer TiN/Ti multilayer coatings)

Table 3. *Cont.*

Ref.	Coating Materials	Coating Technique	Substrate	Study Type	Roughness	Corrosion Resistant Effectiveness	Other Effectiveness
[177]	Ti-Nb	Laser cladding	Cold-rolled NiTi alloy	In vitro (in simulated body fluid)	NA	I_{corr} reduced from 272.4 to 163.7 nA/cm^2 and E_{corr} increased from −0.184 to −0.128 V, which exhibited a good corrosion resistance.	NA
[173]	TiO_2	Sol-gel dip-coating	SS archwires	In vitro (in Ringer's solution)	NA	I_{corr} increased from 0.007 to 39.9 μA/cm^2 and E_{corr} reduced from −162 to −300 mV, which exhibited an unsatisfactory corrosion resistance.	NA
[158]	TiO_2 (NPs)	Atomic layer deposition	316L SS	In vitro (in simulation body fluid solution)	NA	The corrosion rate calculated by current density was 1.913 mpy for an uncoated sample and reduced after coated to 0.203 mpy for titania film, 0.174 mpy for alumina film, and 0.164 mpy for multilayer. The corrosion resistance was effectively enhanced by thin films, multilayer proved to be more corrosion protection than single layers, and Al2O3 had better corrosion resistance than TiO_2.	Antibacterial effectiveness Biocompatibility
[173]	TiO_2/Ag	Sol-gel dip-coating	SS archwires	In vitro (in Ringer's solution)	NA	I_{corr} increased from 0.007 to 30.0 μA/cm^2 and E_{corr} reduced from −162 to −285 mV, which exhibited an unsatisfactory corrosion resistance.	Wear resistance
[159]	ZrO_2	EBPVD	316L SS	In vitro (in artificial saliva and artificial saliva containing 0.2% or 2% of NaF)	R_a reduced from 10 to 3 nm after being coated.	I_{corr} reduced from 0.34–9.27 to 0.04–0.46 μA/cm^2; E_{corr} increased from −0.08775 to −0.0739, from −0.2122 to −0.1267, and from −0.2831 to −0.1714 mV when immersed in artificial saliva, artificial saliva containing 0.2% NaF, and artificial saliva containing 2% of NaF, respectively, which exhibited a good corrosion resistance.	Antibacterial effectiveness Wear resistance Biocompatibility

NA: not applicable; E_{corr}: corrosion potential values; I_{corr}: corrosion current density; aC:H: hydrogenated forms of diamond-like carbon; Cr: chromium; Fe: ferrum; B-HAp: boron-doped hydroxyapatite; NaCl: sodium chloride; GB-PP: grid-blasted plasma-polymerized; AISI: American Iron and Steel Institute; SPM: scanning probe microscope; Ni-Cr: nickel-chromium; Ti-Cr-N: titanium-chromium-nitrogen; RMS: root mean square; TiN/Ti: titanium nitride/titanium; Ti_6Al_4V: titanium-aluminum-vanadium alloy with 6% aluminum and 4% vanadium; Ti-Nb: titanium-niobium; TiO_2/Ag: titanium dioxide/silver.

5. Discussion

It is necessary to modify the surface of orthodontic accessories—especially brackets and archwires—using surface coatings. This article reviews previous surface coatings and modification effects related to friction reduction, antibacterial properties, and corrosion resistance.

From the perspective of a research mindset, there are numerous types of functional coatings for the surface of orthodontic attachments. They have different methods and conditions of coating, but all have certain functions under different experimental conditions. This may seem like another new kind of coating has been discovered to choose from; however, in fact, it makes little sense to keep discovering new materials without comparing a range of materials and coating methods horizontally under the same conditions. Therefore, a detailed comparison of newly discovered coating materials with existing materials should be performed in various aspects in order to tailor the most appropriate coating materials for clinical applications. In addition, different coating methods and conditions have different effects on the properties and effects of the coating. When the coating temperature, duration, and/or raw material ratio are changed, the nature of the coating also changes [133]. With this in mind, the best conditions for preparing a certain functional coating can be found by changing the coating process. Moreover, there have been few relevant animal experiments and clinical trials, making it challenging to determine the clinical value of coatings. The use of animals for validation of the effectiveness and safety of various surface coatings is essential, and clinical trials should also be conducted to document what is really happening in clinical applications in more detail.

In clinical practice, it is expected that a single coating on the surface of a fixed orthodontic appliance will meet all of the performance requirements for friction reduction, antibacterial activity, and corrosion resistance. However, it is difficult to find a single coating that can satisfy all three conditions at the same time. A few coatings, such as TaAgB and PTFE, offer relatively comprehensive performance improvements [68,92,97,147,192]. However, studies on these materials are scarce, and there is not enough clear and stable clinical evidence. Recently, new coating materials such as tantalum nitride-copper (TaCuN) solid solution [195] have also been verified with excellent modification properties, in addition to novel coatings in other field, such as ferrum-aluminum/dizinc magnesium (Fe-Al/$MgZn_2$) multilayer [196] and nanographene piece + carbon dots and nickel-tungsten (Ni-W) co-deposition coatings [197], but they have not been tested on brackets or archwires. Moreover, the durability of the coatings needs to be further improved. As a long-term fixed intraoral orthodontic accessory, the brackets need to maintain their surface properties for months or more than a year, and the coatings are likely to be damaged and peeled off by sliding [62]. This suggests that functional coatings need to be able to remain effective in the mouth for a long period, in addition to the abovementioned properties. Therefore, coatings with high durability are needed.

The antifriction effect of GSEC and epoxy can last up to 30 days [90,97]. For antibacterial coatings, N-doped TiO_2 inhibited *Streptococcus mutans* for up to 90 days in vitro and up to 60 days in clinical trials [132,134], making it the most durable material in the scope of this review. Ag and ZnO have also been validated for a period of 3 months [112]. TiO_2 nanoparticles were effective within 3 weeks [128], and lysosome coating lasted 2 weeks [154]. For titanium sputter coating, corrosion can be reduced within 30 days [175]. Likewise, DLC will have this effect for 14 days [76]. Most of the remaining materials have not been confirmed by long-term experiments and cannot be used in the comparison. Therefore, sustainability aspects should be an important research direction in the future, with considerable significance for the life and validity of the coating.

In a comprehensive view, it is necessary to formulate and compare various surface coating materials and technologies in order to obtain surface coatings that combine multiple functions, are proven to be safe and effective in clinical trials, and have good durability to increase efficiency and reduce risks in the orthodontic treatment process. Among all functional coatings, the best friction-reducing coatings are TiN, TaAgB, and GO, which not only have remarkably low CoF, but also have advantages such as antibacterial properties, wear resistance, corrosion resistance, and good biocompatibility [30,32,58,68,89,181]. The next choices are TiCN, GSEC, CN_x, and ZnO, which do not have as many additional advantages, but also have excellent friction reduction properties [27,29,31,32,46,90,112]. However, CrN [30], epoxy [93], and parylene [97] are least recommended as friction-reducing coatings because they have been shown in recent studies not to reduce surface friction and sometimes even increase it. Among the antimicrobial coatings, AgNP/PTFE coatings [157], CTS/PEG coatings [156], and F-O-Ti coatings [136] are the most recommended because they have the advantages of corrosion resistance, durability, and biocompatibility, on the basis of antimicrobial rates of over 90%. Ag and ZnO may not be as strong as the former coatings in terms of wear resistance, aesthetics, and durability. However, Ag and ZnO, as classic antimicrobial coatings, have also been repeatedly proven to have excellent antimicrobial properties and biocompatibility, and have been put into clinical application for a longer time and with good results [109,139]. Interestingly, Al compound and composite coatings generally perform better in corrosion-resistant coatings, such as Al_2O_3 coatings [30,158], Al_2O_3/TiO_2 multi-layer coatings [158], and Al-SiO_2 coatings [70]. In particular, Al-SiO_2 coatings not only significantly reduce the corrosion on the surface of substrates, but also have the advantages of friction reduction, wear resistance, and biocompatibility [70]. This kind of multilayer and composite material is also gradually becoming a research trend in recent years, which might be suitable for application.

The materials and methods already available are adequate, but there are still directions that need to be studied. Some future research prospects are presented here. First, the control of the thickness of the coatings: on the basis of ensuring the modification effect, the thickness of the coating should be reduced as much as possible, which not only saves material but also lessens the adverse effect of the coating's thickness on the friction [60,78]. Second, it is necessary to verify the clinical effect that the coating can eventually bring—for example, the improvement of the friction-reducing coatings in terms of practical orthodontic efficiency, the contribution of the antimicrobial coatings to avoiding the demineralization of the tooth surface and the inflammation of the periodontal tissues, and the ability of the anticorrosion coatings to reduce patients' metal-ion allergies. However, the biggest problem with animal experiments is the difficulty of controlling the experimental conditions [198]. Therefore, if the safety of the materials can be confirmed, preclinical trials can be considered directly, without animal experiments. Finally, the long-term effects of various coatings are uncertain. For orthodontic treatment, they can be as short as a few months or as long as a few years. Orthodontic appliances—especially brackets—are present in the oral environment from start to finish, so the coatings on the bracket surfaces need to maintain the longest-lasting modified effect. In fact, laser cladding may offer greater advantages than chemical, electrochemical, and physical deposition methods in terms of coating thickness and permanence. These processes have little effect on the dimensions of the substrates and are less susceptible to peeling off [199]. In this regard, more efforts could be made in laser cladding as a coating method in the future. There are still some limitations to this review. Our discussion in this case is directed only to the surface coating of orthodontic fixed appliances, while other surface modification methods have not been discussed extensively. In the case of orthodontic appliances, the main focus is only on the coating of the two main components—brackets and archwires. Other components of orthodontic appliances, such as ligature wires, buccal tubes, and orthodontic band strips, have not been discussed in depth. Of course, functional surface coatings for orthodontic bands have recently emerged, such as AgNPs coatings with enhanced antimicrobial properties [200]. Going forward, each attachment for fixed appliances should be taken into account for future improvements.

6. Conclusions

In order to increase treatment efficiency and reduce side effects, functional surface coatings of orthodontic fixed appliances are receiving increasingly more attention. Over the past decade, research on friction-reducing, antibacterial, and corrosion-resistant coatings has grown, better materials have been discovered, and methods and conditions have been explored to optimize the performance of the materials. However, the current coating materials have not yet achieved the perfect integration of these three functions, and most of the research is limited to in vitro experiments. Thus, there is still a certain distance to go before clinical applications, so further research is needed in the future.

Author Contributions: Conceptualization, B.H., X.L. and R.Z.; validation, X.L. and B.H.; data curation, R.Z.; writing—original draft preparation, R.Z.; writing—review and editing, R.Z., X.L. and B.H.; visualization, R.Z.; supervision, X.L. and B.H.; project administration, B.H. and X.L.; funding acquisition, X.L. and B.H. All authors have read and agreed to the published version of the manuscript.

Funding: This research was funded by the Beijing Municipal Natural Science Foundation—Haidian Original Innovation Joint Fund, Key project, L222001; the Program for New Clinical Techniques and Therapies of Peking University School and Hospital of Stomatology, grant number PKUSSNCT-21B11; and the National Natural Science Foundation of China, grant numbers 51972005, U21A2055, 82101074, and 51672009.

Institutional Review Board Statement: Not applicable.

Informed Consent Statement: Not applicable.

Data Availability Statement: Not applicable.

Conflicts of Interest: The authors declare no conflict of interest.

References

1. Arango, S.; Peláez-Vargas, A.; García, C. Coating and Surface Treatments on Orthodontic Metallic Materials. *Coatings* **2013**, *3*, 1–15. [CrossRef]
2. Lee, S.M.; Hwang, C.-J. A comparative study of frictional force in self-ligating brackets according to the bracket-archwire angulation, bracket material, and wire type. *Korean J. Orthod.* **2015**, *45*, 13–19. [CrossRef] [PubMed]
3. Braun, S.; Bluestein, M.; Moore, B.K.; Benson, G. Friction in perspective. *Am. J. Orthod. Dentofac. Orthop.* **1999**, *115*, 619–627. [CrossRef] [PubMed]
4. Julien, K.C.; Buschang, P.H.; Campbell, P.M. Prevalence of white spot lesion formation during orthodontic treatment. *Angle Orthod.* **2013**, *83*, 641–647. [CrossRef]
5. Øgaard, B.; Rølla, G.; Arends, J. Orthodontic appliances and enamel demineralization: Part 1. Lesion development. *Am. J. Orthod. Dentofac. Orthop.* **1988**, *94*, 68–73. [CrossRef]
6. Featherstone, J.D.B. The Continuum of Dental Caries—Evidence for a Dynamic Disease Process. *J. Dent. Res.* **2004**, *83*, 39–42. [CrossRef]
7. Wu, Y.; Cao, L.; Cong, J. The periodontal status of removable appliances vs fixed appliances: A comparative meta-analysis. *Medicine* **2020**, *99*, e23165. [CrossRef]
8. Beberhold, K.; Sachse-Kulp, A.; Schwestka-Polly, R.; Hornecker, E.; Ziebolz, D. The Orthodontic Plaque Index: An oral hygiene index for patients with multibracket appliances. *Orthod. Art Pract. Dentofac. Enhanc.* **2012**, *13*, 94–99.
9. House, K.; Sernetz, F.; Dymock, D.; Sandy, J.R.; Ireland, A.J. Corrosion of orthodontic appliances—Should we care? *Am. J. Orthod. Dentofac. Orthop.* **2008**, *133*, 584–592. [CrossRef] [PubMed]
10. Chaturvedi, T.P.; Upadhayay, S.N. An overview of orthodontic material degradation in oral cavity. *Indian J. Dent. Res.* **2010**, *21*, 275–284. [CrossRef]
11. Eliades, T.; Athanasiou, A.E. In vivo aging of orthodontic alloys: Implications for corrosion potential, nickel release, and biocompatibility. *Angle Orthod.* **2002**, *72*, 222–237.
12. Chaturvedi, T.P. *Corrosion Behaviour of Orthodontic Alloys—A Review*; Banaras Hindu University: Varanasi, India, 2008.
13. Pulikkottil, V.J.; Chidambaram, S.; Bejoy, P.U.; Femin, P.K.; Paul, P.; Rishad, M. Corrosion resistance of stainless steel, nickel-titanium, titanium molybdenum alloy, and ion-implanted titanium molybdenum alloy archwires in acidic fluoride-containing artificial saliva: An in vitro study. *J. Pharm. Bioallied Sci.* **2016**, *8*, S96–S99. [CrossRef] [PubMed]
14. Martín-Cameán, A.; Jos, A.; Puerto, M.; Calleja, A.; Iglesias-Linares, A.; Solano, E.; Cameán, A.M. In vivo determination of aluminum, cobalt, chromium, copper, nickel, titanium and vanadium in oral mucosa cells from orthodontic patients with mini-implants by Inductively coupled plasma-mass spectrometry (ICP-MS). *J. Trace Elem. Med. Biol.* **2015**, *32*, 13–20. [CrossRef] [PubMed]
15. Kovac, V.; Poljsak, B.; Bergant, M.; Scancar, J.; Mezeg, U.; Primozic, J. Differences in Metal Ions Released from Orthodontic Appliances in an In Vitro and In Vivo Setting. *Coatings* **2022**, *12*, 190. [CrossRef]
16. Rahilly, G.; Price, N. Nickel allergy and orthodontics. *J. Orthod.* **2003**, *30*, 171–174. [CrossRef] [PubMed]
17. Noble, J.; Ahing, S.I.; Karaiskos, N.E.; Wiltshire, W.A. Nickel allergy and orthodontics, a review and report of two cases. *Br. Dent. J.* **2008**, *204*, 297–300. [CrossRef] [PubMed]

18. Gracco, A.; Dandrea, M.; Deflorian, F.; Zanella, C.; De Stefani, A.; Bruno, G.; Stellini, E. Application of a Molybdenum and Tungsten Disulfide Coating to Improve Tribological Properties of Orthodontic Archwires. *Nanomaterials* **2019**, *9*, 753. [CrossRef]
19. Ito, A.; Kitaura, H.; Sugisawa, H.; Noguchi, T.; Ohori, F.; Mizoguchi, I. Titanium Nitride Plating Reduces Nickel Ion Release from Orthodontic Wire. *Appl. Sci.* **2021**, *11*, 9745. [CrossRef]
20. Wang, Y.; Ding, C.; Ge, Z.; Li, Z.; Chen, L.; Guo, X.; Dong, G.; Zhou, P. A novel antibacterial and fluorescent coating composed of polydopamine and carbon dots on the surface of orthodontic brackets. *J. Mater. Sci. Mater. Med.* **2023**, *34*, 10. [CrossRef]
21. Usui, T.; Iwata, T.; Miyake, S.; Otsuka, T.; Koizumi, S.; Shirakawa, N.; Kawata, T. Mechanical and frictional properties of aesthetic orthodontic wires obtained by hard chrome carbide plating. *J. Dent. Sci.* **2018**, *13*, 151–159. [CrossRef] [PubMed]
22. Katz, A.; Redlich, M.; Rapoport, L.; Wagner, H.D.; Tenne, R. Self-lubricating coatings containing fullerene-like WS2 nanoparticles for orthodontic wires and other possible medical applications. *Tribol. Lett.* **2006**, *21*, 135–139. [CrossRef]
23. Kachoei, M.; Eskandarinejad, F.; Divband, B.; Khatamian, M. The effect of zinc oxide nanoparticles deposition for friction reduction on orthodontic wires. *Dent. Res. J.* **2013**, *10*, 499–505.
24. Elhelbawy, N.; Ellaithy, M. Comparative evaluation of Stainless-steel wires and brackets coated with nanoparticles of Chitosan or Zinc oxide upon friction: An in vitro study. *Int. Orthod.* **2021**, *19*, 274–280. [CrossRef] [PubMed]
25. Kachoei, M.; Nourian, A.; Divband, B.; Kachoei, Z.; Shirazi, S. Zinc-oxide nanocoating for improvement of the antibacterial and frictional behavior of nickel-titanium alloy. *Nanomedicine* **2016**, *11*, 2511–2527. [CrossRef] [PubMed]
26. Behroozian, A.; Kachoei, M.; Khatamian, M.; Divband, B. The effect of ZnO nanoparticle coating on the frictional resistance between orthodontic wires and ceramic brackets. *J. Dent. Res. Dent. Clin. Dent. Prospect.* **2016**, *10*, 106. [CrossRef]
27. Hammad, S.M.; El-Wassefy, N.A.; Shamaa, M.S.; Fathy, A. Evaluation of zinc-oxide nanocoating on the characteristics and antibacterial behavior of nickel-titanium alloy. *Dent. Press J. Orthod.* **2020**, *25*, 51–58. [CrossRef]
28. Kao, C.-T.; Guo, J.-U.; Huang, T.-H. Comparison of friction force between corroded and noncorroded titanium nitride plating of metal brackets. *Am. J. Orthod. Dentofac. Orthop.* **2011**, *139*, 594–600. [CrossRef]
29. Zhang, M.; Liu, X.; Shang, H.; Lin, J. Comparison of TiN and CNx coatings on orthodontic stainless steel: Tribological and biological evaluation. *Surf. Coat. Technol.* **2019**, *362*, 381–387. [CrossRef]
30. Arici, N.; Akdeniz, B.S.; Oz, A.A.; Gencer, Y.; Tarakci, M.; Arici, S. Effectiveness of medical coating materials in decreasing friction between orthodontic brackets and archwires. *Korean J. Orthod.* **2021**, *51*, 270–281. [CrossRef]
31. Suciu, V.; Ferreira, A.; Correa, M.A.; Vaz, F.; Munteanu, D. Structural, Mechanical, and Decorative Properties of Sputtered TiN and Ti (N, C) Films for Orthodontic Applications; an In Vitro Study. *Materials* **2021**, *14*, 5175. [CrossRef]
32. Zhang, J.; Lou, J.; He, H.; Xie, Y. Comparative Investigation on the Tribological Performances of TiN, TiCN, and Ti-DLC Film-Coated Stainless Steel. *JOM* **2019**, *71*, 4872–4879. [CrossRef]
33. Sheng, L.; Xiao, Y.; Jiao, C.; Du, B.; Li, Y.; Wu, Z.; Shao, L. Influence of layer number on microstructure, mechanical properties and wear behavior of the TiN/Ti multilayer coatings fabricated by high-power magnetron sputtering deposition. *J. Manuf. Process.* **2021**, *70*, 529–542. [CrossRef]
34. Bewilogua, K.; Hofmann, D. History of diamond-like carbon films—From first experiments to worldwide applications. *Surf. Coat. Technol.* **2014**, *242*, 214–225. [CrossRef]
35. Husmann, P.; Bourauel, C.; Wessinger, M.; Jäger, A. The frictional behavior of coated guiding archwires. *J. Orofac. Orthop. /Fortschr. Kieferorthopädie* **2002**, *63*, 199–211. [CrossRef]
36. Farronato, G.; Maijer, R.; Carìa, M.P.; Esposito, L.; Alberzoni, D.; Cacciatore, G. The effect of Teflon coating on the resistance to sliding of orthodontic archwires. *Eur. J. Orthod.* **2012**, *34*, 410–417. [CrossRef]
37. Katta, A.; Vannala, V.; Navaneethakrishnan, K.K.; Kandasamy, R.; Venkatachalam, B.; Arafath, M. Coated, uncoated stainless steel ligatures versus Self ligation-In the perspective of friction. *Pak. Orthod. J.* **2013**, *5*, 60–66.
38. Sukh, R.; Singh, G.K.; Tandon, P.; Singh, G.P.; Singh, A. A comparative study of frictional resistance during simulated canine retraction on typodont model. *J. Orthod. Sci.* **2013**, *2*, 61–66. [CrossRef]
39. Abbas, A.A.; Alhuwaizi, A.F. The effect of wire dimension, type and thickness of coating layer on friction of coated stainless-steel arch wires. *Int. J. Med. Res. Health Sci.* **2018**, *7*, 115–121.
40. Kawaguchi, K.; Iijima, M.; Muguruma, T.; Endo, K.; Mizoguchi, I. Effects of bioactive glass coating by electrophoretic deposition on esthetical, bending, and frictional performance of orthodontic stainless steel wire. *Dent. Mater. J.* **2020**, *39*, 593–600. [CrossRef] [PubMed]
41. Mhaske, A.R.; Shetty, P.C.; Bhat, N.S.; Ramachandra, C.S.; Laxmikanth, S.M.; Nagarahalli, K.; Tekale, P.D. Antiadherent and antibacterial properties of stainless steel and NiTi orthodontic wires coated with silver against *Lactobacillus acidophilus*—An in vitro study. *Prog. Orthod.* **2015**, *16*, 40. [CrossRef] [PubMed]
42. Redlich, M.; Katz, A.; Rapoport, L.; Wagner, H.D.; Feldman, Y.; Tenne, R. Improved orthodontic stainless steel wires coated with inorganic fullerene-like nanoparticles of WS_2 impregnated in electroless nickel–phosphorous film. *Dent. Mater.* **2008**, *24*, 1640–1646. [CrossRef]
43. Chow, P.K.; Singh, E.; Viana, B.C.; Gao, J.; Luo, J.; Li, J.; Lin, Z.; Elías, A.L.; Shi, Y.; Wang, Z.; et al. Wetting of Mono and Few-Layered WS_2 and MoS_2 Films Supported on Si/SiO_2 Substrates. *ACS Nano* **2015**, *9*, 3023–3031. [CrossRef]
44. Kozbial, A.; Gong, X.; Liu, H.; Li, L. Understanding the Intrinsic Water Wettability of Molybdenum Disulfide (MoS_2). *Langmuir* **2015**, *31*, 8429–8435. [CrossRef]

45. Wang, L.; Jibin, P.U.; Wang, H.; Zeng, C.; Xue, Q. Tantalum-Doped Molybdenum Disulfide/Tungsten Disulfide Multi-Layer Film as Well as Preparation Method and Use Thereof. U.S. Patent Application US20220341023A1, 27 October 2022.
46. Wang, P.; Luo, X.; Qin, J.; Pan, Z.; Zhou, K. Effect of Graphene Sheets Embedded Carbon Films on the Fretting Wear Behaviors of Orthodontic Archwire–Bracket Contacts. *Nanomaterials* **2022**, *12*, 3430. [CrossRef] [PubMed]
47. Schilling, K.; Bradford, B.; Castelli, D.; Dufour, E.; Nash, J.F.; Pape, W.; Schulte, S.; Tooley, I.; van den Bosch, J.; Schellauf, F. Human safety review of "nano" titanium dioxide and zinc oxide. *Photochem. Photobiol. Sci.* **2010**, *9*, 495–509. [CrossRef] [PubMed]
48. Karandish, M.; Pakshir, M.; Moghimi, M.; Jafarpour, D. Evaluating the Mechanical Properties of Zinc-Coated Stainless Steel Orthodontic Wires Using Physical Vapor Deposition. *Int. J. Dent.* **2021**, *2021*, 1–8. [CrossRef] [PubMed]
49. Deshmukh, S.B.; Bari, R.H. Nanostructured ZrO_2 thin films deposited by spray pyrolysis techniques for ammonia gas sensing application. *Int. Lett. Chem. Phys. Astron.* **2015**, *56*, 120–130. [CrossRef]
50. Sollazzo, V.; Pezzetti, F.; Scarano, A.; Piattelli, A.; Bignozzi, C.A.; Massari, L.; Brunelli, G.; Carinci, F. Zirconium oxide coating improves implant osseointegration in vivo. *Dent. Mater.* **2008**, *24*, 357–361. [CrossRef]
51. Manicone, P.F.; Iommetti, P.R.; Raffaelli, L. An overview of zirconia ceramics: Basic properties and clinical applications. *J. Dent.* **2007**, *35*, 819–826. [CrossRef] [PubMed]
52. Golshah, A.; Feyli, S.A. Effect of zirconium oxide nano-coating on frictional resistance of orthodontic wires. *J. Orthod. Sci.* **2022**, *11*, 35. [CrossRef]
53. Davidson, J.A. Zirconium Oxide Coated Prosthesis for Wear and Corrosion Resistance. U.S. Patent Application US5037438A, 6 August 1991.
54. Zhang, Z.X.; Fu, B.F.; Zhang, D.Y.; Cheng, Y.; Sheng, L.Y.; Lai, C.; Xi, T.F. Safety and efficacy of nano lamellar TiN coatings on nitinol atrial septal defect occluders in vivo. *Mater. Sci. Eng. C* **2013**, *33*, 1355–1360. [CrossRef] [PubMed]
55. Kim, H.; Kim, C.Y.; Kim, D.W.; Lee, I.S.; Lee, G.H.; Park, J.C.; Lee, S.J.; Lee, K.Y. Wear performance of self-mating contact pairs of TiN and TiAlN coatings on orthopedic grade Ti-6Al-4V. *Biomed. Mater.* **2010**, *5*, 044108. [CrossRef]
56. Steele, J.G.; McCabe, J.F.; Barnes, I.E. Properties of a titanium nitride coating for dental instruments. *J. Dent.* **1991**, *19*, 226–229. [CrossRef] [PubMed]
57. Yılmaz, H.; Zortuk, F.B.; Özyılmaz, A.T. Evaluation of the surface properties of modified Ni-Ti arch wires. *Int. Dent. Res.* **2021**, *11*, 167–175. [CrossRef]
58. Sugisawa, H.; Kitaura, H.; Ueda, K.; Kimura, K.; Ishida, M.; Ochi, Y.; Kishikawa, A.; Ogawa, S.; Takano-Yamamoto, T. Corrosion resistance and mechanical properties of titanium nitride plating on orthodontic wires. *Dent. Mater. J.* **2018**, *37*, 286–292. [CrossRef] [PubMed]
59. Krishnan, V.; Krishnan, A.; Remya, R.; Ravikumar, K.K.; Nair, S.A.; Shibli, S.M.A.; Varma, H.K.; Sukumaran, K.; Kumar, K.J. Development and evaluation of two PVD-coated β-titanium orthodontic archwires for fluoride-induced corrosion protection. *Acta Biomater.* **2011**, *7*, 1913–1927. [CrossRef]
60. da Silveira, R.E.; Elias, C.N.; do Amaral, F.L.B. Assessment of frictional resistance and surface roughness in orthodontic wires coated with two different nanoparticles. *Microsc. Res. Tech.* **2022**, *85*, 1884–1890. [CrossRef]
61. Maliael, M.T.; Jain, R.K.; Srirengalakshmi, M. Effect of Nanoparticle Coatings on Frictional Resistance of Orthodontic Archwires: A Systematic Review and Meta-analysis. *World J. Dent.* **2022**, *13*, 417–424.
62. Syed, S.S.; Kulkarni, D.; Todkar, R.; Bagul, R.S.; Parekh, K.; Bhujbal, N. A novel method of coating orthodontic archwires with nanoparticles. *J. Int. Oral Health JIOH* **2015**, *7*, 30.
63. Ghasemi, T.; Arash, V.; Rabiee, S.M.; Rajabnia, R.; Pourzare, A.; Rakhshan, V. Antimicrobial effect, frictional resistance, and surface roughness of stainless steel orthodontic brackets coated with nanofilms of silver and titanium oxide: A preliminary study. *Microsc. Res. Tech.* **2017**, *80*, 599–607. [CrossRef]
64. Jung, S.-C.; Kim, S.-J.; Imaishi, N.; Cho, Y.-I. Effect of TiO_2 thin film thickness and specific surface area by low-pressure metal–organic chemical vapor deposition on photocatalytic activities. *Appl. Catal. B Environ.* **2005**, *55*, 253–257. [CrossRef]
65. Zhang, L.-C.; Attar, H. Selective Laser Melting of Titanium Alloys and Titanium Matrix Composites for Biomedical Applications: A Review. *Adv. Eng. Mater.* **2015**, *18*, 463–475. [CrossRef]
66. Krishnan, M.; Seema, S.; Tiwari, B.; Sharma, H.S.; Londhe, S.; Arora, V. Surface characterization of nickel titanium orthodontic arch wires. *Med. J. Armed Forces India* **2015**, *71*, S340–S345. [CrossRef] [PubMed]
67. Shah, P.K.; Sharma, P.; Goje, S.K. Comparative Evaluation of Frictional Resistance of Silver-Coated Stainless Steel Wires with Uncoated Stainless Steel Wires: An In vitro Study. *Contemp. Clin. Dent.* **2018**, *9*, S331–S336. [CrossRef] [PubMed]
68. Wang, J.; Yu, D.; Li, Y.; Miao, K.; Bao, X.; Hu, M.; Zhang, K. Tribological and biological assessments of TaAgB solid solution coatings for orthodontic treatment. *Appl. Surf. Sci.* **2022**, *597*, 153704. [CrossRef]
69. Hasan, K.M.F.; Xiaoyi, L.; Shaoqin, Z.; Horváth, P.G.; Bak, M.; Bejó, L.; Sipos, G.; Alpár, T. Functional silver nanoparticles synthesis from sustainable point of view: 2000 to 2023—A review on game changing materials. *Heliyon* **2022**, *8*, e12322. [CrossRef]
70. Wu, H.; Yang, J.; Yan, Y.; Zheng, B.; Algahefi, A.L.; Ma, S.; Liu, Y. Study of Al–SiO_2 Aesthetic Composite Coating on Orthodontic Metal Archwire. *Coatings* **2022**, *12*, 746. [CrossRef]
71. Nik, T.H.; Ghadirian, H.; Hooshmand, T.; Kharazifard, M.J.; Nasiri, M.; Mahd, M.J. Effect of 0.05% Sodium Fluoride Mouthwash on Surface Roughness and Friction between Ceramic Brackets and Rhodium-Coated and Uncoated Stainless Steel Wires. *Front. Dent.* **2019**, *16*, 121–129. [CrossRef] [PubMed]

72. Choi, S.; Hwang, E.Y.; Park, H.K.; Park, Y.G. Correlation between frictional force and surface roughness of orthodontic archwires. *Scanning* **2015**, *37*, 399–405. [CrossRef] [PubMed]
73. Kusy, R.P.; Whitley, J.Q.; Mayhew, M.J.; Buckthal, J.E. Surface roughness of orthodontic archwires via laser spectroscopy. *Angle Orthod.* **1988**, *58*, 33–45. [PubMed]
74. Kusy, R.P.; Whitley, J.Q. Effects of surface roughness on the coefficients of friction in model orthodontic systems. *J. Biomech.* **1990**, *23*, 913–925. [CrossRef]
75. Aldabagh, D.J.; Alzubaydi, T.L.; Alhuwaizi, A.F. Surface Characterization of Stainless Steel 316L Coated with Various Nanoparticle Types. *Int. J. Biomater.* **2023**, *2023*, 3997281. [CrossRef]
76. Kobayashi, S.; Ohgoe, Y.; Ozeki, K.; Sato, K.; Sumiya, T.; Hirakuri, K.K.; Aoki, H. Diamond-like carbon coatings on orthodontic archwires. *Diam. Relat. Mater.* **2005**, *14*, 1094–1097. [CrossRef]
77. Akaike, S.; Hayakawa, T.; Kobayashi, D.; Aono, Y.; Hirata, A.; Hiratsuka, M.; Nakamura, Y. Reduction in static friction by deposition of a homogeneous diamond-like carbon (DLC) coating on orthodontic brackets. *Dent. Mater. J.* **2015**, *34*, 888–895. [CrossRef]
78. Muguruma, T.; Iijima, M.; Nagano-Takebe, F.; Endo, K.; Mizoguchi, I. Frictional Properties and Characterization of a Diamond-Like Carbon Coating Formed on Orthodontic Stainless Steel. *J. Biomater. Tissue Eng.* **2017**, *7*, 119–126. [CrossRef]
79. Huang, T.-H.; Guo, J.-U.; Kao, C.-T. A comparison of the friction associated with diamond-like carbon (DLC) or titanium nitride (TiN) plating metal brackets. *Surf. Coat. Technol.* **2010**, *205*, 1917–1921. [CrossRef]
80. Kang, T.; Huang, S.-Y.; Huang, J.-J.; Li, Q.-H.; Diao, D.-F.; Duan, Y.-Z. The effects of diamond-like carbon films on fretting wear behavior of orthodontic archwire-bracket contacts. *J. Nanosci. Nanotechnol.* **2015**, *15*, 4641–4647. [CrossRef]
81. Muguruma, T.; Iijima, M.; Brantley, W.A.; Mizoguchi, I. Effects of a diamond-like carbon coating on the frictional properties of orthodontic wires. *Angle Orthod.* **2011**, *81*, 141–148. [CrossRef] [PubMed]
82. Zhang, H.; Guo, S.; Wang, D.; Zhou, T.; Wang, L.; Ma, J. Effects of nanostructured, diamondlike, carbon coating and nitrocarburizing on the frictional properties and biocompatibility of orthodontic stainless steel wires. *Angle Orthod.* **2016**, *86*, 782–788. [CrossRef]
83. Huang, S.Y.; Huang, J.J.; Kang, T.; Diao, D.F.; Duan, Y.Z. Coating NiTi archwires with diamond-like carbon films: Reducing fluoride-induced corrosion and improving frictional properties. *J. Mater. Sci. Mater. Med.* **2013**, *24*, 2287–2292. [CrossRef] [PubMed]
84. Tantiwinyupong, N.; Chintavalakorn, R.; Santiwong, P.; Khantachawana, A. Frictional and Mechanical Properties of Surface Modified Nickel-Titanium Orthodontic Archwires. *Key Eng. Mater.* **2019**, *801*, 39–43. [CrossRef]
85. Akaike, S.; Kobayashi, D.; Aono, Y.; Hiratsuka, M.; Hirata, A.; Hayakawa, T.; Nakamura, Y. Relationship between static friction and surface wettability of orthodontic brackets coated with diamond-like carbon (DLC), fluorine-or silicone-doped DLC coatings. *Diam. Relat. Mater.* **2016**, *61*, 109–114. [CrossRef]
86. Muguruma, T.; Iijima, M.; Brantley, W.A.; Nakagaki, S.; Endo, K.; Mizoguchi, I. Frictional and mechanical properties of diamond-like carbon-coated orthodontic brackets. *Eur. J. Orthod.* **2011**, *35*, 216–222. [CrossRef] [PubMed]
87. Muguruma, T.; Iijima, M.; Kawaguchi, M.; Mizoguchi, I. Effects of sp2/sp3 Ratio and Hydrogen Content on In Vitro Bending and Frictional Performance of DLC-Coated Orthodontic Stainless Steels. *Coatings* **2018**, *8*, 199. [CrossRef]
88. Rokaya, D.; Srimaneepong, V.; Qin, J.; Siraleartmukul, K.; Siriwongrungson, V. Graphene oxide/silver nanoparticle coating produced by electrophoretic deposition improved the mechanical and tribological properties of NiTi alloy for biomedical applications. *J. Nanosci. Nanotechnol.* **2019**, *19*, 3804–3810. [CrossRef] [PubMed]
89. Dai, D.; Zhou, D.; He, L.; Wang, C.; Zhang, C. Graphene oxide nanocoating for enhanced corrosion resistance, wear resistance and antibacterial activity of nickel-titanium shape memory alloy. *Surf. Coat. Technol.* **2022**, *431*, 128012. [CrossRef]
90. Pan, Z.; Zhou, Q.; Wang, P.; Diao, D. Robust low friction performance of graphene sheets embedded carbon films coated orthodontic stainless steel archwires. *Friction* **2021**, *10*, 142–158. [CrossRef]
91. Mousavi, S.M.; Shamohammadi, M.; Rastegaar, Z.; Skini, M.; Rakhshan, V. Effect of esthetic coating on surface roughness of orthodontic archwires. *Int. Orthod.* **2017**, *15*, 312–321. [CrossRef]
92. Bhat, K.R.R.; Ahmed, N.; Joseph, R.; Abrar, A.Y. Comparative Evaluation of Frictional Resistance between Different Types of Ceramic Brackets and Stainless Steel Brackets with Teflon-Coated Stainless Steel and Stainless Steel Archwires: An In-Vitro Study. *Cureus* **2022**, *14*, e24161. [CrossRef] [PubMed]
93. Al-Ghroosh, D.H.; Basim, A.; Nahidh, M.; Ghazi, A. Assessment of Static Friction Generated from Different Aesthetic Archwires (In-Vitro Study). *J. Pharm. Sci. Res.* **2018**, *10*, 3310.
94. Joseph, J.N.; Ammayappan, P.; Sankar, H.; Yashwant, V.A.; Alexander, L. Comparison of frictional resistance of different esthetic archwires in different esthetic brackets in dry and wet fields: An in vitro study. *J. Dent. Res. Rev.* **2021**, *8*, 176. [CrossRef]
95. Kameda, T.; Sato, H.; Oka, S.; Miyazaki, A.; Ohkuma, K.; Terada, K. Low temperature polytetrafluoroethylene (PTFE) coating improves the appearance of orthodontic wires without changing their mechanical properties. *Dent. Mater. J.* **2020**, *39*, 721–734. [CrossRef]
96. Kunimatsu, R.; Tsuka, Y.; Nakajima, K.; Sumi, K.; Yoshimi, Y.; Kado, I.; Inada, A.; Kiritoshi, Y.; Tanimoto, K. The Influence of 2-Methacryloyloxyethyl Phosphorylcholine Polymer Materials on Orthodontic Friction and Attachment of Oral Bacteria. *Materials* **2022**, *15*, 5770. [CrossRef] [PubMed]

97. Lin, C.Y.; Kim, J.S.; Lin, E.T.; Lin, E.T. Prolonged water immersion alters resistance to sliding of aesthetic orthodontic coated wires. *Orthod. Craniofac. Res.* **2021**, *24*, 111–120. [CrossRef]
98. Moradpoor, H.; Safaei, M.; Mozaffari, H.R.; Sharifi, R.; Imani, M.M.; Golshah, A.; Bashardoust, N. An overview of recent progress in dental applications of zinc oxide nanoparticles. *RSC Adv.* **2021**, *11*, 21189–21206. [CrossRef] [PubMed]
99. Yin, I.X.; Zhang, J.; Zhao, I.S.; Mei, M.L.; Li, Q.; Chu, C.H. The Antibacterial Mechanism of Silver Nanoparticles and Its Application in Dentistry. *Int. J. Nanomed.* **2020**, *15*, 2555–2562. [CrossRef]
100. Monteiro, D.R.; Gorup, L.F.; Takamiya, A.S.; Ruvollo-Filho, A.C.; de Camargo, E.R.; Barbosa, D.B. The growing importance of materials that prevent microbial adhesion: Antimicrobial effect of medical devices containing silver. *Int. J. Antimicrob. Agents* **2009**, *34*, 103–110. [CrossRef] [PubMed]
101. Rivero, P.J.; Urrutia, A.; Goicoechea, J.; Zamarreño, C.R.; Arregui, F.J.; Matías, I.R. An antibacterial coating based on a polymer/sol-gel hybrid matrix loaded with silver nanoparticles. *Nanoscale Res. Lett.* **2011**, *6*, 305. [CrossRef]
102. Sun, Q.; Zhang, L.; Bai, R.; Zhuang, Z.; Zhang, Y.; Yu, T.; Peng, L.; Xin, T.; Chen, S.; Han, B. Recent Progress in Antimicrobial Strategies for Resin-Based Restoratives. *Polymers* **2021**, *13*, 1590. [CrossRef] [PubMed]
103. Dakal, T.C.; Kumar, A.; Majumdar, R.S.; Yadav, V. Mechanistic Basis of Antimicrobial Actions of Silver Nanoparticles. *Front. Microbiol.* **2016**, *7*, 1831. [CrossRef] [PubMed]
104. Gonçalves, I.S.; Viale, A.B.; Sormani, N.N.; Pizzol, K.E.D.C.; Araujo-Nobre, A.R.d.; Oliveira, P.C.S.d.; Barud, H.G.d.O.; Antonio, S.G.; Barud, H.d.S. Antimicrobial orthodontic wires coated with silver nanoparticles. *Braz. Arch. Biol. Technol.* **2020**, *63*. [CrossRef]
105. Arash, V.; Keikhaee, F.; Rabiee, S.M.; Rajabnia, R.; Khafri, S.; Tavanafar, S. Evaluation of antibacterial effects of silver-coated stainless steel orthodontic brackets. *J. Dent. Tehran Univ. Med. Sci.* **2016**, *13*, 49–54.
106. Jasso-Ruiz, I.; Velazquez-Enriquez, U.; Scougall-Vilchis, R.J.; Morales-Luckie, R.A.; Sawada, T.; Yamaguchi, R. Silver nanoparticles in orthodontics, a new alternative in bacterial inhibition: In vitro study. *Prog. Orthod.* **2020**, *21*, 24. [CrossRef] [PubMed]
107. Jasso-Ruiz, I.; Velazquez-Enriquez, U.; Scougall-Vilchis, R.J.; Lara-Carrillo, E.; Toral-Rizo, V.H.; López-Castañares, R.; Morales-Luckie, R.A. Synthesis and characterization of silver nanoparticles on orthodontic brackets: A new alternative in the prevention of white spots. *Coatings* **2019**, *9*, 480. [CrossRef]
108. Metin-Gürsoy, G.; Taner, L.; Akca, G. Nanosilver coated orthodontic brackets: In vivo antibacterial properties and ion release. *Eur. J. Orthod.* **2017**, *39*, 9–16. [CrossRef] [PubMed]
109. Meyer-Kobbe, V.; Doll, K.; Stiesch, M.; Schwestka-Polly, R.; Demling, A. Comparison of intraoral biofilm reduction on silver-coated and silver ion-implanted stainless steel bracket material: Biofilm reduction on silver ion-implanted bracket material. *J. Orofac. Orthop./Fortschr. Kieferorthopädie* **2018**, *80*, 32–43. [CrossRef]
110. Ryu, H.-S.; Bae, I.-H.; Lee, K.-G.; Hwang, H.-S.; Lee, K.-H.; Koh, J.-T.; Cho, J.-H. Antibacterial effect of silver-platinum coating for orthodontic appliances. *Angle Orthod.* **2012**, *82*, 151–157. [CrossRef]
111. Fatani, E.J.; Almutairi, H.H.; Alharbi, A.O.; Alnakhli, Y.O.; Divakar, D.D.; Muzaheed; Alkheraif, A.A.; Khan, A.A. In vitro assessment of stainless steel orthodontic brackets coated with titanium oxide mixed Ag for anti-adherent and antibacterial properties against *Streptococcus mutans* and *Porphyromonas gingivalis*. *Microb. Pathog.* **2017**, *112*, 190–194. [CrossRef] [PubMed]
112. Zeidan, N.K.; Enany, N.M.; Mohamed, G.G.; Marzouk, E.S. The antibacterial effect of silver, zinc-oxide and combination of silver/zinc oxide nanoparticles coating of orthodontic brackets (an in vitro study). *BMC Oral Health* **2022**, *22*, 1–14. [CrossRef] [PubMed]
113. Oliveira, D.C.; Thomson, J.J.; Alhabeil, J.A.; Toma, J.M.; Plecha, S.C.; Pacheco, R.R.; Cuevas-Suárez, C.E.; Piva, E.; Lund, R.G. In vitro *Streptococcus mutans* adhesion and biofilm formation on different esthetic orthodontic archwires. *Angle Orthod.* **2021**, *91*, 786–793. [CrossRef]
114. Lee, B.-S.; Lin, Y.-C.; Hsu, W.-C.; Hou, C.-H.; Shyue, J.-J.; Hsiao, S.-Y.; Wu, P.-J.; Lee, Y.-T.; Luo, S.-C. Engineering Antifouling and Antibacterial Stainless Steel for Orthodontic Appliances through Layer-by-Layer Deposition of Nanocomposite Coatings. *ACS Appl. Bio Mater.* **2020**, *3*, 486–494. [CrossRef] [PubMed]
115. Chen, H.; Zhang, Y.; Yu, T.; Song, G.; Xu, T.; Xin, T.; Lin, Y.; Han, B. Nano-Based Drug Delivery Systems for Periodontal Tissue Regeneration. *Pharmaceutics* **2022**, *14*, 2250. [CrossRef] [PubMed]
116. Cho, M.; Chung, H.; Choi, W.; Yoon, J. Linear correlation between inactivation of *E. coli* and OH radical concentration in TiO_2 photocatalytic disinfection. *Water Res.* **2004**, *38*, 1069–1077. [CrossRef] [PubMed]
117. Bai, R.; Peng, L.; Sun, Q.; Zhang, Y.; Zhang, L.; Wei, Y.; Han, B. Metallic Antibacterial Surface Treatments of Dental and Orthopedic Materials. *Materials* **2020**, *13*, 4594. [CrossRef]
118. Chhattani, S.; Shetty, P.C.; Laxmikant, S.M.; Ramachandra, C.S. In vitro assessment of photocatalytic titanium oxide surface-modified stainless steel and nickel titanium orthodontic wires for its antiadherent and antibacterial properties against *Streptococcus mutans*. *J. Indian Orthod. Soc.* **2014**, *48*, 82–87. [CrossRef]
119. Chun, M.-J.; Shim, E.; Kho, E.-H.; Park, K.-J.; Jung, J.; Kim, J.-M.; Kim, B.; Lee, K.-H.; Cho, D.-L.; Bai, D.-H.; et al. Surface modification of orthodontic wires with photocatalytic titanium oxide for its antiadherent and antibacterial properties. *Angle Orthod.* **2007**, *77*, 483–488. [CrossRef]
120. Özyildiz, F.; Uzel, A.; Hazar, A.S.; Güden, M.; Ölmez, S.; Aras, I.; Karaboz, I. Photocatalytic antimicrobial effect of TiO_2 anatase thin-film-coated orthodontic arch wires on 3 oral pathogens. *Turk. J. Biol.* **2014**, *38*, 289–295. [CrossRef]
121. Liu, J.; Lou, Y.; Zhang, C.; Yin, S.; Li, H.; Sun, D.; Sun, X. Improved corrosion resistance and antibacterial properties of composite arch-wires by N-doped TiO_2 coating. *RSC Adv.* **2017**, *7*, 43938–43949. [CrossRef]

122. Özyıldız, F.; Güden, M.; Uzel, A.; Karaboz, I.; Akil, O.; Bulut, H. Antimicrobial activity of TiO_2-coated orthodontic ceramic brackets against *Streptococcus mutans* and Candida albicans. *Biotechnol. Bioprocess Eng.* **2010**, *15*, 680–685. [CrossRef]
123. Math, M.; Shah, A.G.; Gangurde, P.; Karandikar, A.G.; Gheware, A.; Jadhav, B.S. In-vitro Comparative Assessment of Antibacterial and Anti-adherent Effect of Two Types of Surface Modificants on Stainless Steel Orthodontic Brackets against *Streptococcus mutans. J. Indian Orthod. Soc.* **2021**, *56*, 282–289. [CrossRef]
124. Shah, A.G.; Shetty, P.C.; Ramachandra, C.S.; Bhat, N.S.; Laxmikanth, S.M. In vitro assessment of photocatalytic titanium oxide surface modified stainless steel orthodontic brackets for antiadherent and antibacterial properties against *Lactobacillus acidophilus. Angle Orthod.* **2011**, *81*, 1028–1035. [CrossRef] [PubMed]
125. Choi, J.-Y.; Chung, C.J.; Oh, K.-T.; Choi, Y.-J.; Kim, K.-H. Photocatalytic antibacterial effect of TiO_2 film of TiAg on *Streptococcus mutans. Angle Orthod.* **2009**, *79*, 528–532. [CrossRef] [PubMed]
126. Baby, R.D.; Subramaniam, S.; Arumugam, I.; Padmanabhan, S. Assessment of antibacterial and cytotoxic effects of orthodontic stainless steel brackets coated with different phases of titanium oxide: An in-vitro study. *Am. J. Orthod. Dentofac. Orthop.* **2017**, *151*, 678–684. [CrossRef]
127. Venkatesan, K.; Kailasam, V.; Padmanabhan, S. Evaluation of titanium dioxide coating on surface roughness of nickel-titanium archwires and its influence on *Streptococcus mutans* adhesion and enamel mineralization: A prospective clinical study. *Am. J. Orthod. Dentofac. Orthop.* **2020**, *158*, 199–208. [CrossRef] [PubMed]
128. Mollabashi, V.; Farmany, A.; Alikhani, M.Y.; Sattari, M.; Soltanian, A.R.; Kahvand, P.; Banisafar, Z. Effects of TiO_2-coated stainless steel orthodontic wires on *Streptococcus mutans* bacteria: A clinical study. *Int. J. Nanomed.* **2020**, *15*, 8759–8766. [CrossRef]
129. Choi, J.Y.; Kim, K.H.; Choy, K.C.; Oh, K.T.; Kim, K.N. Photocatalytic antibacterial effect of TiO_2 film formed on Ti and TiAg exposed to *Lactobacillus acidophilus. J. Biomed. Mater. Res.* **2006**, *80*, 353–359.
130. Domínguez-Espíndola, R.B.; Bruguera-Casamada, C.; Silva-Martínez, S.; Araujo, R.M.; Brillas, E.; Sirés, I. Photoelectrocatalytic inactivation of *Pseudomonas aeruginosa* using an Ag-decorated TiO_2 photoanode. *Sep. Purif. Technol.* **2019**, *208*, 83–91. [CrossRef]
131. Cao, B.; Wang, Y.; Li, N.; Liu, B.; Zhang, Y. Preparation of an orthodontic bracket coated with an nitrogen-doped TiO_2-xNy thin film and examination of its antimicrobial performance. *Dent. Mater. J.* **2013**, *32*, 311–316. [CrossRef]
132. Salehi, P.; Babanouri, N.; Roein-Peikar, M.; Zare, F. Long-term antimicrobial assessment of orthodontic brackets coated with nitrogen-doped titanium dioxide against *Streptococcus mutans. Prog. Orthod.* **2018**, *19*, 35. [CrossRef]
133. Cao, S.; Liu, B.; Fan, L.; Yue, Z.; Liu, B.; Cao, B. Highly antibacterial activity of N-doped TiO_2 thin films coated on stainless steel brackets under visible light irradiation. *Appl. Surf. Sci.* **2014**, *309*, 119–127. [CrossRef]
134. Monica, A.; Padmanabhan, S. The effect of nitrogen-doped titanium dioxide–modified stainless steel brackets on *Streptococcus mutans*: A randomized clinical trial. *Angle Orthod.* **2022**, *92*, 396–401. [CrossRef] [PubMed]
135. Teixeira, L.P.; Gontijo, L.C.; Franco Júnior, A.R.; Pereira, M.F.; Schuenck, R.P.; Malacarne-Zanon, J. Evaluation of antimicrobial potential and surface morphology in thin films of titanium nitride and calcium phosphate on orthodontic brackets. *Am. J. Orthod. Dentofac. Orthop.* **2021**, *160*, 209–214. [CrossRef] [PubMed]
136. Chen, M.; Li, H.; Wang, X.; Qin, G.; Zhang, E. Improvement in antibacterial properties and cytocompatibility of titanium by fluorine and oxygen dual plasma-based surface modification. *Appl. Surf. Sci.* **2019**, *463*, 261–274. [CrossRef]
137. Mobeen, N.; Duraisamy, S.; Ravi, K. Evaluation of the Ion release from nanoparticles coated orthodontic brackets-In vitro Study. *Int. J. Orthod. Rehabil.* **2022**, *13*, 10–21. [CrossRef]
138. Ramazanzadeh, B.; Jahanbin, A.; Yaghoubi, M.; Shahtahmassbi, N.; Ghazvini, K.; Shakeri, M.; Shafaee, H. Comparison of antibacterial effects of ZnO and CuO nanoparticles coated brackets against *Streptococcus mutans. J. Dent.* **2015**, *16*, 200–205.
139. Gholami, M.; Esmaeilzadeh, M.; Kachoei, Z.; Kachoei, M.; Divband, B. Influence of Physical Dimension and Morphological-Dependent Antibacterial Characteristics of ZnO Nanoparticles Coated on Orthodontic NiTi Wires. *BioMed Res. Int.* **2021**, *2021*, 6397698. [CrossRef]
140. Espinosa-Cristóbal, L.F.; López-Ruiz, N.; Cabada-Tarín, D.; Reyes-López, S.Y.; Zaragoza-Contreras, A.; Constandse-Cortéz, D.; Donohué-Cornejo, A.; Tovar-Carrillo, K.; Cuevas-González, J.C.; Kobayashi, T. Antiadherence and Antimicrobial Properties of Silver Nanoparticles against *Streptococcus mutans* on Brackets and Wires Used for Orthodontic Treatments. *J. Nanomater.* **2018**, *2018*, 9248527. [CrossRef]
141. Kim, I.-H.; Park, H.-S.; Kim, Y.K.; Kim, K.-H.; Kwon, T.-Y. Comparative short-term in vitro analysis of *mutans streptococci* adhesion on esthetic, nickel-titanium, and stainless-steel arch wires. *Angle Orthod.* **2014**, *84*, 680–686. [CrossRef]
142. Asiry, M.A.; AlShahrani, I.; Almoammar, S.; Durgesh, B.H.; Al Kheraif, A.A.; Hashem, M.I. Influence of epoxy, polytetrafluoroethylene (PTFE) and rhodium surface coatings on surface roughness, nano-mechanical properties and biofilm adhesion of nickel titanium (Ni-Ti) archwires. *Mater. Res. Express* **2018**, *5*, 026511. [CrossRef]
143. Vellwock, A.E.; Yao, H. Biomimetic and bioinspired surface topographies as a green strategy for combating biofouling: A review. *Bioinspir. Biomim.* **2021**, *16*, 041003. [CrossRef]
144. Vellwock, A.E.; Su, P.; Zhang, Z.; Feng, D.; Yao, H. Reconciling the Conflict between Optical Transparency and Fouling Resistance with a Nanowrinkled Surface Inspired by Zebrafish's Cornea. *ACS Appl. Mater. Interfaces* **2022**, *14*, 7617–7625. [CrossRef]
145. Zhang, M.; Liu, X.; Xie, Y.; Zhang, Q.; Zhang, W.; Jiang, X.; Lin, J. Biological safe gold nanoparticle-modified dental aligner prevents the *Porphyromonas gingivalis* biofilm formation. *ACS Omega* **2020**, *5*, 18685–18692. [CrossRef]
146. Xie, Y.; Zhang, M.; Zhang, W.; Liu, X.; Zheng, W.; Jiang, X. Gold nanoclusters-coated orthodontic devices can inhibit the formation of *Streptococcus mutans* biofilm. *ACS Biomater. Sci. Eng.* **2020**, *6*, 1239–1246. [CrossRef]

147. Raji, S.H.; Shojaei, H.; Ghorani, P.S.; Rafiei, E. Bacterial colonization on coated and uncoated orthodontic wires: A prospective clinical trial. *Dent. Res. J.* **2014**, *11*, 680–683.
148. Al-Lami, A.A.; Al-Sheakli, I.I. Quantitative assessment of *Mutans Streptococci* adhesion to coated and uncoated orthodontic archwires: In vitro study. *J. Baghdad Coll. Dent.* **2014**, *26*, 156–162.
149. Demling, A.; Elter, C.; Heidenblut, T.; Bach, F.-W.; Hahn, A.; Schwestka-Polly, R.; Stiesch, M.; Heuer, W. Reduction of biofilm on orthodontic brackets with the use of a polytetrafluoroethylene coating. *Eur. J. Orthod.* **2010**, *32*, 414–418. [CrossRef] [PubMed]
150. Fuchslocher Hellemann, C.; Grade, S.; Heuer, W.; Dittmer, M.P.; Stiesch, M.; Schwestka-Polly, R.; Demling, A.P. Three-dimensional analysis of initial biofilm formation on polytetrafluoroethylene in the oral cavity. *J. Orofac. Orthop./Fortschr. Kieferorthopädie* **2013**, *74*, 458–467. [CrossRef]
151. Oliveira, A.S.; Kaizer, M.R.; Azevedo, M.S.; Ogliari, F.A.; Cenci, M.S.; Moraes, R.R. (Super) hydrophobic coating of orthodontic dental devices and reduction of early oral biofilm retention. *Biomed. Mater.* **2015**, *10*, 065004. [CrossRef] [PubMed]
152. Zhang, X.; Bai, R.; Sun, Q.; Zhuang, Z.; Zhang, Y.; Chen, S.; Han, B. Bio-inspired special wettability in oral antibacterial applications. *Front. Bioeng. Biotechnol.* **2022**, *10*, 1001616. [CrossRef] [PubMed]
153. Peng, L.; Chang, L.; Liu, X.; Lin, J.; Liu, H.; Han, B.; Wang, S. Antibacterial property of a polyethylene glycol-grafted dental material. *ACS Appl. Mater. Interfaces* **2017**, *9*, 17688–17692. [CrossRef]
154. He, L.; Cui, Y.; Zhang, C. The corrosion resistance, cytotoxicity, and antibacterial properties of lysozyme coatings on orthodontic composite arch wires. *RSC Adv.* **2020**, *10*, 18131–18137. [CrossRef] [PubMed]
155. Liu, X.; Peng, L.; Meng, J.; Zhu, Z.; Han, B.; Wang, S. Protein-mediated anti-adhesion surface against oral bacteria. *Nanoscale* **2018**, *10*, 2711–2714. [CrossRef]
156. Peng, L.; Chang, L.; Si, M.; Lin, J.; Wei, Y.; Wang, S.; Liu, H.; Han, B.; Jiang, L. Hydrogel-Coated Dental Device with Adhesion-Inhibiting and Colony-Suppressing Properties. *ACS Appl. Mater. Interfaces* **2020**, *12*, 9718–9725. [CrossRef]
157. Zhang, S.; Liang, X.; Gadd, G.M.; Zhao, Q. A sol–gel based silver nanoparticle/polytetrafluorethylene (AgNP/PTFE) coating with enhanced antibacterial and anti-corrosive properties. *Appl. Surf. Sci.* **2021**, *535*, 147675. [CrossRef]
158. Abbass, M.K.; Ajeel, S.A.; Wadullah, H.M. Biocompatibility, Bioactivity and Corrosion Resistance of Stainless Steel 316L Nanocoated with TiO_2 and Al_2O_3 by Atomic Layer Deposition Method. *J. Phys. Conf. Ser.* **2018**, *1032*, 012017. [CrossRef]
159. Kaliaraj, G.S.; Vishwakarma, V.; Kirubaharan, K.; Dharini, T.; Ramachandran, D.; Muthaiah, B. Corrosion and biocompatibility behaviour of zirconia coating by EBPVD for biomedical applications. *Surf. Coat. Technol.* **2018**, *334*, 336–343. [CrossRef]
160. Li, X.; Wang, J.; Han, E.-h.; Ke, W. Influence of fluoride and chloride on corrosion behavior of NiTi orthodontic wires. *Acta Biomater.* **2007**, *3*, 807–815. [CrossRef]
161. da Silva, D.L.; Mattos, C.T.; Simão, R.A.; de Oliveira Ruellas, A.C. Coating stability and surface characteristics of esthetic orthodontic coated archwires. *Angle Orthod.* **2013**, *83*, 994–1001. [CrossRef] [PubMed]
162. Kao, C.-T.; Ding, S.-J.; Chen, Y.-C.; Huang, T.-H. The anticorrosion ability of titanium nitride (TiN) plating on an orthodontic metal bracket and its biocompatibility. *J. Biomed. Mater. Res.* **2002**, *63*, 786–792. [CrossRef] [PubMed]
163. Zuo, J.; Xie, Y.; Zhang, J.; Wei, Q.; Zhou, B.; Luo, J.; Wang, Y.; Yu, Z.M.; Tang, Z.G. TiN coated stainless steel bracket: Tribological, corrosion resistance, biocompatibility and mechanical performance. *Surf. Coat. Technol.* **2015**, *277*, 227–233. [CrossRef]
164. Yuan, X.-S.; Wang, Y.; Cao, L.; Cao, B.-C.; Liang, J. Effects of N-Doped TiO_2 Thin Films on Corrosion Resistance of Stainless Steel Orthodontic Brackets in Artificial Saliva. *Corrosion* **2015**, *71*, 784–794. [CrossRef] [PubMed]
165. Ito, A.; Kitaura, H.; Noguchi, T.; Ohori, F.; Mizoguchi, I. Analysis of Coating Loss from Coated Stainless Steel Orthodontic Wire. *Appl. Sci.* **2022**, *12*, 9497. [CrossRef]
166. Lou, J.; Gao, Z.; Zhang, J.; He, H.; Wang, X. Comparative Investigation on Corrosion Resistance of Stainless Steels Coated with Titanium Nitride, Nitrogen Titanium Carbide and Titanium-Diamond-like Carbon Films. *Coatings* **2021**, *11*, 1543. [CrossRef]
167. Banaszek, K.; Klimek, L.; Dąbrowski, J.R.; Jastrzębski, W. Fretting Wear in Orthodontic and Prosthetic Alloys with Ti(C,N) Coatings. *Processes* **2019**, *7*, 874. [CrossRef]
168. Banaszek, K.; Maślanka, M.; Semenov, M.; Klimek, L. Corrosive Studies of a Prosthetic Ni-Cr Alloy Coated with Ti(C,N) Type Layers. *Materials* **2022**, *15*, 2471. [CrossRef] [PubMed]
169. Liu, C.; Chu, P.K.; Lin, G.; Yang, D. Effects of Ti/TiN multilayer on corrosion resistance of nickel–titanium orthodontic brackets in artificial saliva. *Corros. Sci.* **2007**, *49*, 3783–3796. [CrossRef]
170. Katić, V.; Ćurković, H.O.; Semenski, D.; Baršić, G.; Marušić, K.; Špalj, S. Influence of surface layer on mechanical and corrosion properties of nickel-titanium orthodontic wires. *Angle Orthod.* **2014**, *84*, 1041–1048. [CrossRef]
171. Gil, F.J.; Solano, E.; Mendoza, A.; Peña, J. Inhibition of Ni Release from NiTi and NiTiCu Orthodontic Archwires by Nitrogen Diffusion Treatment. *J. Appl. Biomater. Biomech.* **2004**, *2*, 151–155.
172. Liu, J.-K.; Liu, I.-H.; Liu, C.; Chang, C.-J.; Kung, K.-C.; Liu, Y.-T.; Lee, T.-M.; Jou, J.-L. Effect of titanium nitride/titanium coatings on the stress corrosion of nickel–titanium orthodontic archwires in artificial saliva. *Appl. Surf. Sci.* **2014**, *317*, 974–981. [CrossRef]
173. Kielan-Grabowska, Z.; Bącela, J.; Zięty, A.; Seremak, W.; Gawlik-Maj, M.; Kawala, B.; Borak, B.; Detyna, J.; Sarul, M. Improvement of Properties of Stainless Steel Orthodontic Archwire Using TiO_2:Ag Coating. *Symmetry* **2021**, *13*, 1734. [CrossRef]
174. Ohtsu, N.; Yamasaki, K.; Taniho, H.; Konaka, Y.; Tate, K. Pulsed anodization of NiTi alloy to form a biofunctional Ni-free oxide layer for corrosion protection and hydrophilicity. *Surf. Coat. Technol.* **2021**, *412*, 127039. [CrossRef]
175. Anuradha, P.; Varma, N.K.S.; Balakrishnan, A. Reliability performance of titanium sputter coated Ni–Ti arch wires: Mechanical performance and nickel release evaluation. *Bio-Med. Mater. Eng.* **2015**, *26*, 67–77. [CrossRef]

176. Kosari Mehr, A.; Kosari Mehr, A.; Babaei, R. Performance of reactively co-sputtered titanium chromium nitride films in artificial saliva: Corrosion protection and reduction in the release of potentially toxic elements. *Surf. Coat. Technol.* **2021**, *427*, 127855. [CrossRef]
177. Hu, J.; Ren, Y.; Huang, Q.; He, H.; Liang, L.; Liu, J.; Li, R.; Wu, H. Microstructure and Corrosion Behavior of Ti-Nb Coatings on NiTi Substrate Fabricated by Laser Cladding. *Coatings* **2021**, *11*, 597. [CrossRef]
178. Ohgoe, Y.; Kobayashi, S.; Ozeki, K.; Aoki, H.; Nakamori, H.; Hirakuri, K.K.; Miyashita, O. Reduction effect of nickel ion release on a diamond-like carbon film coated onto an orthodontic archwire. *Thin Solid Film.* **2006**, *497*, 218–222. [CrossRef]
179. Kobayashi, S.; Ohgoe, Y.; Ozeki, K.; Hirakuri, K.; Aoki, H. Dissolution effect and cytotoxicity of diamond-like carbon coatings on orthodontic archwires. *J. Mater. Sci. Mater. Med.* **2007**, *18*, 2263–2268. [CrossRef] [PubMed]
180. Peng, J.; Liao, J.; Peng, Y.; Xiao, Y.; Huang, J.; Li, L. Enhancement of sp3 C Fraction in Diamond-like Carbon Coatings by Cryogenic Treatment. *Coatings* **2022**, *12*, 42. [CrossRef]
181. Srimaneepong, V.; Rokaya, D.; Thunyakitpisal, P.; Qin, J.; Saengkiettiyut, K. Corrosion Resistance of Graphene oxide/Silver Coatings on Ni–Ti alloy and Expression of IL-6 and IL-8 in Human Oral Fibroblasts. *Sci. Rep.* **2020**, *10*, 3247. [CrossRef]
182. Fróis, A.; Evaristo, M.; Santos, A.C.; Louro, C.S. Salivary pH Effect on Orthodontic Appliances: In Vitro Study of the SS/DLC System. *Coatings* **2021**, *11*, 1302. [CrossRef]
183. Redlich, M.; Gorodnev, A.; Feldman, Y.; Kaplan-Ashiri, I.; Tenne, R.; Fleischer, N.; Genut, M.; Feuerstein, N. Friction reduction and wear resistance of electro-co-deposited inorganic fullerene-like WS_2 coating for improved stainless steel orthodontic wires. *J. Mater. Res.* **2008**, *23*, 2909–2915. [CrossRef]
184. de Amorim, M.C.; da Rocha Gomes, S.; da Silva, B.P.; Aoki, I.V.; Basting, R.T. Surface Micromorphology, Ion Release and Resistance to Corrosion of Orthodontic Wires Aesthetic Coating Subject to Degradation. *J. Bio- Tribo-Corros.* **2021**, *8*, 22. [CrossRef]
185. Escobar, L.M.; Rivera, J.R.; Arbelaez, E.; Torres, L.F.; Villafañe, A.; Díaz-Báez, D.; Mora, I.; Lafaurie, G.I.; Tanaka, M. Comparison of Cell Viability and Chemical Composition of Six Latest Generation Orthodontic Wires. *Int. J. Biomater.* **2021**, *2021*, 8885290. [CrossRef] [PubMed]
186. Khaneh Masjedi, M.; Niknam, O.; Haghighat Jahromi, N.; Javidi, P.; Rakhshan, V. Effects of Fixed Orthodontic Treatment Using Conventional, Copper-Included, and Epoxy-Coated Nickel-Titanium Archwires on Salivary Nickel Levels: A Double-Blind Randomized Clinical Trial. *Biol. Trace Elem. Res.* **2016**, *174*, 27–31. [CrossRef] [PubMed]
187. Trino, L.D.; Dias, L.F.G.; Albano, L.G.S.; Bronze-Uhle, E.S.; Rangel, E.C.; Graeff, C.F.O.; Lisboa-Filho, P.N. Zinc oxide surface functionalization and related effects on corrosion resistance of titanium implants. *Ceram. Int.* **2018**, *44*, 4000–4008. [CrossRef]
188. Aksoy, M.E.; Aksakal, B.; Aslan, N.; Dikici, B. Boron-Doped Hydroxyapatite Coatings on NiTi Alloys Using the Electrophoretic Deposition Method: Enhanced Corrosion and Adhesion Performances. *J. Mater. Eng. Perform.* **2021**, *30*, 7365–7375. [CrossRef]
189. Mehrvarz, A.; Khalil-Allafi, J.; Etminanfar, M.; Mahdavi, S. The study of morphological evolution, biocorrosion resistance, and bioactivity of pulse electrochemically deposited Hydroxyapatite/ZnO composite on NiTi superelastic alloy. *Surf. Coat. Technol.* **2021**, *423*, 127628. [CrossRef]
190. Catauro, M.; Barrino, F.; Bononi, M.; Colombini, E.; Giovanardi, R.; Veronesi, P.; Tranquillo, E. Coating of Titanium Substrates with ZrO_2 and ZrO_2-SiO_2 Composites by Sol-Gel Synthesis for Biomedical Applications: Structural Characterization, Mechanical and Corrosive Behavior. *Coatings* **2019**, *9*, 200. [CrossRef]
191. Krishnan, M.; Seema, S.; Kumar, A.V.; Varthini, N.P.; Sukumaran, K.; Pawar, V.R.; Arora, V. Corrosion resistance of surface modified nickel titanium archwires. *Angle Orthod.* **2014**, *84*, 358–367. [CrossRef]
192. Mareci, D.; Earar, K.; Zetu, I.; Bolat, G.; Crimu, C.; Istrate, B.; Munteanu, C.; Matei, M.N. Comparative electrochemical behaviour, of uncoated and coated Ni Ti, for dental orthodontic wires. *Mater. Plast.* **2015**, *52*, 150–153.
193. Lin, C.-W.; Chung, C.-J.; Chou, C.-M.; He, J.-L. In vitro wear tests of the dual-layer grid blasting-plasma polymerized superhydrophobic coatings on stainless steel orthodontic substrates. *Thin Solid Film.* **2019**, *687*, 137464. [CrossRef]
194. Ahmed, R.A.; Farghali, R.A.; Alshahrani, W.A. Influence of Fluoride and/or Bovine Albumin on Electrochemical Properties of Bare and ionic liquid-Coated Ni47Ti49Co4 Orthodontic archwires in artificial saliva solution. *Int. J. Electrochem. Sci.* **2021**, *16*, 2. [CrossRef]
195. Yu, D.; Miao, K.; Li, Y.; Bao, X.; Hu, M.; Zhang, K. Sputter-deposited TaCuN films: Structure, tribological and biomedical properties. *Appl. Surf. Sci.* **2021**, *567*, 150796. [CrossRef]
196. Il-Jeong, P.; Kim, M.-S.; Suk-Kyu, L.E.E. Highly Corrosion-Resistant Plated Steel Sheet Having Excellent Plating Adhesion and Resistance to Liquid Metal Embrittlement. U.S. Patent Application US11530470B2, 27 September 2018.
197. Cui, H.-Z.; Zhang, G.-S.; Tian, S.-C.; Wei, N.; Tian, J.; Wang, X.-Z. A Kind of Layered Bionic Wear-and Corrosion-Resistant Anti-Friction Coating and Preparation Method and Application. CN Patent Application CN109208044B, 9 June 2020.
198. Singh, V.P.; Pratap, K.; Sinha, J.; Desiraju, K.; Bahal, D.; Kukreti, R. Critical evaluation of challenges and future use of animals in experimentation for biomedical research. *Int. J. Immunopathol. Pharmacol.* **2016**, *29*, 551–561. [CrossRef]

199. Zhao, Y.; Zhang, T.; Chen, L.; Yu, T.; Sun, J.; Guan, C. Microstructure and mechanical properties of Ti–C–TiN-reinforced Ni204-based laser-cladding composite coating. *Ceram. Int.* **2021**, *47*, 5918–5928. [CrossRef]
200. Bahrami, R.; Pourhajibagher, M.; Badiei, A.; Masaeli, R.; Tanbakuchi, B. Evaluation of the cell viability and antimicrobial effects of orthodontic bands coated with silver or zinc oxide nanoparticles: An in vitro study. *Korean J. Orthod.* **2023**, *53*, 16–25. [CrossRef] [PubMed]

Disclaimer/Publisher's Note: The statements, opinions and data contained in all publications are solely those of the individual author(s) and contributor(s) and not of MDPI and/or the editor(s). MDPI and/or the editor(s) disclaim responsibility for any injury to people or property resulting from any ideas, methods, instructions or products referred to in the content.

International Journal of
Molecular Sciences

Article

Extensive Investigation on the Effect of Niobium Insertion on the Physical and Biological Properties of 45S5 Bioactive Glass for Dental Implant

Imen Hammami [1], Sílvia Rodrigues Gavinho [1], Ana Sofia Pádua [2], Maria do Carmo Lança [3], João Paulo Borges [3], Jorge Carvalho Silva [2], Isabel Sá-Nogueira [4,5], Suresh Kumar Jakka [1,*] and Manuel Pedro Fernandes Graça [1,*]

1 I3N and Physics Department, University of Aveiro, 3810-193 Aveiro, Portugal
2 Physics Department and CENIMAT-I3N, NOVA School of Science and Technology, 2829-516 Caparica, Portugal
3 Materials Science Department and CENIMAT-I3N, NOVA School of Science and Technology, 2829-516 Caparica, Portugal
4 UCIBIO, Applied Molecular Biosciences Unit, Department of Life Sciences, NOVA School of Science and Technology, NOVA University Lisbon, 2819-516 Caparica, Portugal
5 Associate Laboratory i4HB—Institute for Health and Bioeconomy, NOVA School of Science and Technology, NOVA University Lisbon, 2819-516 Caparica, Portugal
* Correspondence: suresh@ua.pt (S.K.J.); mpfg@ua.pt (M.P.F.G.); Tel.: +351-964236731 (M.P.F.G.)

Abstract: Dental implants have emerged as one of the most consistent and predictable treatments in the oral surgery field. However, the placement of the implant is sometimes associated with bacterial infection leading to its loss. In this work, we intend to solve this problem through the development of a biomaterial for implant coatings based on 45S5 Bioglass® modified with different amounts of niobium pentoxide (Nb_2O_5). The structural feature of the glasses, assessed by XRD and FTIR, did not change in spite of Nb_2O_5 incorporation. The Raman spectra reveal the Nb_2O_5 incorporation related to the appearance of NbO_4 and NbO_6 structural units. Since the electrical characteristics of these biomaterials influence their osseointegration ability, AC and DC electrical conductivity were studied by impedance spectroscopy, in the frequency range of 10^2–10^6 Hz and temperature range of 200–400 K. The cytotoxicity of glasses was evaluated using the osteosarcoma Saos-2 cells line. The in vitro bioactivity studies and the antibacterial tests against Gram-positive and Gram-negative bacteria revealed that the samples loaded with 2 mol% Nb_2O_5 had the highest bioactivity and greatest antibacterial effect. Overall, the results showed that the modified 45S5 bioactive glasses can be used as an antibacterial coating material for implants, with high bioactivity, being also non-cytotoxic to mammalian cells.

Keywords: Bioglass®; biomaterial; niobium oxide; osseointegration; antibacterial properties; cytotoxicity; electrical properties; bioactivity; bone regeneration

check for updates

Citation: Hammami, I.; Gavinho, S.R.; Pádua, A.S.; Lança, M.d.C.; Borges, J.P.; Silva, J.C.; Sá-Nogueira, I.; Jakka, S.K.; Graça, M.P.F. Extensive Investigation on the Effect of Niobium Insertion on the Physical and Biological Properties of 45S5 Bioactive Glass for Dental Implant. *Int. J. Mol. Sci.* **2023**, *24*, 5244. https://doi.org/10.3390/ijms24065244

Academic Editor: Mary Anne Melo

Received: 31 December 2022
Revised: 1 March 2023
Accepted: 8 March 2023
Published: 9 March 2023

Copyright: © 2023 by the authors. Licensee MDPI, Basel, Switzerland. This article is an open access article distributed under the terms and conditions of the Creative Commons Attribution (CC BY) license (https://creativecommons.org/licenses/by/4.0/).

1. Introduction

The loss of one or more teeth has become a common problem resulting from trauma, bone or dental pathology, cancer, or simply ageing. According to statistics reported by the American Association of Oral and Maxillofacial Surgeons [1], 69% of adults aged between 35 and 44 have lost at least one permanent tooth due to an accident, gum disease, a failed root canal, or tooth decay. In addition, by the age of 74, approximately 26% of those persons have lost all their permanent teeth. Therefore, the demand for dental implants has increased rapidly. The statistics in 2010 revealed more than 300,000 dental implants placed per year [2]. The clinical success of dental implants is related to their ability to ensure rapid osteointegration and prevent the development of peri-implantitis [3]. These conditions are essential to ensure the long-term success of the implant. Peri-implantitis disease is an inflammatory condition that affects the tissues around dental implants. Just like a natural tooth, bacteria can grow at the implant neck and over time irritate gums, which remain in a chronic inflamed state and can lead to bone loss around the implant(s). Initially, the implant(s) remain clinically stable, without any sign of mobility being evidenced. In the absence of effective treatment, bone lysis continues, ultimately leading to implant loss. Therefore, peri-implantitis is considered a secondary implant failure.

To mitigate bacterial contamination, various measures have been used such as careful disinfection and rigorous aseptic surgical protocols [4]. Infection, which can also occur after surgery, is characterized by bacterial colonization and biofilm formation on implant surfaces. The biofilm is considered the primary etiologic reason for the inflammation of peri-implant tissues [5]. Peri-implantitis-associated biofilms are caused by a plethora of microbial species including anaerobes and facultative aerobes, and *Staphylococcus aureus* plays a predominant role in the development of this pathology [6]. To ensure the long-term success of the dental implant, the selection of the biomaterial to be used is a critical factor. Research on the development of new biomaterials or the manipulation of the structure and composition of existing biomaterials has been carried out to improve the properties of biomedical devices [7–9].

Typically, biomaterials for medical prostheses are made of metallic materials, mainly stainless steel and titanium alloys [10–12]. However, metal prostheses show dramatic failure due to their higher mechanical properties than those of bone tissue, resulting in necrosis and stress shielding of the tissue in contact. It is known that the long-term use of these materials can lead to an excessive release of metal ions which can promote local inflammation, pain, and even clinical failure [13–15]. Besides this issue, the formation of the biofilm problem can also lead to implant failure. Thus, an optimal implant should be able to prevent bacterial adhesion and enhance osteointegration [16,17]. To address these challenges, new progress in biomaterial science has led to the development of bioactive materials for implant coatings [4,18,19].

It has been reported that modifying the surface topography of an implant, by depositing a bioactive coating, improves the mechanical interlocking of the bone with the implant and increases the processes of proliferation and adhesion of osteoblasts [20–22]. Among the bioactive materials, silicate-based bioglasses containing calcium and phosphorus are mainly used in dental and orthopaedic surgery [23,24]. The first bioactive glasses were discovered in 1969 by L.L. Hench, called 45S5 Bioglass®, with the weight composition of 45% SiO_2, 24.5% Na_2O, 24.5% CaO, and 6% P_2O_5 [25]. This bioactive glass promotes the nucleation of calcium phosphate hydroxyapatite (HA) and carbonate (HAC) during ion exchanges that occur through contact/reaction with physiological fluids [26–28]. The formation of HAC on the surface of bioactive glasses promotes osteoinduction, osteoconduction, osseointegration, and angiogenesis [29–31]. A further benefit of using bioglass is the ability to alter its composition by incorporating extra ions that can improve its functionality without being toxic [32,33].

Several reports have shown that metallic ions (including magnesium, zinc, copper, silver, etc.) could be employed to generate antibacterial activity [34–37]. However, the information concerning their long-term effects on human health is limited. Loading these ions into bioactive systems is an effective strategy to control their release over a long period [38,39]. Moreover, bioactive materials can optimize the response of the biological system by interacting with the adjacent tissues, inducing reactions that promote their development and regeneration [40]. The presence of niobium (Nb) species in biomaterials has been reported to improve their bioactive and mechanical properties [41,42]. Some investigations reported that niobium ions promote the differentiation and mineralization of osteogenic cells [41,43]. The insertion of niobium in metallic alloys and ceramic matrices has shown superior corrosion resistance and low cytotoxicity [44,45]. Although niobium-containing bioactive glasses have recently attracted great interest, the effect of their incorporation on the physical and biological response of the 45S5 bioactive glasses is, according to our knowledge, very low.

For this purpose, 45S5 bioactive glasses containing Nb_2O_5 were prepared by a melt-quenching technique. X-ray diffraction, IR, and Raman spectroscopies were used to evaluate the influence of Nb_2O_5 on the structure of 45S5 Bioglass®. The electrical characteristics of the glasses were studied due to their potential to be electrically charged/polarized, therefore optimizing the osseointegration. Cytotoxicity tests using human osteosarcoma Saos-2 cells were used to assess the biocompatibility of the 45S5 bioactive glasses modified by the Nb insertion. The in vitro investigation of bioactivity was conducted by immersing samples (pellets made of powdered bioactive glass) in simulated body fluid (SBF), which is a solution with the same ionic strength and mineral content as the blood plasma. The changes in bioglasses, in terms of dissolution and precipitation of the protective layer containing calcium phosphates, were evaluated using scanning electron microscopy (SEM) with energy-dispersive X-ray analysis (EDX). The antibacterial activity of the bioglasses was evaluated from their ability to inhibit the growth of the *S. aureus* bacteria, which is commonly associated with implant infections, as well as other bacteria involved in the formation of the pathogenic biofilm.

2. Results

2.1. Structural Characterization

The XRD patterns, illustrated in Figure 1, consist of typical broad bands, without traces of crystalline phases, which confirm the amorphous character of the glasses. These results match well with the ones reported in the literature for 45S5 glasses [3,46]. The insertion of niobium, up to 8% mole of Nb_2O_5, did not promote the formation of any crystalline phase but enlarged the band at smaller angles. This phenomenon was already observed in phosphate glasses with niobium [47].

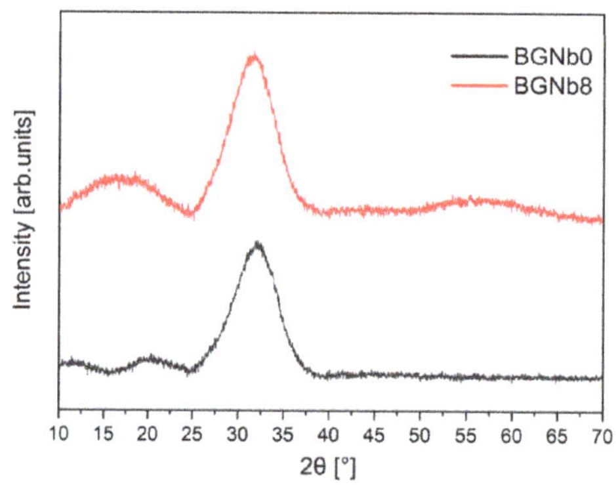

Figure 1. XRD patterns of 45S5 bioglass samples modified by the insertion of 0 and 8 mol% of Nb_2O_5.

Figure 2 depicts the features revealed by the FTIR spectra of the glasses. The Si-O-Si stretching modes are ascribed to the bands detected at around 1010 cm^{-1} and 721 cm^{-1} [46,48–54]. The presence of the non-bridging oxygen ions is shown by the appearance of a band at 912 cm^{-1} that is attributed to the Si-O_{NBO} stretching mode [46,51,53–55]. The presence of a shoulder at 596 cm^{-1} is attributed to the bending vibration of the P-O molecule [46,50,53–56]. The band that appeared at around 497 cm^{-1} is associated with the Si-O-Si bending mode [46,48–54].

Figure 3 shows the Raman spectra of the bioactive glasses prepared. For a more detailed analysis, the Raman spectra of BGNb1, BGNb2, BGNb4, and BGNb8 were deconvolved by a Gaussian fitting (Figure 4).

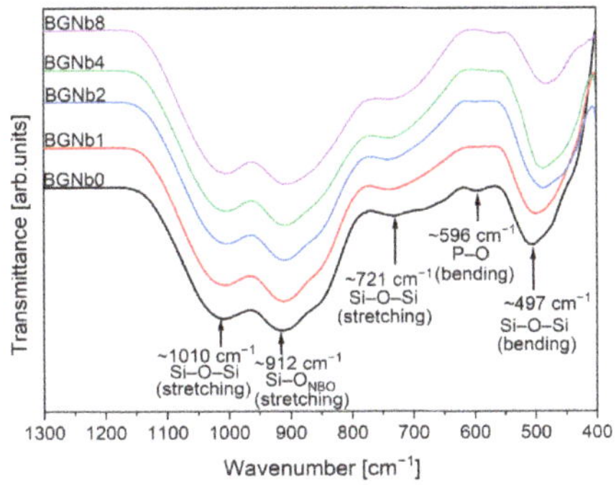

Figure 2. FTIR spectra of bioactive glasses modified by Nb_2O_5 insertion.

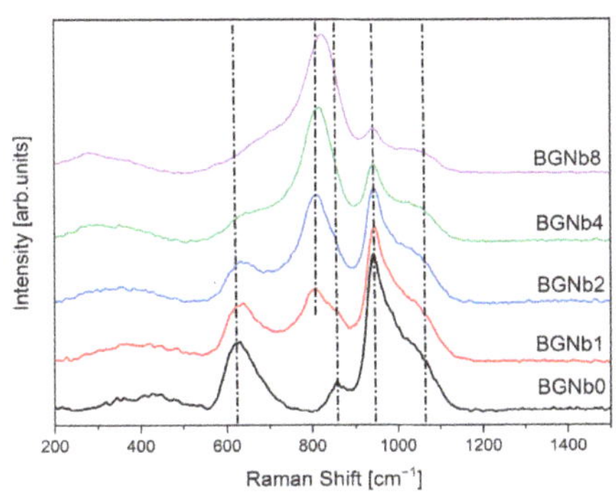

Figure 3. Raman spectra of bioactive glasses modified by Nb_2O_5 insertion.

The whole range of Raman spectra can be separated into two regions: low wavenumber (500–900 cm^{-1}) and high wavenumber (>900 cm^{-1}). In the low-frequency region, a broad band at about 630 cm^{-1} is observed for BGNb0 which can be associated with the rocking motion of bridging oxygen in structural units that contain non-bridging oxygen ions (NBO) [57–59]. With the insertion of Nb_2O_5 in the glass network, it is noted that the band is shifted towards a higher frequency. By fitting and deconvoluting this peak (Figure 4), two bands can be differentiated, one located at 630–680 cm^{-1} and the other at 710–750 cm^{-1}, and attributed to the vibrational mode of the Si-O-Si bond and the vibration of Nb-O, respectively, in NbO_6 octahedra, with a low degree of distortion [60–62]. Moreover, an additional broad peak at 780–890 cm^{-1} is observed for the modified bioglass. The deconvoluted Raman spectra (Figure 4), show that this peak could be distinguished as three different bands located at 782–804 cm^{-1}, 806–829 cm^{-1}, and 857–886 cm^{-1}, which are assigned to vibrations of NbO_6 octahedrons with different degrees of distortion, to the vibration of NbO_4 units, and to the symmetric stretching of Q_0 units in the silicate network (SiO_4^{4-}), respectively [63–65]. The vibrational bands of NbO_6 units increased with increasing Nb_2O_5 concentration up to 2 mol%, whereas the NbO_4 vibrational band decreased. A shoulder is observed for the BGNb0 sample at around 857 cm^{-1}, corresponding to symmetric stretching of Q_0 Si units [54,65]. At the high-frequency region, a vibration band can be observed at around 945 cm^{-1} and a shoulder at 1050 cm^{-1}. From the result obtained by a deconvolution process (Figure 4), five vibrational modes located at 900–923 cm^{-1}, 938–944 cm^{-1}, 967–975 cm^{-1}, 997–1023 cm^{-1}, and 1051–1078 cm^{-1} can be distinguished, which correspond to symmetric stretching of Q_1 Si, Q_2 Si, Q_0 P, and Q_1 P units, and asymmetric stretching of bridging oxygen in all Q species, respectively [54,57,65].

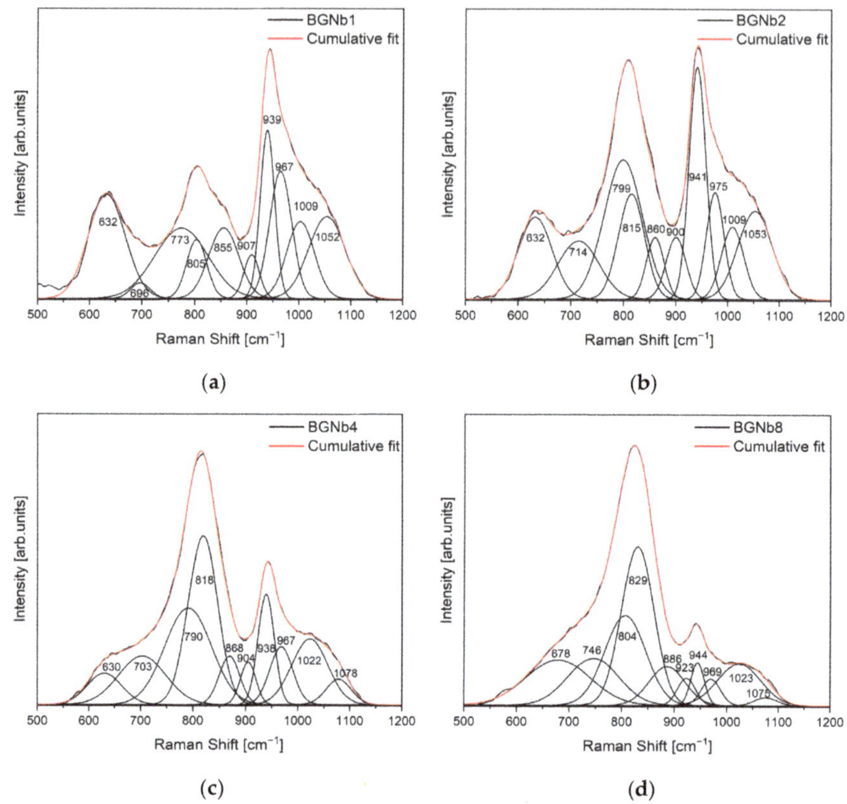

Figure 4. Deconvoluted Raman spectra of (**a**) BGNb1; (**b**) BGNb2; (**c**) BGNb4; and (**d**) BGNb8.

The sum of the area of Raman vibration bands associated with non-bridging oxygen ions is plotted in Figure 5 as a function of Nb_2O_5 concentration. An increase in the number of NBOs for the bioglass with Nb_2O_5 content up to 2 mol% can be observed, indicating the existence of an optimal amount of Nb_2O_5 that can be added to the base bioglass.

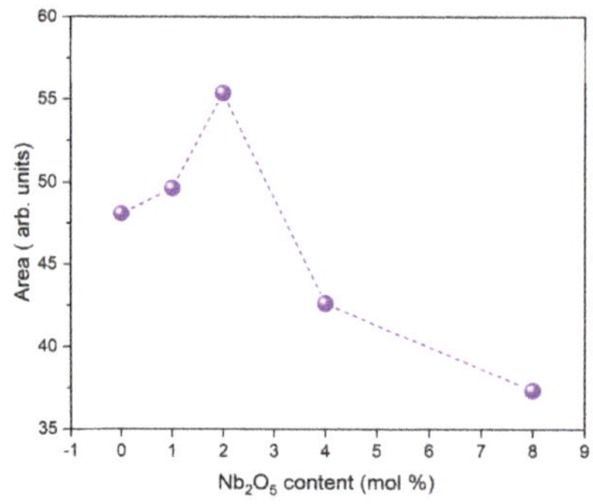

Figure 5. Sum of the areas of the bands associated with NBOs vibrations.

2.2. Thermal Analysis

The differential thermal analysis (DTA) spectra of BGNb2 and BGNb8 are shown in Figure 6. The thermograms of both samples demonstrate the presence of a glass transition temperature, T_g, followed by an exothermic peak, T_c, ascribed to structural alteration associated with the formation of crystalline phases, and an endothermic peak, T_m, attributed to the melting point of bioglass. The T_g values were determined from the thermograms using the tangent method on the measured DTA curve, above and below the glass transition region. The value presented in Table 1 is the mean value of the abscissas of the tangents' interceptions. The uncertainty in these measurements is 2%. In a prior study, a comparable thermal response was observed for the 45S5 bioglass [3]. The critical temperatures of the bioglasses modified by Nb_2O_5 are reported in Table 1.

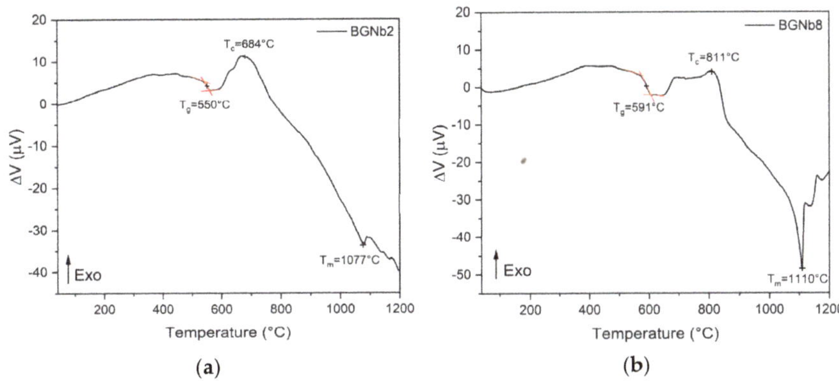

Figure 6. DTA spectra of (**a**) BGNb2; (**b**) BGNb8.

Table 1. The characteristic temperatures for BGNb0, BGNb2, and BGNb8.

Sample	T_g (°C)	T_c (°C)	T_m (°C)
BGNb0 [3]	552	728	1175
BGNb2	550	684	1077
BGNb8	591	811	1110

2.3. Electrical Characterization

Figure 7 depicts the DC conductivity versus 1000/T on a logarithmic scale. It is visible that the conductivity rises with the increase in the temperature, which should be related to the increased mobility of the charge carriers. For temperatures above 300 K, approximately, this variation becomes linear, indicating that the Arrhenius formalism may be used to evaluate the activation energy associated with this thermally activated process. The calculated activation energy at the high-temperature region (inset in Figure 7) for all samples is registered in Table 2. The activation energy (E_a (DC)) decreases with the increase in Nb_2O_5 concentration, reaching a minimum value at the value of 2 mol%. After, a reversing trend is observed. This behavior is similar to the one observed for the NBO ions amount (Figure 5). On the contrary, an opposite trend was observed for the DC conductivity (σ_{DC}), which shows a maximum value for the sample with 2 mol% of Nb_2O_5.

The dielectric properties of the samples were analyzed using the modulus formalism, ($M^* = 1/\varepsilon^*$), to minimize the effect of electrode polarization and conductivity. Figure 8 shows the presence of one dielectric relaxation which is shifted to higher frequencies with increasing the temperature. It should be noted that other formalisms, such as permittivity or impedance, did not exhibit this dielectric relaxation behavior. Therefore, the observed relaxation behavior should be associated with an intrinsic characteristic related to dipole formation between the network modifier ions and the non-bridging oxygen ions.

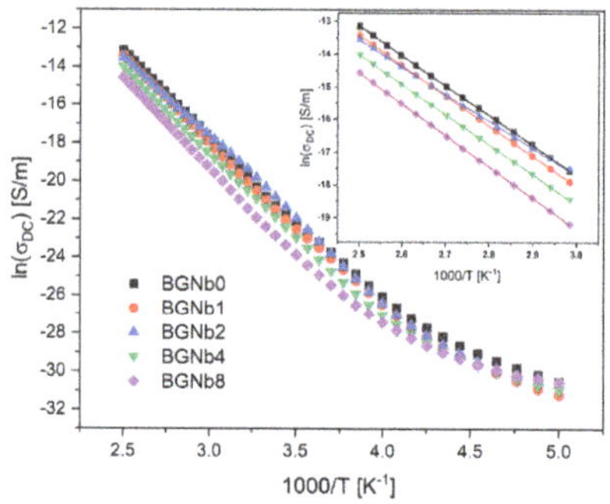

Figure 7. DC conductivity versus 1000/T (inset: a magnification of the high-temperature measurement zone is shown in the inset; the lines reflect the Arrhenius fit).

Table 2. The DC conductivity (σ_{DC}), DC activation energy E_a (DC), AC conductivity (σ_{AC}), and AC activation energy E_a (AC), for all bioglass samples.

Sample	σ_{DC} ($\times 10^{-9}$) [S/m] (At 300 K)	E_a (DC) [kJ/mol]	σ_{AC} ($\times 10^{-7}$) [S/m] (At 350 K, 1 kHz)	E_a (AC) [kJ/mol] (At 1 kHz)	σ_{AC} ($\times 10^{-7}$) [S/m] (At 350 K, 10 kHz)	E_a (AC) [kJ/mol] (At 10 kHz)
BGNb0	0.91 ± 0.01	75.42 ± 0.08	1.50 ± 0.01	49.39 ± 0.98	4.91 ± 0.05	37.95 ± 0.97
BGNb1	0.78 ± 0.04	76.79 ± 0.06	1.84 ± 0.04	50.21 ± 0.96	5.65 ± 0.11	38.49 ± 0.79
BGNb2	1.52 ± 0.10	69.35 ± 0.20	2.36 ± 0.03	48.61 ± 0.87	6.54 ± 0.13	37.83 ± 0.61
BGNb4	0.51 ± 0.02	76.47 ± 0.15	1.17 ± 0.01	47.55 ± 0.86	3.69 ± 0.06	36.37 ± 0.66
BGNb8	0.19 ± 0.01	79.47 ± 0.16	1.23 ± 0.02	44.25 ± 0.92	4.60 ± 0.21	34.11 ± 0.69

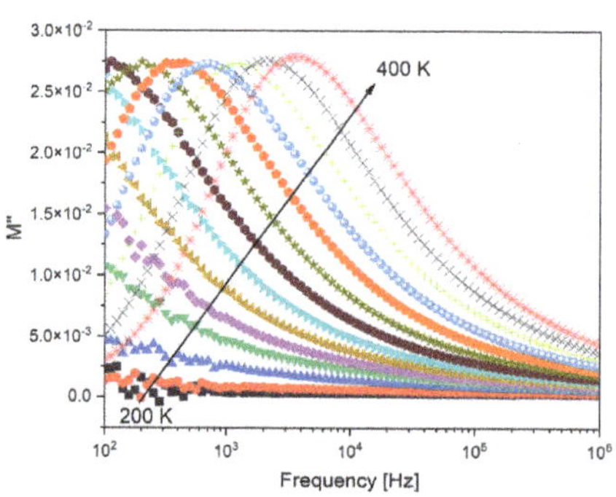

Figure 8. The imaginary part of the dielectric modulus (M") versus frequency for BGNb2 glass, between 200 K and 400 K.

The AC conductivity (Figure 9) showed the same feature, in resonance with DC conductivity, where the variation increases linearly in the high-temperature region. The results obtained for AC conductivity are registered in Table 2 and show an agreement with DC conductivity results. The bioglass with 2 mol% Nb_2O_5 content exhibits the highest AC conductivity. The activation energy value decreases with increasing frequency for all bioglass samples. This can be explained due to the increase in the applied frequency which enhances the electronic jumps between the localized states [66,67].

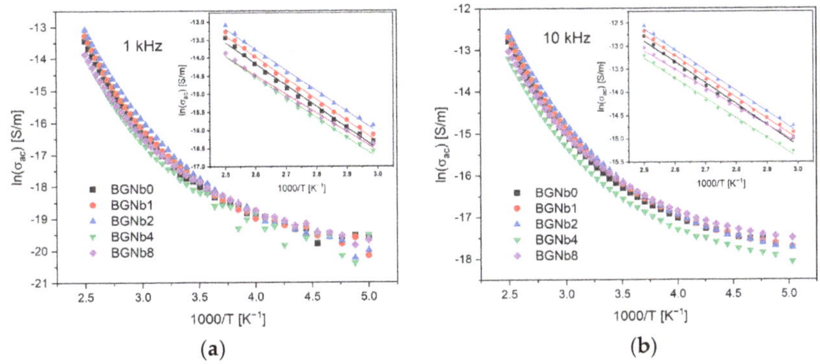

Figure 9. AC conductivity versus 1000/T at (**a**) 1 kHz; (**b**) 10 kHz (inset: magnification of the high-temperature measurement zone; the lines reflect the Arrhenius fit).

2.4. In Vitro Evaluation of the Biocompatibility

The human osteosarcoma Saos-2 cells were used to evaluate the cytotoxicity of the bioactive glasses prepared. Two sets of samples were tested, which we call the non-passivated and the passivated ones. The difference between the two sets is that the non-passivated is the filtered solution of the pristine bioglass powders that were in contact with the McCoy 5A medium for 24 h. The passivated solution was achieved by changing the medium after 24 h to a new medium, which was kept in contact with the powders for a new 24 h period. The filtered solution was then called the passivated sample. Figure 10 displays the results obtained for the passivated and non-passivated samples. All samples showed an improvement in cell viability when the concentration decreased from 100 mg/mL to 6.75 mg/ mL, as expected. Moreover, it can be seen that the passivated samples exhibit higher cell viability compared to the non-passivated samples. The passivated samples, which better simulate physiological conditions [68], revealed cell viability higher than 80%, except for the BGNb0 and BGNb1 samples that showed toxicity at the 100 mg/mL concentration and BGNb0 at 50 mg/mL, respectively.

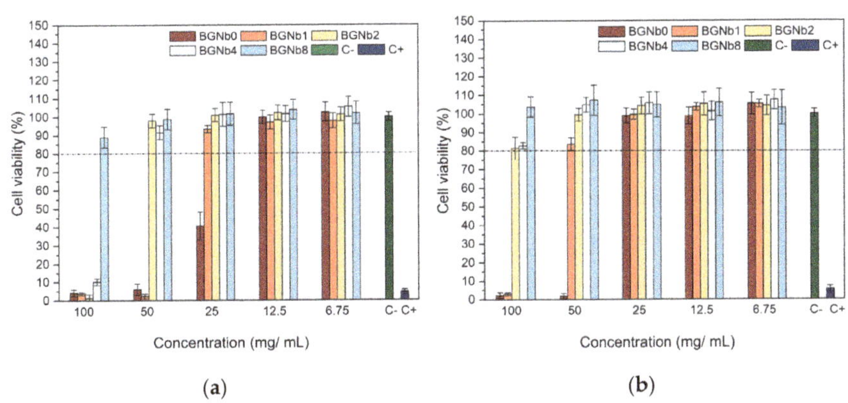

Figure 10. Relative viability of (**a**) non-passivated and (**b**) passivated bioactive glass samples modified by Nb_2O_5 insertion, in culture with Saos-2 cells.

2.5. Antibacterial Activity

Figure 11 depicts the antibacterial activity of the bioglasses modified by Nb_2O_5 insertion evaluated by the observed inhibition halo against the Gram-positive *Staphylococcus aureus* and *Streptococcus mutans* and the Gram-negative *Escherichia coli* bacteria. By the rise of the inhibition halo, we can conclude that the samples have antibacterial activity. An increase in the inhibition halo with increasing the content of Nb_2O_5 up to 2 mol% is noted. A decrease in the antibacterial effect was observed with increasing the Nb_2O_5 from 2 to 8 mol%.

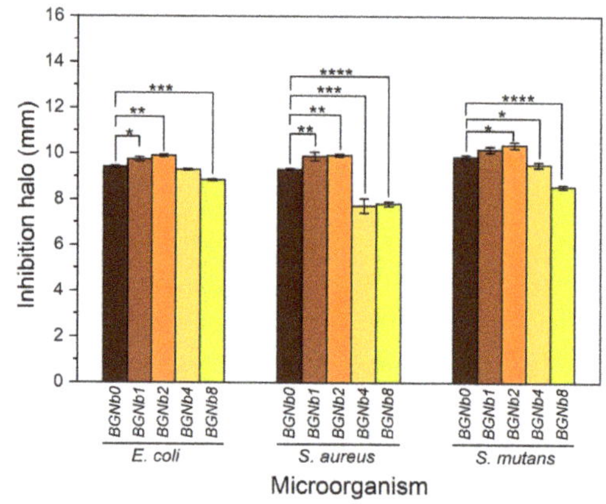

Figure 11. Antibacterial activity of the bioglass samples against *E. coli*, *S. aureus*, and *S. mutans* based on inhibition halo. Results are reported as mean ± SD. The asterisks indicate significance in an unpaired *t*-test; * $p \leq 0.05$; ** $p \leq 0.01$; *** $p \leq 0.001$; **** $p \leq 0.0001$.

2.6. In Vitro Degradation and Bioactivity Assay

The in vitro bioactivity, i.e., the ability to induce an apatite-like layer formation, of the bioglasses prepared, was investigated using an immersion test in simulated body fluid (SBF). Figure 12 shows the SEM micrographs and the EDS data of the surface of bioglass pellets after immersion in SBF.

The SEM micrographs confirmed the bioactivity of the glasses. The surface morphologies of pellets show the presence of spherical (cauliflower-like) particles characteristic of an apatitic layer formation. With the increase in the immersion time in SBF, this layer becomes thicker. It can be seen that for the samples with an immersion time of 28 days, the size of the particles decreases with increasing the percentage of niobium beyond 2 mol%, and in the case of the sample with the highest load percentage, BGNb8, only small amounts of apatite particles start to appear. The EDS graphs (Figure 12(d1–d5)) confirm the change in the amount of the chemical elements with the soaking time in SBF. The amount of Na and Si atoms tends to decrease with the immersion time, while the amount of P and Ca tends to increase.

The variation of Si and Na atomic percentage and the Ca/P ratio, determined using the data from EDS analysis on the surface of the bioglass samples before and after immersion in SBF solution, are illustrated in Figure 13. An abrupt drop in the atomic percentage of the elements is evident during the initial 2 days of immersion in all samples while no significant change is observed afterward. The samples with lower percentages of niobium, BGNb1, and BGNb2, show a faster decrease in Na and Si percentage in the first 2 d compared to the other glasses, and the value of the Ca/P ratio decreases reaching a value of 1.60–1.80, comparable to that of hydroxyapatite (Ca/P = 1.67) [69–71]. The decrease in the atomic percentage of Na and Si in the sample modified with 8 mol% of Nb_2O_5 is not significant, in contrast with the other samples, and the Ca/P ratio is considered high.

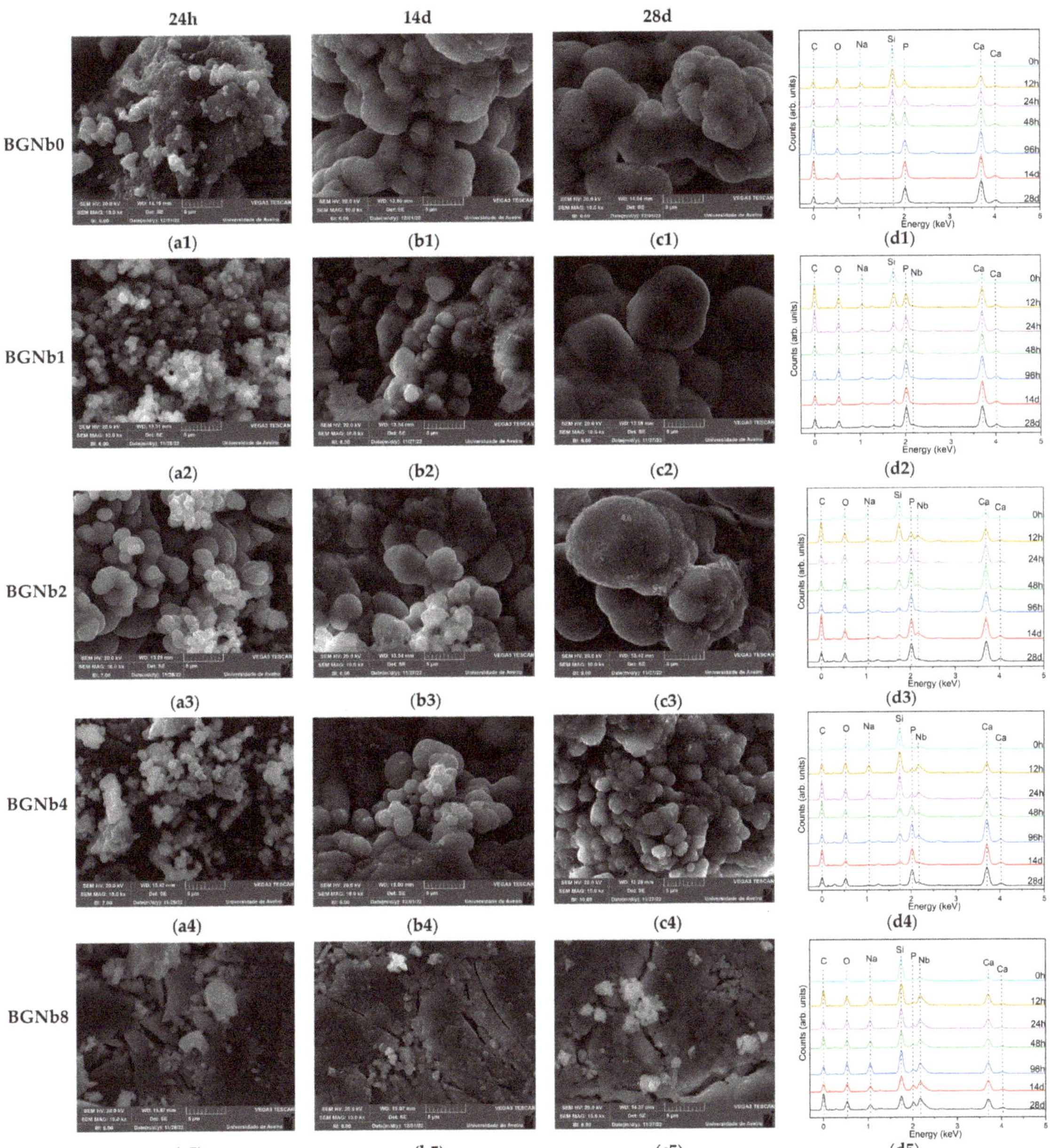

Figure 12. SEM micrographs of five groups of bioactive glass samples modified by the insertion of different percentages of Nb_2O_5 after immersion in SBF for (**a1–a5**) 24 h; (**b1–b5**) 14 d; (**c1–c5**) 28 d; (**d1–d5**) plots of the EDS analysis for all immersion times. (The magnification of SEM images is 10 kX).

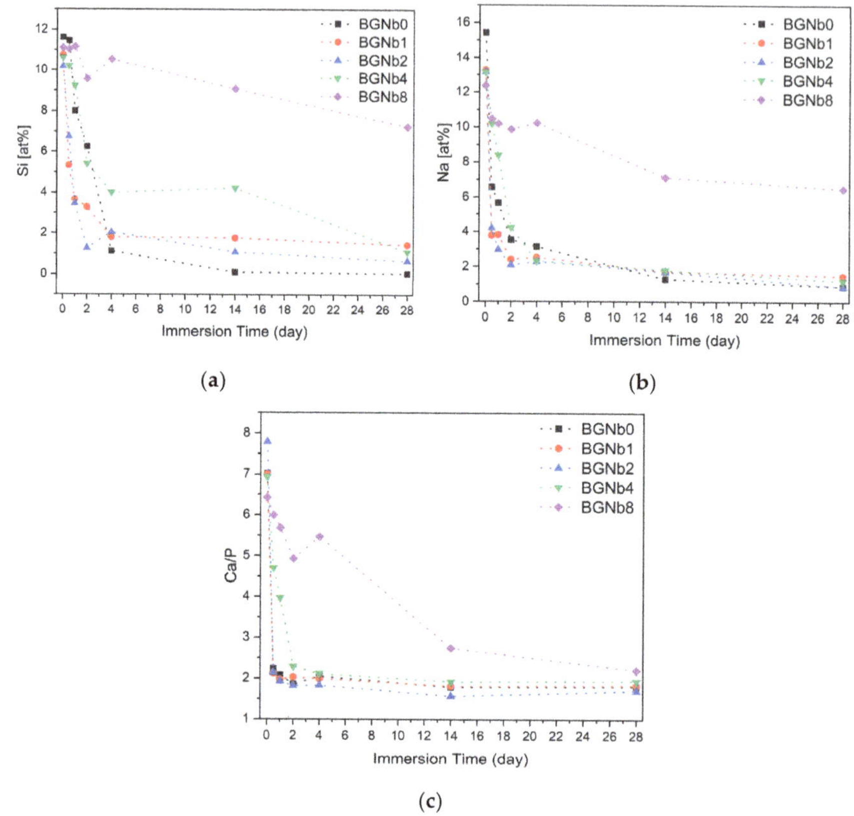

Figure 13. The variation of elemental concentration of (**a**) Si; (**b**) Na; (**c**) Ca/P ratio on the surface of bioactive glasses after immersion in SBF as a function of immersion time.

Figure 14 depicts the variation of pH of the bioactive glass samples after soaking in SBF for different periods, from 12 h up to 28 days. For all samples, the pH initially increases compared to the initial pH of the SBF solution, 7.4, and then, after 2 days of sample immersion, it starts to decrease.

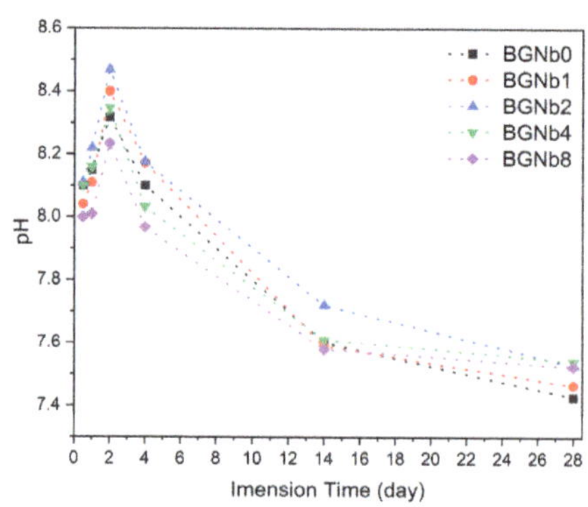

Figure 14. Variation of pH of the SBF solution with the immersion time.

3. Discussion

The results obtained by XRD and FTIR do not reciprocate any type of modification in the structure of the glass matrix with the insertion of Nb_2O_5. However, Raman spectra clearly showed the appearance of NbO_4 and NbO_6 structural groups, located at 800–810 cm^{-1} and 827–862 cm^{-1}, for the glasses modified with Nb_2O_5 (Figures 3 and 4). As the concentration of Nb_2O_5 is raised up to 2 mol%, the quantity of NbO_6 units, related to niobium ions acting as modifiers in the glass matrix, increases progressively. Exceeding the 2 mol% of Nb_2O_5, there is a fractional conversion of NbO_6 units into NbO_4 units, which contributes to the formation of the glass network. Similar behavior was already found in other glasses containing niobium [72–74]. The presence of NbO_6 units is predicted to act as a network modifier through the depolymerization of Si-O-Si bonds. These octahedrons can form a chain structure with different degrees of distortion by sharing their vertices with at least two silicon octahedrons. With the increasing NbO_6 concentration, a higher concentration of NBO is expected [59,72]. The BGNb2 sample exhibits a high concentration of NbO_6 structural groups (Figure 4), and thus a high concentration of NBO as justified in Figure 5.

The presence of NbO_6 and NbO_4 plays a critical role in the thermal response of the bioglass. From Figure 7 and Table 1, it can be observed that the Tg increases with the insertion of a high concentration of niobium, which indicates that an increase in network connectivity occurs due to the presence of a high concentration of NbO_4, the network former, leading to a decrease in the amount of NBO ions. For the BGNb2 sample, a slight decrease in the characteristic temperatures is observed compared to the bioglass base, which is attributed to the increase in the amount of NBO ions and the weakness in the glass network, facilitating the ion mobility. Similar behaviors were reported in the literature for niobium glass systems [75,76].

The existence of NBO ions could affect the electrical properties of the prepared glasses. It is known that in the bioactive glass system, the conductivity is mainly related to the energy transported by Na^+ and Ca^+ ions moving through the glass network [3,77]. For the glasses modified with Nb_2O_5, the presence of NbO_6 units leads to the degradation of the bioglass network by the formation of more NBO ions. Such structural alterations may have a significant impact on Na^+ and Ca^+ mobility. The mobility of such network modifier ions increases with the rise in NBO ions. One can notice in Table 2 that the sample with 2 mol% content of Nb_2O_5 exhibits the highest DC conductivity and therefore the lowest activation energy since the decrease in activation energy suggests an increase in the charge carrier's mobility. The same happens to the AC conductivity, the increase in the conductivity with samples loaded with Nb_2O_5 up to 2 mol% should be essentially related to the increase in Na^+ ions' mobility.

From the biomedical applications point of view, it is critical to evaluate the cytotoxicity of the bioactive glasses given their intended use as implant coating materials. The evaluation of the bioactive glasses' cytotoxicity against the human osteosarcoma Saos-2 cells, represented in Figure 10, shows low cell viability at high extract concentration, especially in the case of the samples with no or low content of Nb_2O_5. It was found that increasing the content of Nb_2O_5 inserted into the bioactive glass network can increase the biocompatibility of materials, which is consistent with the results of previous studies [75,78,79]. The non-passivated samples show lower cell viability compared to the passivated samples, which could be explained by the high rate of ion-exchange reactions that take place during the first 24 h of material–cell culture medium interactions leading to an increase in local pH [52]. Based on the results of passivated samples, which better mimic physiological circumstances, all modified bioactive glasses by niobium insertion can be used in a biomedical application where the extracellular fluid is exposed to bioactive glasses in circumstances corresponding to the 100 mg/mL extract production. The exceptions are the glasses with Nb_2O_5 content lower than 2 mol%, which show cell viability less than 80% at that concentration.

An important factor in reducing the risk of implant infection is preventing bacterial colonization. The 45S5 glass particles promote a considerable antibacterial effect against certain oral bacteria mainly due to the change in pH and the osmotic pressure effect [80,81]. The increase in pH to a more alkaline range creates an unfriendly environment for bacteria, resulting in morphological alterations. Additionally, variations in ion concentration in the bacterial environment result in a reduction in pressure across the bacterial cell membrane, which causes bacteria to shrink and therefore damages the cell membrane [81,82]. The antibacterial test results obtained using the agar diffusion method demonstrate the antibacterial properties against Gram-positive and Gram-negative bacteria (Figure 11). The inhibition halo tends to increase with increasing the Nb_2O_5 concentration up to 2mol%, which can be due to the high ionic strength of Nb^{5+} ions preventing bacteria from growing by creating a hyperosmotic environment [72,74]. Additionally, the presence of a disintegrated glass network due to the high concentration of NbO_6 units in those glasses offers a suitable environment for the leaching of alkali metals ions, Na^+, and Ca^{2+}, which could lead to an increase in pH and thus promote the death of bacteria. Beyond 2 mol% of Nb_2O_5 insertion, the antibacterial activity of the samples decreased, which could be related to the conversion of certain NbO_6 structural units into NbO_4 units, as proved by the Raman analysis.

To assess the in vitro bioactivity of the synthesized bioactive glasses in a biological medium, the SBF immersion test was chosen. This method serves a better understanding of the physicochemical reactions taking place on bioglass in physiological fluids. When evaluating bioglass in vitro, it is critical to take into account changes in the surface chemistry, mainly the formation of an apatitic layer, since it has a significant impact on osteoblast cell adhesion and proliferation [83]. The formation of an apatitic layer is purely physicochemical and does not involve cells. This was observed in the SEM images of bioglasses after the immersion in SBF which is a cell-free solution (Figure 12). The apatite particles are formed on the surface of the bioglass, and with increasing immersing time the particles tend to aggregate, and a dense layer is formed. However, the size of the formed apatite seems to be smaller for the samples with the highest concentration of Nb_2O_5 (above 2 mol%) compared to the other samples. To better understand the effect of niobium on the formation of the hydroxyapatite (HA) layer on the bioglass surface, we refer to the mechanism proposed by L. Hench for the original Bioglass® which is adaptable to the majority of bioactive materials [25]. Similarly to the 45S5 bioactive glass, the SiO_4 tetrahedron serves as the main structural unit of the bioactive glasses containing niobium, which may form bonds with other SiO_4 tetrahedrons through Si-O-Si bonding, also known as bridging oxygen (BO) ions. Network modifiers alter the structure of the glass by converting BO into NBO, resulting in a drop in the glass network connectivity and thus a rise in the dissolution rate and consequently the release of ions. In fact, when the glass comes into contact with SBF, the alkali and alkaline earth ions (Na^+ and Ca^{2+}) present on the surface of the glass are exchanged with the H^+ and H_3O^+ ions of the medium. This leads to an increase in the pH of the medium as the H^+ ions are replaced by cations. The rise in pH values promotes the breaking, in the glass network, of the Si-O-Si bonds leading to a faster dissolution of the glass and the formation of silanol units ($Si(OH)_4$). These silanol units then condense on the surface of the glass, forming a hydrated silica layer which promotes the nucleation of the carbonated hydroxyapatite. From the results obtained (Figure 13b), a faster drop can be seen in the atomic percentage of Na on the bioglass surface in the first days of SBF immersion. The presence of niobium acting as a network modifier, at high concentrations in the samples with a low percentage of Nb_2O_5 ($\leq$2 mol%), creates a higher number of NBO ions and therefore promotes a faster release of the Na^+ ions. For the samples with Nb_2O_5 > 2 mol%, the amount of NbO_4 units increases, reducing the amount of NBO ions, and therefore the mobility or release of the network modifiers, namely the Na ions, becomes more difficult leading to a reduction in the glass dissolution rate. The same trend is observed for Si, which exhibits faster release from the bioglass with Nb_2O_5 concentration below 4 mol%, as indicated in Figure 13a by the faster decrease in the atomic percentage of this element. This proves the enhanced surface reactivity for the bioglasses with niobium load up to 2 mol%. The SEM micrographs in Figure 12 show that after 24 h of SBF immersion, the amount and the size of spheroidal apatite particles are more pronounced in the sample BGNb2, suggesting the existence of an optimal composition for the apatite formation. With longer soaking times in SBF, the concentration of the ions stabilizes and the Ca/P ratio decreases, reaching a value between 1.6 and 1.8 for the sample with niobium content up to 2 mol%, close to that of the HA of the human bones [69,70]. For the samples with higher niobium amount, the value of the Ca/P ratio is above 2, higher than the ideal value of 1.67. This higher value of Ca/P suggests that those samples are less bioactive [84,85], meaning that the apatite structure does not develop easily on the glass surface when exposed to SBF. After 28 days of soaking in SBF, the apatite particles grow and become denser, which indicates the formation of a dense layer of crystalline HA on the bioglass surface, while the growth was reduced for the samples with niobium content of 4 mol% and only the beginning of the formation of small apatite particles was observed for the sample with 8 mol% of niobium loads.

The variation in the pH is indicated in Figure 14. The increase in the pH in the first 48 h was due to the high release of the alkaline metal ions from the bioglass surface, and then it decreases due to the formation of silanol. The incorporation of 2 mol % of niobium into the bioglass resulted in a higher value of pH, which can be explained by the highest degradation rate of this sample and this is consistent with the condition of the creation of an HA-like layer on the surface of the bioactive glass with higher crystallinity [59,72].

4. Materials and Methods

4.1. Bioglass Preparation

A series of 45S5 bioactive glasses (45% SiO_2, 24.5% Na_2O, 24.5% CaO, and 6% P_2O_5 (wt%)) modified by the insertion of different amounts of Nb_2O_5, from 0 to 8 mol% (designed by BGNb0, BGNb1 ... BGNb8) was prepared using the melt-quenching technique. In brief, high-purity grade SiO_2, P_2O_5, $CaCO_3$, Na_2CO_3, and Nb_2O_5 (>99.99%) were mixed and ground in a planetary ball mill at 300 rpm for 1 h to homogenize, before undergoing calcination at 800 °C for 8 h. The powder was then placed in platinum crucibles and melted at 1300 °C for 1 h. The melt was quenched between the casting plates to obtain bulk glass samples.

4.2. Structural Characterization

Malvern Panalytical Aeris powder diffractometer (CuK_α radiation, λ = 1.54056 Å) was used to collect the X-ray diffraction (XRD) patterns at room temperature. The acquisition was performed using a scan step of 0.02° in 1 s in a 2θ angle range of 10–70°.

The FTIR spectra were performed on FT Perkin-Elmer Spectrum BX Spectrometer in the ATR crystal (Golden Gate Diamond ATR Accessory), in the range of 400 and 1300 cm^{-1}. The measurements were obtained from bioglass powder dispersed in KBr pellets. During acquisition, the room temperature and humidity were kept at approximately 23 °C and 35%, respectively.

Raman spectra were recorded at room temperature using a Horiba Jobin Yvon HR 800 spectrometer with an Ar^+ laser (λ = 532 nm). Spectra were collected in a back-scattering geometry between 200 and 1500 cm^{-1} with a 50x lens focused on the sample.

4.3. Thermal Analysis

The thermal properties of the glasses were investigated using a simultaneous differential thermal analysis (DTA)/thermogravimetric (TG) measurement. A Hitachi STA 7300 system was employed, under Nitrogen N50 (99.999%) flowing at 200 mL/min and heating at 10 °C/min.

4.4. Electrical Characterization

Bulk glass samples were used for the electrical measurements. The samples were polished until parallel surfaces with a thickness of about 1 mm were obtained and then painted with silver conductive paste to form the electrodes. The direct current conductivity (σ_{DC}) of the samples was measured with a Keithley 617 electrometer, capable of measuring currents down to 10^{-14} A. This measurement was performed in a temperature range between 200 and 400 K where a voltage of 100 V was applied across the bulk glass. The AC electrical conductivity (σ_{AC}) and impedance measurements were also performed in the temperature range of 200 to 400 K, using a precision impedance meter, Agilent 4294, operating in a broad frequency window from 100 Hz to 1 MHz and in the C_p–R_p configuration. The temperature of the samples was regulated by an Oxford Research IT-C4 and monitored using a platinum sensor in both DC and AC measurements.

The complex electric permittivity ε^* was calculated with the following equation [86,87]:

$$\varepsilon^* = \varepsilon' - j\,\varepsilon'' = C_p\,(d/\varepsilon_0\,A) - j\,d\,(\omega\,R_p\,\varepsilon_0\,A) \quad (1)$$

At the high-temperature range, the activation energy (E_a) of the DC and AC conductivities was determined by fitting the data to the Arrhenius model [88–91].

$$\sigma = \sigma_0\,\exp(-E_A/(k_B\,T)), \quad (2)$$

where σ_0 is a pre-exponential factor, E_A is the activation energy, K_B is the Boltzmann constant, and T the temperature.

4.5. In Vitro Evaluation of the Biocompatibility

In accordance with International Standard "ISO 10993-5 Biological evaluation of medical devices—Part 5: Tests for in vitro cytotoxicity" [92], the cytotoxicity of the bioglasses was assessed at various concentrations against the human osteosarcoma cell line (Saos-2 cells, ATCC® HTB-85™). Extracts were produced by placing the bioglass powder in contact with a culture medium, McCoy 5A medium (Merck KGaA, Darmstadt, Germany) at a concentration of 100 mg/mL. For non-passivated extract, the medium in contact with the powder was kept in an incubator for 24 h at 37 °C then filtered with a 0.22 µm millipore filter and stored at 37 °C. For the passivated extract, a new McCoy 5A medium was added to the same bioglass powder and then placed in the incubator for another 24 h at 37 °C. The 96-well plates were seeded with a Saos-2 cell line at a density of 30,000 cells/cm^2 and placed in an incubator with 5% CO_2 atmosphere for 24 h at 37 °C. After, the culture medium was removed and on the same plate, negative controls (C-) (viable cells), positive controls (C+) (in which the toxic compound dimethyl sulphoxide was added), non-passivated and passivated extracts with appropriate dilutions (50 mg/mL, 25 mg/mL, 12.5 mg/mL, and 6.75 mg/mL) were placed in an incubator for 48 h. The resazurin cell viability indicator was used to assess cell populations [93]. Using a Biotek ELX800 microplate reader, the optical absorbances of each well were measured at 570 and 600 nm. To confirm the assay's reproducibility, two biological replicates, with six statistical replicates each, were carried out in this test for each sample.

4.6. Antibacterial Activity

The bacterial strains *Escherichia coli* K12 DSM498, *Staphylococcus aureus* COL MRSA (methicillin-resistant strain), and *Streptococcus mutans* DSM20523, were used as models to assess the antibacterial activity of the glasses, as previously described [46]. The method of agar diffusion assay plates, using the two-layer bioassay, was performed with TSB medium with the molten seeded overlay containing approximately 10^8 CFU/mL of the appropriate indicator bacteria. Bioglass pellets with a diameter of 6 mm were placed in the center of the plates, left for 4 h at room temperature, and then incubated for 24 h at 37 °C. In the case of *S. mutans* the plates were placed in a 5% CO_2 incubator.

Images of the pellets were taken, and the diameter of the inhibition halo was measured with ImageJ software; each pellet was measured 30 times in several orientations [94]. The results of the eight independent assays for each bacterium were statistically analyzed with an unpaired *t*-test, comparing the bioactive glass base composition with each of the different samples using GraphPad Prism 8.0 software.

4.7. In Vitro Bioactivity Assay

The bioactivity evaluation of the bioglasses was conducted by immersing samples (pellets made of powdered bioglass) in simulated body fluid SBF, according to ISO 23317:2017 Standards. The samples were placed in different flasks, soaked in SBF, and remained inside an incubator at 37 °C with continuous oscillating support for 12, 24, 48, 96 h, 14, and 28 days. The SBF solutions were refreshed every 48 h to mimic the biological environment.

To calculate the volume of SBF used for each sample, we used the following formula:

$$V_s = 100 \text{ mm} \times S_a \quad (3)$$

where V_s is the volume of SBF in mm^3, and S_a is the surface area of the pellet in mm^2.

After soaking, the pellets were collected and gently rinsed with deionized water and then dried at room temperature.

This assay aimed to determine the change in ion concentration and the formation of an apatite-like layer on the bioglass surface over 28 days with the presence of a different concentration of Nb_2O_5. For that, TESCAN Vega 3 scanning electron microscopy (SEM) (TESCAN ORSAY HOLDING, a.s., Brno-Kohoutovice, Czech Republic) equipped with energy-dispersive X-ray spectroscopy (EDS) (Bruker EDS) was performed on the glass surface to determine the morphological and compositional changes resulting from the reaction in SBF.

5. Conclusions

Bioactive 45S5 glasses modified by the insertion of different amounts of niobium pentoxide, Nb_2O_5, were successfully synthesized using the melt-quenching technique. XRD and FTIR results show that there was no alteration in the glass matrix with the addition of Nb. The characterization using Raman spectroscopy showed the appearance of additional bands for the bioactive glass containing niobium attributed to the distortion of NbO_6 units and the vibration of NbO_4. The fractional conversion of the network modifier units of NbO_6 into the NbO_4 network former affects the electrical properties of the sample and causes a decrease in the bioactivity and antibacterial effect. The sample with 2 mol% of Nb_2O_5 content presented the highest percentage of NbO_6 units and showed a higher dissolution rate and maximal growth of the HA layer on its surface in the in vitro immersion tests in SBF. Moreover, the evaluation of the antibacterial activity against *E. coli*, *S. aureus*, and *S. mutans* revealed that glass loaded with 2 mol% of Nb_2O_5 had the greatest antibacterial effect. We can conclude that the 45S5 bioactive glass modified by the insertion of 2 mol% of Nb_2O_5 is more suitable for biomedical applications and can be employed as a coating material for a dental implant without being harmful to osteoblasts cells.

Author Contributions: Conceptualization, I.H., S.R.G., A.S.P., I.S.-N. and M.P.F.G.; methodology, I.H., S.R.G., A.S.P., I.S.-N. and M.P.F.G.; software, I.H.; validation, I.H. and M.P.F.G.; formal analysis, I.H.; investigation, I.H., S.R.G., A.S.P., S.K.J., I.S.-N. and M.P.F.G.; resources, J.C.S., J.P.B., I.S.-N. and M.P.F.G.; data curation, I.H.; writing—original draft preparation, I.H.; writing—review and editing, M.P.F.G., I.S.-N., J.C.S., J.P.B., S.K.J. and M.d.C.L.; visualization, I.H. and M.P.F.G.; supervision, M.P.F.G., J.C.S. and J.P.B. All authors have read and agreed to the published version of the manuscript.

Funding: This research was funded by FEDER funds through the COMPETE 2020 Program and National Funds through FCT—Portuguese Foundation for Science and Technology under the project LISBOA-01-0247-FEDER-039985/POCI-01-0247-FEDER-039985, LA/P/0037/2020, UIDP/50025/2020, and UIDB/50025/2020 of the Associate Laboratory Institute of Nanostructures, Nanomodelling and Nanofabrication—i3N, UIDP/04378/2020 and UIDB/04378/2020 of the Research Unit on Applied Molecular Biosciences—UCIBIO, and LA/P/0140/2020 of the Associate Laboratory Institute for Health and Bioeconomy—i4HB. S.R. Gavinho and A. Sofia Pádua acknowledges FCT—Portuguese Foundation for Science and Technology for the PhD grant (SFRH/BD/148233/2019 and UI/DB/151287/2021, respectively). S.K. Jakka acknowledges FCT—Fundaçao para a Ciência e a Tecnologia, Portugal, I.P., in the scope of the framework contract foreseen in the numbers 4, 5, and 6 of article 23 of the Decree Law 57/2016 of 29 August, changed by Law 57/2017 of 19 July.

Institutional Review Board Statement: Not applicable.

Informed Consent Statement: Not applicable.

Data Availability Statement: The data presented in this study are available on request from the corresponding author.

Conflicts of Interest: The authors declare no conflict of interest.

References

1. Gaviria, L.; Salcido, J.P.; Guda, T.; Ong, J.L. Current trends in dental implants. *J. Korean Assoc. Oral Maxillofac. Surg.* **2014**, *40*, 50. [CrossRef] [PubMed]
2. Gupta, A.; Dhanraj, M.; Sivagami, G. Status of surface treatment in endosseous implant: A literary overview. *Indian J. Dental Res.* **2010**, *21*, 433. [CrossRef]
3. Gavinho, S.R.; Graça, M.P.F.; Prezas, P.R.; Kumar, J.S.; Melo, B.M.G.; Sales, A.J.M.; Almeida, A.F.; Valente, M.A. Structural, thermal, morphological and dielectric investigations on 45S5 glass and glass-ceramics. *J. Non-Cryst. Solids* **2021**, *562*, 120780. [CrossRef]
4. Zhao, L.; Wang, H.; Huo, K.; Cui, L.; Zhang, W.; Ni, H.; Zhang, Y.; Wu, Z.; Chu, P.K. Antibacterial nano-structured titania coating incorporated with silver nanoparticles. *Biomaterials* **2011**, *32*, 5706–5716. [CrossRef]
5. Wang, Z.; Shen, Y.; Haapasalo, M. Dent. Mater. with antibiofilm properties. *Dent. Mater.* **2014**, *30*, e1–e16. [CrossRef]
6. Charalampakis, G.; Leonhardt, Å.; Rabe, P.; Dahlén, G. Clinical and microbiological characteristics of peri-implantitis cases: A retrospective multicentre study. *Clin. Oral Implants Res.* **2012**, *23*, 1045–1054. [CrossRef] [PubMed]
7. Tibbitt, M.W.; Rodell, C.B.; Burdick, J.A.; Anseth, K.S. Progress in material design for biomedical applications. *Proc. Natl. Acad. Sci. USA* **2015**, *112*, 14444–14451. [CrossRef] [PubMed]
8. Festas, A.J.; Ramos, A.; Davim, J.P. Medical devices biomaterials—A review. *Proc. Instit. Mech. Eng. Part L* **2020**, *234*, 218–228. [CrossRef]
9. Chen, F.-M.; Liu, X. Advancing biomaterials of human origin for tissue engineering. *Prog. Polym. Sci.* **2016**, *53*, 86–168. [CrossRef]
10. Becerikli, M.; Jaurich, H.; Wallner, C.; Wagner, J.M.; Dadras, M.; Jettkant, B.; Pöhl, F.; Seifert, M.; Jung, O.; Mitevski, B. P2000-A high-nitrogen austenitic steel for application in bone surgery. *PLoS ONE* **2019**, *14*, e0214384. [CrossRef]
11. Ramakrishna, S.; Ramalingam, M.; Kumar, T.S.; Soboyejo, W.O. *Biomaterials: A nano Approach*; CRC Press: Boca Raton, FL, USA, 2016.
12. Singh, N.; Hameed, P.; Ummethala, R.; Manivasagam, G.; Prashanth, K.G.; Eckert, J. Selective laser manufacturing of Ti-based alloys and composites: Impact of process parameters, application trends, and future prospects. *Mater. Today Adv.* **2020**, *8*, 100097. [CrossRef]
13. Boccaccini, A.R.; Brauer, D.S.; Hupa, L. *Bioactive Glasses: Fundamentals, Technology and Applications*; Royal Society of Chemistry: London, UK, 2016.
14. Ylänen, H. *Bioactive Glasses: Materials, Properties and Applications*; Woodhead Publishing: Sawston, UK, 2017.
15. Shukla, A.A.; Etzel, M.R.; Gadam, S. *Process Scale Bioseparations for the Biopharmaceutical Industry*; CRC Press: Boca Raton, FL, USA, 2006.
16. Zhang, B.G.; Myers, D.E.; Wallace, G.G.; Brandt, M.; Choong, P.F. Bioactive coatings for orthopaedic implants—Recent trends in development of implant coatings. *Int. J. Mol. Sci.* **2014**, *15*, 11878–11921. [CrossRef]
17. Davidson, D.J.; Spratt, D.; Liddle, A.D. Implant materials and prosthetic joint infection: The battle with the biofilm. *EFORT Open Rev.* **2019**, *4*, 633–639. [CrossRef] [PubMed]
18. Trisi, P.; Marcato, C.; Todisco, M. Bone-to-implant apposition with machined and MTX microtextured implant surfaces in human sinus grafts. *Int. J. Period. Restor. Dent.* **2003**, *23*, 426–437.
19. Bornstein, M.M.; Lussi, A.; Schmid, B.; Belser, U.C.; Buser, D. Early loading of nonsubmerged titanium implants with a sandblasted and acid-etched (SLA) surface: 3-year results of a prospective study in partially edentulous patients. *Int. J. Oral Maxillofac. Implants* **2003**, *18*, 659–666.
20. Al Mugeiren, O.M.; Baseer, M.A. Dental implant bioactive surface modifiers: An update. *J. Int. Soc. Prevent. Commun. Dent.* **2019**, *9*, 1. [CrossRef]
21. Damiati, L.; Eales, M.G.; Nobbs, A.H.; Su, B.; Tsimbouri, P.M.; Salmeron-Sanchez, M.; Dalby, M.J. Impact of surface topography and coating on osteogenesis and bacterial attachment on titanium implants. *J. Tissue Eng.* **2018**, *9*, 2041731418790694. [CrossRef] [PubMed]
22. Priyadarshini, B.; Rama, M.; Chetan; Vijayalakshmi, U. Bioactive coating as a surface modification technique for biocompatible metallic implants: A review. *J. Asian Ceramic Soc.* **2019**, *7*, 397–406. [CrossRef]
23. Islam, M.T.; Felfel, R.M.; Abou Neel, E.A.; Grant, D.M.; Ahmed, I.; Hossain, K.M.Z. Bioactive calcium phosphate–based glasses and ceramics and their biomedical applications: A review. *J. Tissue Eng.* **2017**, *8*, 2041731417719170. [CrossRef]
24. Alamri, A.; Salloot, Z.; Alshaia, A.; Ibrahim, M.S. The effect of bioactive glass-enhanced orthodontic bonding resins on prevention of demineralization: A systematic review. *Molecules* **2020**, *25*, 2495. [CrossRef]
25. Hench, L.L. The story of Bioglass®. *J. Mater. Sci. Mater. Med.* **2006**, *17*, 967–978. [CrossRef]
26. Hench, L.L. (Ed.) *An Introduction to Bioceramics*, 2nd ed.; Imperial College Press: London, UK, 2013; ISBN 978-1-908977-15-1.

27. Boccaccini, A.R. (Ed.) *Tissue Engineering Using Ceramics and Polymers*, 2nd ed.; Elsevier: Amsterdam, The Netherlands, 2014; ISBN 978-0-85709-716-3.
28. Zheng, K.; Sui, B.; Ilyas, K.; Boccaccini, A.R. Porous bioactive glass micro-and nanospheres with controlled morphology: Developments, properties and emerging biomedical applications. *Mater. Horiz.* **2021**, *8*, 300–335. [CrossRef]
29. Xynos, I.D.; Edgar, A.J.; Buttery, L.D.; Hench, L.L.; Polak, J.M. Ionic products of bioactive glass dissolution increase proliferation of human osteoblasts and induce insulin-like growth factor II mRNA expression and protein synthesis. *Biochem. Biophys. Res. Commun.* **2000**, *276*, 461–465. [CrossRef]
30. Vollenweider, M.; Brunner, T.J.; Knecht, S.; Grass, R.N.; Zehnder, M.; Imfeld, T.; Stark, W.J. Remineralization of human dentin using ultrafine bioactive glass particles. *Acta Biomater.* **2007**, *3*, 936–943. [CrossRef]
31. Rahaman, M.N.; Liu, X.; Bal, B.S.; Day, D.E.; Bi, L.; Bonewald, L.F. Bioactive glass in bone tissue engineering. *Biomater. Sci. Process. Prop. Appl. II* **2012**, *237*, 73–82.
32. Pantulap, U.; Arango-Ospina, M.; Boccaccini, A.R. Bioactive glasses incorporating less-common ions to improve biological and physical properties. *J. Mater. Sci. Mater. Med.* **2022**, *33*, 1–41. [CrossRef] [PubMed]
33. Joy-anne, N.O.; Akande, O.; Ecker, M. Incorporation of Novel Elements in Bioactive Glass Compositions to Enhance Implant Performance. In *Current Concepts in Dental Implantology-From Science to Clinical Research*; IntechOpen: London, UK, 2021.
34. Leung, Y.H.; Ng, A.M.; Xu, X.; Shen, Z.; Gethings, L.A.; Wong, M.T.; Chan, C.M.; Guo, M.Y.; Ng, Y.H.; Djurišić, A.B. Mechanisms of antibacterial activity of MgO: Non-ROS mediated toxicity of MgO nanoparticles towards Escherichia coli. *Small* **2014**, *10*, 1171–1183. [CrossRef]
35. Tian, T.; Wu, C.; Chang, J. Preparation and in vitro osteogenic, angiogenic and antibacterial properties of cuprorivaite (CaCuSi 4 O 10, Cup) bioceramics. *RSC Adv.* **2016**, *6*, 45840–45849. [CrossRef]
36. Palanikumar, L.; Ramasamy, S.N.; Balachandran, C. Size-dependent antimicrobial response of zinc oxide nanoparticles. *IET Nanobiotechnol.* **2014**, *8*, 111–117. [CrossRef]
37. Oliveira, R.L.; Barbosa, L.; Hurtado, C.R.; Ramos, L. de P.; Montanheiro, T.L.; Oliveira, L.D.; Tada, D.B.; Triches, E. de S. Bioglass-based scaffolds coated with silver nanoparticles: Synthesis, processing and antimicrobial activity. *J. Biomed. Mater. Res. Part A* **2020**, *108*, 2447–2459. [CrossRef] [PubMed]
38. Mourino, V.; Cattalini, J.P.; Boccaccini, A.R. Metallic ions as therapeutic agents in tissue engineering scaffolds: An overview of their biological applications and strategies for new developments. *J. R. Soc. Interface* **2012**, *9*, 401–419. [CrossRef]
39. Hoppe, A.; Güldal, N.S.; Boccaccini, A.R. A review of the biological response to ionic dissolution products from bioactive glasses and glass-ceramics. *Biomaterials* **2011**, *32*, 2757–2774. [CrossRef] [PubMed]
40. Nour, S.; Baheiraei, N.; Imani, R.; Rabiee, N.; Khodaei, M.; Alizadeh, A.; Moazzeni, S.M. Bioactive materials: A comprehensive review on interactions with biological microenvironment based on the immune response. *J. Bionic Eng.* **2019**, *16*, 563–581. [CrossRef]
41. Fernandes, G.V.D.O.; Alves, G.; Linhares, A.B.R.; Prado da Silva, M.H.; Granjeiro, J.M. Evaluation of cytocompatibility of bioglass-niobium granules with human primary osteoblasts: A multiparametric approach. In *Proceedings of the Key Engineering Materials*; Trans. Tech. Publ.: Stafa-Zurich, Switzerland, 2012; Volume 493, pp. 37–42.
42. Holloway, E.M.; Wu, J.H.; Czerwinski, M.; Sweet, C.W.; Wu, A.; Tsai, Y.-H.; Huang, S.; Stoddard, A.E.; Capeling, M.M.; Glass, I. Differentiation of human intestinal organoids with endogenous vascular endothelial cells. *Dev. Cell* **2020**, *54*, 516–528. [CrossRef]
43. Denry, I.L.; Holloway, J.A.; Nakkula, R.J.; Walters, J.D. Effect of niobium content on the microstructure and thermal properties of fluorapatite glass-ceramics. *J. Biomed. Mater. Res. Part B* **2005**, *75*, 18–24. [CrossRef] [PubMed]
44. Takahashi, K.; Shiraishi, N.; Ishiko-Uzuka, R.; Anada, T.; Suzuki, O.; Masumoto, H.; Sasaki, K. Biomechanical evaluation of Ti-Nb-Sn alloy implants with a low Young's modulus. *Int. J. Mol. Sci.* **2015**, *16*, 5779–5788. [CrossRef] [PubMed]
45. Prado da Silva, M.H.; Ramirez, C.M.; Granjeiro, J.M.; Rossi, A.M. In vitro assessment of new niobium phosphate glasses and glass ceramics. In Proceedings of the Key Engineering Materials; Trans. Tech. Publ.: Stafa-Zurich, Switzerland, 2008; Volume 361, pp. 229–232.
46. Gavinho, S.R.; Pádua, A.S.; Sá-Nogueira, I.; Silva, J.C.; Borges, J.P.; Costa, L.C.; Graça, M.P.F. Biocompatibility, Bioactivity, and Antibacterial Behaviour of Cerium-Containing Bioglass®. *Nanomaterials* **2022**, *12*, 4479. [CrossRef]
47. Senapati, A.; Barik, S.K.; Venkata Krishnan, R.; Chakraborty, S.; Jena, H. Studies on synthesis, structural and thermal properties of sodium niobium phosphate glasses for nuclear waste immobilization applications. *J. Ther. Anal. Calorim.* **2022**, *148*, 1–15. [CrossRef]
48. Gupta, N.; Santhiya, D.; Murugavel, S.; Kumar, A.; Aditya, A.; Ganguli, M.; Gupta, S. Effects of transition metal ion dopants (Ag, Cu and Fe) on the structural, mechanical and antibacterial properties of bioactive glass. *Colloids Surf. A Physicochem. Eng. Asp.* **2018**, *538*, 393–403. [CrossRef]
49. Miola, M.; Verné, E.; Ciraldo, F.E.; Cordero-Arias, L.; Boccaccini, A.R. Electrophoretic deposition of chitosan/45S5 bioactive glass composite coatings doped with Zn and Sr. *Front. Bioeng. Biotechnol.* **2015**, *3*, 159. [CrossRef]
50. Ibrahim, N.F.; Mohamad, H.; Noor, S.N.F.M. Characterization on melt-derived bioactive glass powder from SiO2-CaO-Na2O-P2O5 system. *J. Non-Cryst. Solids* **2017**, *462*, 23–31. [CrossRef]
51. Ibrahim, N.F.; Mohamad, H.; Noor, S.N.F.M.; Ahmad, N. Melt-derived bioactive glass based on SiO2-CaO-Na2O-P2O5 system fabricated at lower melting temperature. *J. Alloys Compd.* **2018**, *732*, 603–612. [CrossRef]

52. El-Rashidy, A.A.; Roether, J.A.; Harhaus, L.; Kneser, U.; Boccaccini, A.R. Regenerating bone with bioactive glass scaffolds: A review of in vivo studies in bone defect models. *Acta Biomater.* **2017**, *62*, 1–28. [CrossRef] [PubMed]
53. Boccaccini, A.R.; Chen, Q.; Lefebvre, L.; Gremillard, L.; Chevalier, J. Sintering, crystallisation and biodegradation behaviour of Bioglass®-derived glass–ceramics. *Farad. Discuss.* **2007**, *136*, 27–44. [CrossRef]
54. Dziadek, M.; Zagrajczuk, B.; Jelen, P.; Olejniczak, Z.; Cholewa-Kowalska, K. Structural variations of bioactive glasses obtained by different synthesis routes. *Ceram. Int.* **2016**, *42*, 14700–14709. [CrossRef]
55. Akhtach, S.; Tabia, Z.; El Mabrouk, K.; Bricha, M.; Belkhou, R. A comprehensive study on copper incorporated bio-glass matrix for its potential antimicrobial applications. *Ceram. Int.* **2021**, *47*, 424–433. [CrossRef]
56. Dunne, C.F.; Twomey, B.; Stanton, K.T. Effect of a blast coating process on the macro-and microstructure of Grade 5 titanium foam. *Mater. Lett.* **2015**, *147*, 75–78. [CrossRef]
57. Aguiar, H.; Serra, J.; González, P.; León, B. Structural study of sol–gel silicate glasses by IR and Raman spectroscopies. *J. Non-Cryst. Solids* **2009**, *355*, 475–480. [CrossRef]
58. Aguiar, H.; Solla, E.L.; Serra, J.; González, P.; León, B.; Almeida, N.; Cachinho, S.; Davim, E.J.C.; Correia, R.; Oliveira, J.M. Orthophosphate nanostructures in SiO2–P2O5–CaO–Na2O–MgO bioactive glasses. *J. Non-Cryst. Solids* **2008**, *354*, 4075–4080. [CrossRef]
59. Lopes, J.H.; Magalhães, A.; Mazali, I.O.; Bertran, C.A. Effect of niobium oxide on the structure and properties of melt-derived bioactive glasses. *J. Am. Ceram. Soc.* **2014**, *97*, 3843–3852. [CrossRef]
60. Aronne, A.; Sigaev, V.N.; Champagnon, B.; Fanelli, E.; Califano, V.; Usmanova, L.Z.; Pernice, P. The origin of nanostructuring in potassium niobiosilicate glasses by Raman and FTIR spectroscopy. *J. Non-Cryst. Solids* **2005**, *351*, 3610–3618. [CrossRef]
61. Santos, L.F.; Wondraczek, L.; Deubener, J.; Almeida, R.M. Vibrational spectroscopy study of niobium germanosilicate glasses. *J. Non-Cryst. Solids* **2007**, *353*, 1875–1881. [CrossRef]
62. Fukumi, K.; Sakka, S. Coordination state of Nb5+ ions in silicate and gallate glasses as studied by Raman spectroscopy. *J. Mater. Sci.* **1988**, *23*, 2819–2823. [CrossRef]
63. Sene, F.F.; Martinelli, J.R.; Gomes, L. Synthesis and characterization of niobium phosphate glasses containing barium and potassium. *J. Non-Cryst. Solids* **2004**, *348*, 30–37. [CrossRef]
64. Graça, M.P.F.; da Silva, M.F.; Valente, M.A. NaNbO3 crystals dispersed in a B2O3 glass matrix–Structural characteristics versus electrical and dielectrical properties. *Solid State Sci.* **2009**, *11*, 570–577. [CrossRef]
65. Araujo, M.S.; Silva, A.C.; Bartolomé, J.F.; Mello-Castanho, S. Structural and thermal behavior of 45S5 Bioglass®-based compositions containing alumina and strontium. *J. Am. Ceram. Soc.* **2020**, *103*, 3620–3630. [CrossRef]
66. Langar, A.; Sdiri, N.; Elhouichet, H.; Ferid, M. Structure and electrical characterization of ZnO-Ag phosphate glasses. *Results Phys.* **2017**, *7*, 1022–1029. [CrossRef]
67. Farid, A.M.; Bekheet, A.E. AC conductivity and dielectric properties of Sb2S3 films. *Vacuum* **2000**, *59*, 932–939. [CrossRef]
68. Hohenbild, F.; Arango-Ospina, M.; Moghaddam, A.; Boccaccini, A.R.; Westhauser, F. Preconditioning of bioactive glasses before introduction to static cell culture: What is really necessary? *Methods Protoc.* **2020**, *3*, 38. [CrossRef]
69. Beaufils, S.; Rouillon, T.; Millet, P.; Le Bideau, J.; Weiss, P.; Chopart, J.-P.; Daltin, A.-L. Synthesis of calcium-deficient hydroxyapatite nanowires and nanotubes performed by template-assisted electrodeposition. *Mater. Sci. Eng. C* **2019**, *98*, 333–346. [CrossRef]
70. Legeros, R.Z. Apatites in biological systems. *Prog. Cryst. Growth Charact.* **1981**, *4*, 1–45. [CrossRef]
71. Silva, C.C.; Graça, M.P.F.; Valente, M.A.; Sombra, A.S.B. Crystallite size study of nanocrystalline hydroxyapatite and ceramic system with titanium oxide obtained by dry ball milling. *J. Mater. Sci.* **2007**, *42*, 3851–3855. [CrossRef]
72. Madhavi, B.; Reddy, A.S.S.; Prasad, P.S.; Prasad, A.; Devi, P.P.K.; Kumar, V.R.; Veeraiah, N. The impact of Nb2O5 on in-vitro bioactivity and antibacterial activity of CaF2–CaO–B2O3–P2O5–SrO glass system. *Ceram. Int.* **2021**, *47*, 28328–28337. [CrossRef]
73. Chu, C.M.; Wu, J.J.; Yung, S.W.; Chin, T.S.; Zhang, T.; Wu, F.B. Optical and structural properties of Sr–Nb–phosphate glasses. *J. Non-Cryst. Solids* **2011**, *357*, 939–945. [CrossRef]
74. Maeda, H.; Lee, S.; Miyajima, T.; Obata, A.; Ueda, K.; Narushima, T.; Kasuga, T. Structure and physicochemical properties of CaO–P2O5–Nb2O5–Na2O glasses. *J. Non-Cryst. Solids* **2016**, *432*, 60–64. [CrossRef]
75. de Siqueira, L.; Campos, T.M.; Camargo, S.E.; Thim, G.P.; Triches, E.S. Structural, crystallization and cytocompatibility evaluation of the 45S5 bioglass-derived glass-ceramic containing niobium. *J. Non-Cryst. Solids* **2021**, *555*, 120629. [CrossRef]
76. Sanghi, S.; Rani, S.; Agarwal, A.; Bhatnagar, V. Influence of Nb2O5 on the structure, optical and electrical properties of alkaline borate glasses. *Mater. Chem. Phys.* **2010**, *120*, 381–386. [CrossRef]
77. Obata, A.; Nakamura, S.; Moriyoshi, Y.; Yamashita, K. Electrical polarization of bioactive glass and assessment of their in vitro apatite deposition. *J. Biomed. Mater. Res. Part A* **2003**, *67*, 413–420. [CrossRef]
78. Altmann, A.S.P.; Collares, F.M.; Leitune, V.C.B.; Arthur, R.A.; Takimi, A.S.; Samuel, S.M.W. In vitro antibacterial and remineralizing effect of adhesive containing triazine and niobium pentoxide phosphate inverted glass. *Clin. Oral Investig.* **2017**, *21*, 93–103. [CrossRef]
79. Obata, A.; Takahashi, Y.; Miyajima, T.; Ueda, K.; Narushima, T.; Kasuga, T. Effects of niobium ions released from calcium phosphate invert glasses containing Nb2O5 on osteoblast-like cell functions. *ACS Appl. Mater. Interfaces* **2012**, *4*, 5684–5690. [CrossRef]
80. Allan, I.; Newman, H.; Wilson, M. Antibacterial activity of particulate Bioglass® against supra-and subgingival bacteria. *Biomaterials* **2001**, *22*, 1683–1687. [CrossRef]

81. Drago, L.; Toscano, M.; Bottagisio, M. Recent evidence on bioactive glass antimicrobial and antibiofilm activity: A mini-review. *Materials* **2018**, *11*, 326. [CrossRef] [PubMed]
82. Hu, S.; Chang, J.; Liu, M.; Ning, C. Study on antibacterial effect of 45S5 Bioglass®. *J. Mater. Sci. Mater. Med.* **2009**, *20*, 281–286. [CrossRef]
83. Dorozhkin, S.V.; Epple, M. Biological and medical significance of calcium phosphates. *Angew. Chem. Int. Ed.* **2002**, *41*, 3130–3146. [CrossRef]
84. Chien, C.-S.; Liao, T.-Y.; Hong, T.-F.; Kuo, T.-Y.; Chang, C.-H.; Yeh, M.-L.; Lee, T.-M. Surface microstructure and bioactivity of hydroxyapatite and fluorapatite coatings deposited on Ti-6Al-4V substrates using Nd-YAG laser. *J. Med. Biol. Eng* **2014**, *34*, 109–115. [CrossRef]
85. Bano, S.; Romero, A.R.; Grant, D.M.; Nommeots-Nomm, A.; Scotchford, C.; Ahmed, I.; Hussain, T. In-vitro cell interaction and apatite forming ability in simulated body fluid of ICIE16 and 13–93 bioactive glass coatings deposited by an emerging suspension high velocity oxy fuel (SHVOF) thermal spray. *Surf. Coat. Technol.* **2021**, *407*, 126764. [CrossRef]
86. El-Mallawany, R.A. *Tellurite Glasses Handbook: Physical Properties and Data*; CRC Press: Boca Raton, FL, USA, 2014.
87. Feroci, M. Investigation of the role of electrogenerated N-heterocyclic carbene in the Staudinger synthesis in ionic liquid. *Int. J. Org. Chem.* **2011**, *1*, 191. [CrossRef]
88. Graça, M.P.F.; Ferreira da Silva, M.G.; Valente, M.A. Preparation, structure, morphology, and dc and ac conductivity of the 88SiO2-6Li2O-6Nb2O5 (% mole) sol-gel derived glass-ceramics. *J. Sol-Gel Sci. Technol.* **2007**, *42*, 1–8. [CrossRef]
89. Macdonald, J. Emphasizing solid materials and systems. In *Impedance Spectroscopy*; John Wiley & Sons Inc.: New York, NY, USA, 1987.
90. Graça, M.P.F.; da Silva, M.F.; Sombra, A.S.B.; Valente, M.A. Electric and dielectric properties of a SiO2–Na2O–Nb2O5 glass subject to a controlled heat-treatment process. *Phys. B Condens. Matter* **2007**, *396*, 62–69. [CrossRef]
91. Graça, M.P.F.; da Silva, M.F.; Sombra, A.S.B.; Valente, M.A. Electrical characterization of SiO2: LiNbO3 glass and glass–ceramics using dc conductivity, TSDC measurements and dielectric spectroscopy. *J. Non-Cryst. Solids* **2007**, *353*, 4390–4394. [CrossRef]
92. *ISO 10993-5:2009*; Biological Evaluation of Medical Devices—Part 5: Tests for In Vitro Cytotoxicity. International Organization for Standardization: Geneva, Switzerland, 2009.
93. Vieira, T.; Silva, J.C.; do Rego, A.B.; Borges, J.P.; Henriques, C. Electrospun biodegradable chitosan based-poly (urethane urea) scaffolds for soft tissue engineering. *Mater. Sci. Eng. C* **2019**, *103*, 109819. [CrossRef] [PubMed]
94. Schneider, C.A.; Rasband, W.S.; Eliceiri, K.W. NIH Image to ImageJ: 25 years of image analysis. *Nat. Methods* **2012**, *9*, 671–675. [CrossRef] [PubMed]

Disclaimer/Publisher's Note: The statements, opinions and data contained in all publications are solely those of the individual author(s) and contributor(s) and not of MDPI and/or the editor(s). MDPI and/or the editor(s) disclaim responsibility for any injury to people or property resulting from any ideas, methods, instructions or products referred to in the content.

International Journal of
Molecular Sciences

MDPI

Article

An Experimental Anodized and Low-Pressure Oxygen Plasma-Treated Titanium Dental Implant Surface—Preliminary Report

Jakub Hadzik [1,*], Kamil Jurczyszyn [1], Tomasz Gębarowski [2], Andrzej Trytek [3], Tomasz Gedrange [1,4], Marcin Kozakiewicz [5], Marzena Dominiak [1], Paweł Kubasiewicz-Ross [1], Anna Trzcionka-Szajna [6], Ernest Szajna [7] and Wojciech Simka [8,*]

1 Department of Dental Surgery Medical, Faculty of Medicine and Dentistry, University of Wrocław, 50-425 Wrocław, Poland
2 Department of Biostructure and Animal Physiology, Wrocław University of Environmental and Life Sciences, 51-631 Wroclaw, Poland
3 Faculty of Mechanics and Technology, Rzeszow University of Technology, 37-450 Stalowa Wola, Poland
4 Division of Orthodontics, Technical University Dresden, 01307 Dresden, Germany
5 Department of Maxillofacial Surgery, Faculty of Military Medicine, Medical University of Łódź, 90-419 Łódź, Poland
6 NanoPrime, 39-200 Dębica, Poland
7 WEA Techlab, 41-301 Dąbrowa Górnicza, Poland
8 Faculty of Chemistry, Silesian University of Technology, 44-100 Gliwice, Poland
* Correspondence: jakub.hadzik@umw.edu.pl (J.H.); wojciech.simka@polsl.pl (W.S.)

check for updates

Citation: Hadzik, J.; Jurczyszyn, K.; Gębarowski, T.; Trytek, A.; Gedrange, T.; Kozakiewicz, M.; Dominiak, M.; Kubasiewicz-Ross, P.; Trzcionka-Szajna, A.; Szajna, E.; et al. An Experimental Anodized and Low-Pressure Oxygen Plasma-Treated Titanium Dental Implant Surface—Preliminary Report. *Int. J. Mol. Sci.* **2023**, *24*, 3603. https://doi.org/10.3390/ijms24043603

Academic Editor: Mary Anne Melo

Received: 27 December 2022
Revised: 6 February 2023
Accepted: 7 February 2023
Published: 10 February 2023

Copyright: © 2023 by the authors. Licensee MDPI, Basel, Switzerland. This article is an open access article distributed under the terms and conditions of the Creative Commons Attribution (CC BY) license (https://creativecommons.org/licenses/by/4.0/).

Abstract: Chemical composition and physical parameters of the implant surface, such as roughness, regulate the cellular response leading to implant bone osseointegration. Possible implant surface modifications include anodization or the plasma electrolytic oxidation (PEO) treatment process that produces a thick and dense oxide coating superior to normal anodic oxidation. Experimental modifications with Plasma Electrolytic Oxidation (PEO) titanium and titanium alloy Ti6Al4V plates and PEO additionally treated with low-pressure oxygen plasma (PEO-S) were used in this study to evaluate their physical and chemical properties. Cytotoxicity of experimental titanium samples as well as cell adhesion to their surface were assessed using normal human dermal fibroblasts (NHDF) or L929 cell line. Moreover, the surface roughness, fractal dimension analysis, and texture analysis were calculated. Samples after surface treatment have substantially improved properties compared to the reference SLA (sandblasted and acid-etched) surface. The surface roughness (Sa) was 0.59–2.38 µm, and none of the tested surfaces had cytotoxic effect on NHDF and L929 cell lines. A greater cell growth of NHDF was observed on the tested PEO and PEO-S samples compared to reference SLA sample titanium.

Keywords: fractal dimension analysis; implant topography; oxygen plasma; plasma electrolytic oxidation; texture analysis

1. Introduction

Titanium (Ti) and titanium alloys have found their way into dentistry as a material for dental implants. Titanium is biologically inert, able to bond with osteoblasts, and has excellent biocompatibility [1]. Moreover, titanium shows excellent corrosion resistance due to titanium oxide, which forms a passive film on its surface. The titanium oxide (TiO_2) film is typically around 3–10 nm, so it can stably stay on the Ti surface [2].

Implant surface chemical composition and physical surface topography parameters such as the roughness of dental implant surfaces have all been linked to biological regulation in cell interactions leading to osseointegration [3].

The physical, chemical, and biological assessment of the implant surface need to be performed before being approved for clinical trials. To compare the surface topography of different implants, appropriate mathematical and physical descriptive methods can be used. A typical physical description of the implant surface topography is described by the Ra, Rz, and Sa. The degree of roughness, marked as Ra (Average roughness), is an average measure of the deviation value of individual surface points in relation to the selected reference plane [4]. Rz is referred to as the maximum vertical roughness, and Sa is a parameter expressing the absolute value of the difference in height of each point compared to the arithmetic mean of the surface [4]. Sa is often used to describe implant surface roughness, and its values between 1 μm and 2 μm appear to be optimal for dental implants, however, the mechanisms behind an optimal bone response to this Sa value remain largely unknown [5]. Among the mathematical methods, a fractal dimension (FD) analysis can be listed together with the texture analysis (TA). FD analysis can be applied in the description of irregular or complex surfaces or shapes. Fractal dimension analysis (FDA) has been reported in surface testing of various dental materials; among them are xenogeneic bone substitutes, lithium disilicate-based crowns, zirconia dental implants, dental restorative composite, or orthodontic wires [6–9]. Since the fractal architecture concept is particularly interesting in surface and materials science, it also can be adapted to assess the surface of titanium dental implants. Texture analysis is used in dentistry or medicine to analyze images of computed tomography or X-ray [10–13]

Numerous implant surface modifications have been presented in the literature, most commonly used for dental implant sandblasting and etching with acid (SLA) and nanostructure-modified surfaces [14]. These modifications, by affecting the biological tissue response, undoubtedly affect both the possibility of earlier loading of the implant as well as shortening implant length [15]. Plasma electrolytic oxidation, used in our study, was used to improve the properties of medical implants. Plasma can increase the adhesiveness of the surface, therefore, can modify the surface cell adhesions.

The first aim of the study was to compare the reference SLA titanium surface to the plasma electrolytic oxidated (PEO) and PEO treated with low-pressure oxygen plasma (PEO-S) titanium plates. The second aim was to check if the texture and fractal dimension analysis can be used to evaluate dental implant surfaces with complex geometry.

The research null hypotheses were raised:

1. None of the experimental samples express cytotoxicity.
2. There is no difference between the cell growth and cell adherence of the experimental surface modified with plasma electrolytic oxidation (PEO) titanium plates and PEO treated with low-pressure oxygen plasma (PEO-S) compared to standard SLA surface.
3. There are no correlations between a fractal dimension (FD) and texture analysis (TA), nor between implant surface roughness, Sa, and cell growth.
4. There are no differences between evaluated surfaces in an aspect of FD and TA.

2. Results

The titanium (Ti) and Ti6Al4V alloy have been sandblasted (Al_2O_3) and then etched to increase surface roughness. As a result of these processes, characteristic craters and pores formed on the surface (Figure 1). After the etching process, all embedded sand grains were removed, and the surface was chemically homogeneous (Figure 1). Only substrate elements (Ti or Ti, Al, and V) and oxygen were detected on the surface. The presence of oxygen is related to the naturally formed oxide layer on titanium and its alloys in the presence of air [16].

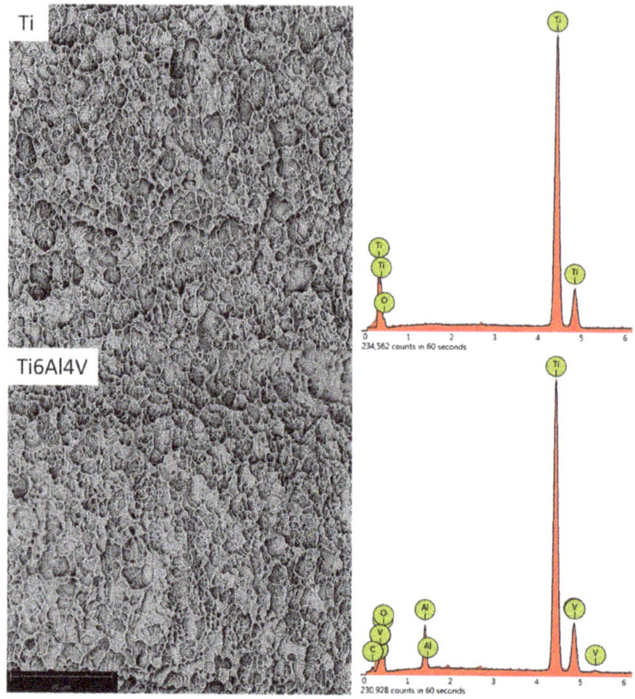

Figure 1. The SEM (scanning electron microscope) images and EDX (energy-dispersive X-ray spectrometer) spectra of Ti and Ti6Al4V samples after sandblasting and etching.

The sandblasted and etched samples are subjected to the plasma electrochemical oxidation (PEO) process in a solution containing calcium and phosphorus compounds [17,18]. During this process, the natural oxide layer on the titanium is thickened. There is an electric discharge of the forming oxide layer, and plasma with a temperature of several thousand degrees Celsius is formed in the puncture channels [19]. This melts and then solidifies the oxide layer. At the same time, the components of the electrolyte, calcium, and phosphorus compounds are incorporated into it [20,21]. The surface morphology of the modified samples is typical for the PEO process (Figure 2). The SEM (scanning electron microscope) images show typical pores resulting from discharges. Analysis of the chemical composition of EDX (energy-dispersive X-ray spectrometer) confirmed the incorporation of calcium and phosphorus into the oxide layer, both in titanium and the Ti6Al4V alloy (Ti-PEO and Ti6Al4V-PEO samples; Figure 2 and Table 1). The EDX spectra also show peaks from the gold that was sputtered on the samples before the SEM observations. Calcium and phosphorus incorporated into the oxide layer are present in the phosphate form [22,23]. It can also be seen that more calcium was incorporated into the Ti6Al4V alloy than in the case of titanium. Additional treatment of the samples with oxygen plasma did not change the surface morphology of the samples (Ti-PEO-S and Ti6Al4V-PEO-S samples; Figure 2). However, their chemical composition changed slightly. In the case of modified titanium and alloy in the oxide layer, the oxygen content increased slightly—around 1% (Ti-PEO-S and Ti6Al4V-PEO-S samples; Table 1). This means that titanium and alloy were oxygenated, which may have a positive effect on their biological response [24].

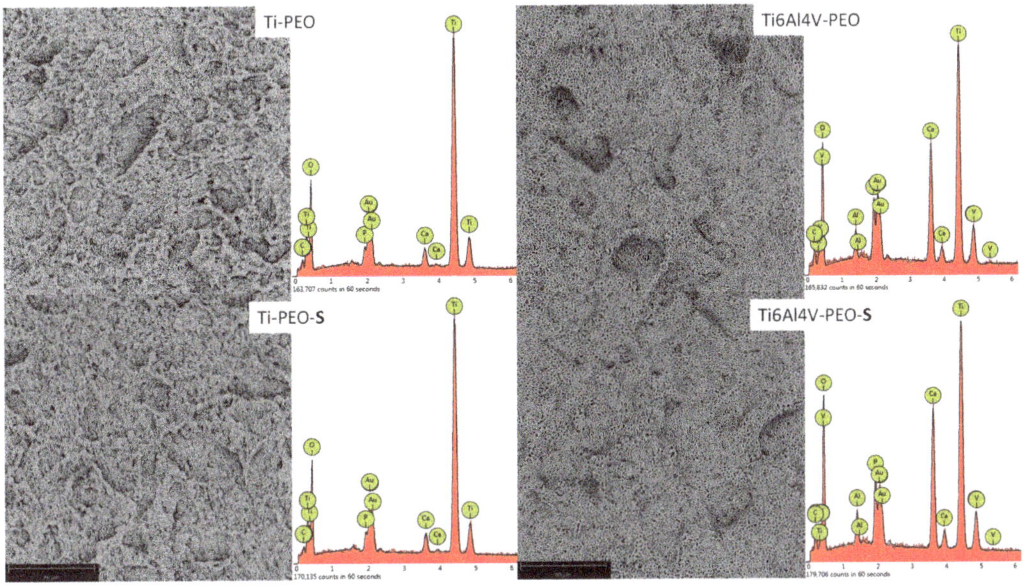

Figure 2. The SEM images of etched Ti and Ti6Al4V samples after the PEO process (Ti-PEO, Ti6Al4V-PEO) and after the PEO process with oxygen plasma treatment (Ti-PEO-S, Ti6Al4V-PEO-S).

Table 1. The semi-quantitative EDX analysis of treated samples, composition in percent (%).

	Ti	Al	V	O *	Ca	P
Ti-PEO	35	-	-	60	3	2
Ti6Al4V-PEO	23	2	1	58	12	3
Ti-PEO-S	35	-	-	61	3	1
Ti6Al4V-PEO-S	21	2	1	61	12	3

* the values must be regarded as only informative.

2.1. Surface Roughness Outcomes

Sa average surface height deviation amplitude was highest for the Ti-PEO surface and reached 2.38 μm; the lowest value of Sa 0.59 μm was noted for the Ti6Al4V-PEO-S sample. Surface roughness (Sa) for all the examined surfaces is presented in Table 2.

Table 2. Mean values of surface roughness (Sa) for each examined surface in μm.

Surface Name	Sa	SD
Ti-PEO	2.38	0.14
Ti6Al4V-PEO	1.74	0.21
Ti-PEO-S	0.86	0.02
Ti6Al4V-PEO-S	0.59	0.02
TI6Al4V	0.72	0.01
Ti	1.42	0.01

SD—standard deviation.

2.2. Biological Analysis Outcomes

2.2.1. In Vitro Cytotoxicity Assessment

The test was performed according to the protocol described in the standard for the cytotoxicity of medical devices. The study assessed the effect of extracts obtained from biomaterials on the vitality of two cell lines used for this purpose. No toxicity of the tested extracts was found. The results are shown in the graphs (Figure 3). No significant effect on the vitality of the tested cultures was found in the tests performed.

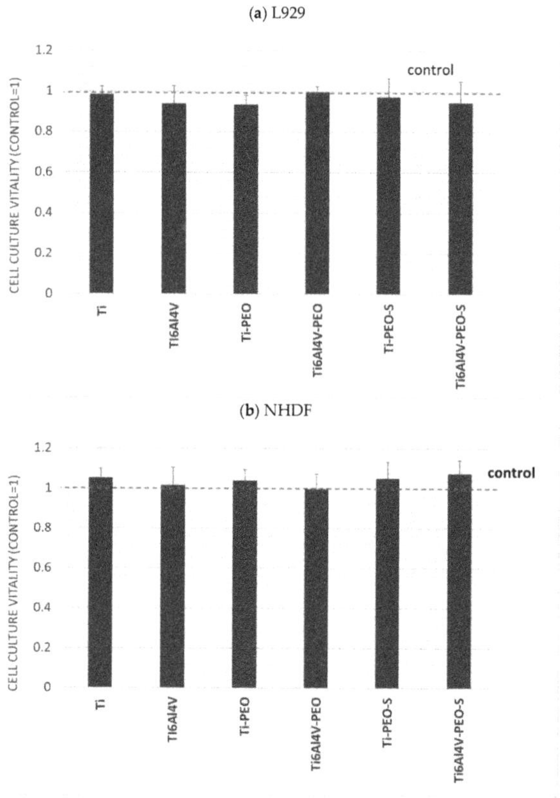

Figure 3. (**a**) cell culture vitality of the L929, and (**b**) cell culture vitality of the NHDF. Evaluation of the cytotoxicity of the tested biomaterials. Results are averages from 5 independent experiments, presented as a ratio of the value obtained in the test culture to the control. Red line—control. There was no statistically significant decrease in the viability of the culture compared to the control ($p < 0.05$).

In the MTT assay, the L929 consensus reference line and the normal human dermal fibroblasts (NHDF) line were used. For both lines, cell morphology assessment was performed according to the guidelines described in the standard and the laboratory procedures. The assessment of cell morphology was performed by comparison with the specifications for the lines and control cultures. No cytopathic changes were observed in the cultures.

2.2.2. Co-Culture of Cells with Test Materials

To assess the effect of the tested materials on cells, cell cultures were performed in their presence. During the study, cell growth was assessed in the immediate vicinity of the material and underneath, at the edge of the material. No changes in cell morphology or other effects on the culture were observed. To confirm the results, staining was performed to assess the number of live and dead cells. The number of dead cells in culture was similar for all test materials and controls. How the test was performed for each material is shown in Figure 4.

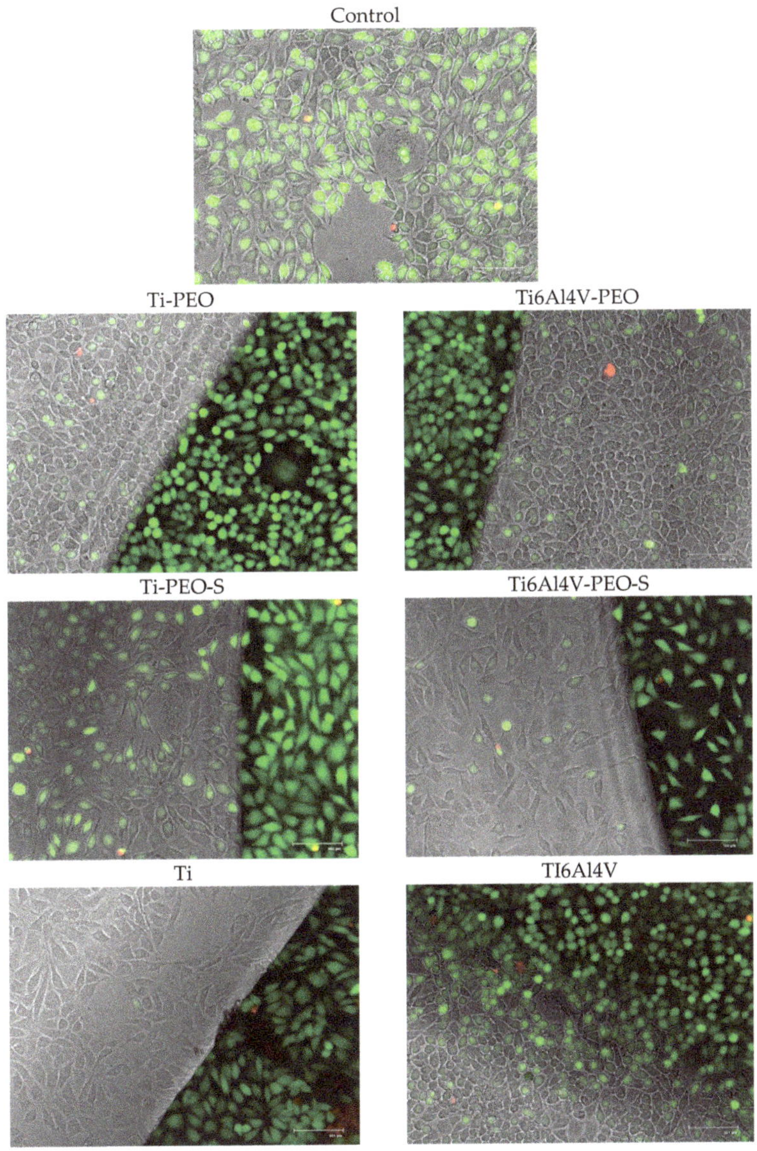

Figure 4. Co-culture of L929 cells and test materials. Live dead (green/red) staining under 20× objective magnification, FLOID microscope. Microphotographs show cells growing close to and directly under the material.

2.2.3. Cell Attachment

The most important bioassay performed was the evaluation of cell adhesion and growth directly on the test surfaces. Adherent cells can grow on a variety of surfaces, e.g., glass, but most commonly on surfaces made of different plastics. These surfaces are subjected to various modifications that make the surface in the culture dish more hydrophilic for maximum cell adhesion. In the study conducted, the control surface was modified polystyrene (TPP, Trasadingen, Switzerland). The results obtained are presented as number of cells per test surface and the fluorescence value obtained as the average value of the tested surface. The results shown in Figure 5 indicate that the introduced modifications significantly improved cell adhesion to the test surfaces compared to the original Ti and Ti6Al4V surfaces. For the PEO and PEO-S surfaces, a significant increase in cell number and adhesion fusion is observed, which is similar to or significantly better than the surface standardly used for cell culture, providing optimal cell growth conditions. Comparing PEO and PEO-S surfaces, the latter provides better conditions for cell growth.

(**a**) Number of cells per mm² of surface tested

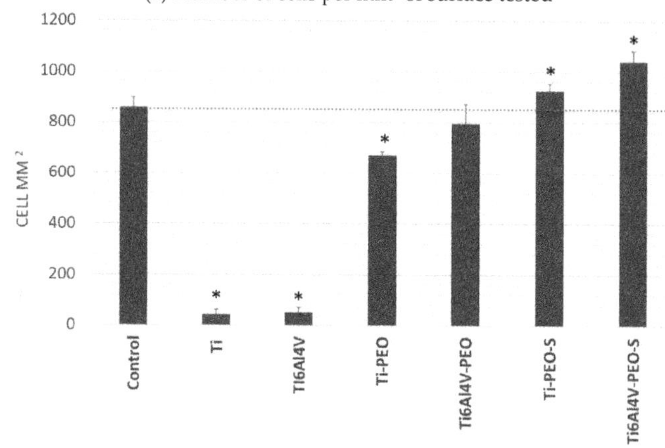

(**b**) Average fluorescence value measured from an area of 1 mm²

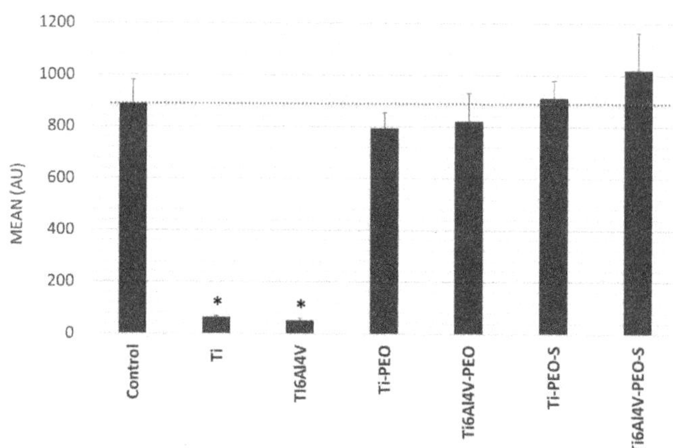

Figure 5. Adhesion and proliferation of NHDF on test surfaces modified after 72 h: (**a**) number of cells per mm^2 of test surface, (**b**) mean fluorescence value measured from an area of 1 mm^2. * Statistically significant difference ($p < 0.05$) compared with control.

2.3. Fractal Dimension Analysis

Table 3 presents mean values of fractal dimension in 100 µm × 100 µm scale. The lowest value of FD was seen in Ti6Al4V-PEO (1.782000) and the highest value in Ti6Al4V (1.888920).

Table 3. Post hoc ANOVA results (least significant difference) for comparison of fractal dimension for ROI size 100 μm × μm between each examined surface.

	Surface Name	FD (ROI = 100 μm × 100 μm)		$p < 0.05$
		Mean	SD	
1	Ti-PEO	1.854240	0.004839	2,3,4,5,6
2	Ti6Al4V-PEO	1.782000	0.007372	1,3,5,6
3	Ti-PEO-S	1.816000	0.014739	1,2,3,5
4	Ti6Al4V-PEO-S	1.782740	0.011618	1,3,5,6
5	Ti	1.804660	0.007630	1,2,4,6
6	Ti6Al4V	1.888920	0.008397	1,2,3,4,5

fractal dimension (FD), standard deviation (SD).

Table 4 shows mean values of fractal dimension in 5 μm × 5 μm scale. The lowest value of FD was seen in Ti (1.6931) and the highest value in Ti6Al4V (1.7731).

Table 4. Post hoc ANOVA results (least significant difference) for comparison of fractal dimension for ROI size 5 μm × μm between each examined surface.

	Surface	FD (ROI = 5 μm × 5 μm)		$p < 0.05$
		Mean	SD	
1	Ti-PEO	1.775360	0.012263	2,3,4,5
2	Ti6Al4V-PEO	1.746280	0.019934	1,3,5,6
3	Ti-PEO-S	1.728280	0.010314	1,2,4,5,6
4	Ti6Al4V-PEO-S	1.751200	0.009727	1,3,5,6
5	Ti	1.693080	0.006080	1,2,3,4,6
6	Ti6Al4V	1.773160	0.016584	2,3,4,5

fractal dimension (FD), standard deviation (SD).

Table 5 shows Pearson correlation coefficients between FD and Sa and amounts of cells in mm^2, medium au. We have revealed almost no linear correlation ($r = -0.02$) between Sa and amounts of cells in mm^2. A very weak negative correlation ($r = -0.52/r = -0.56$) was revealed between the value of FD (in scale 100 μm × 100 μm) and amounts of cells in mm^2, medium au.

Table 5. The values of the Pearson correlation coefficient between the value of fractal dimension calculated in different scale (100 μm × 100 μm and 5 μm × 5μm) and the Sa and number of cells per mm^2 and medium Au and differential entropy.

Feature	Versus (vs.)	Feature	r
FD (100 μm × 100 μm)	vs.	Sa	0.045
FD (5 μm × 5 μm)	vs.	Sa	0.126
FD (100 μm ×100 μm)	vs.	cells [mm^2]	−0.561
FD (5 μm × 5 μm)	vs.	cells [mm^2]	0.194
FD (100 μm ×100 μm)	vs.	medium Au	−0.523
FD (5 μm × 5 μm)	vs.	medium Au	0.239
Sa	vs.	cells [mm^2]	−0.028
Sa	vs.	medium Au	0.084

Pearson correlation coefficient (r), fractal dimension (FD), differential entropy (DifEntrp).

2.4. Texture Analysis

An example of the results of investigating the surface structure of dental implants by SEM image texture analysis is shown in Figure 6.

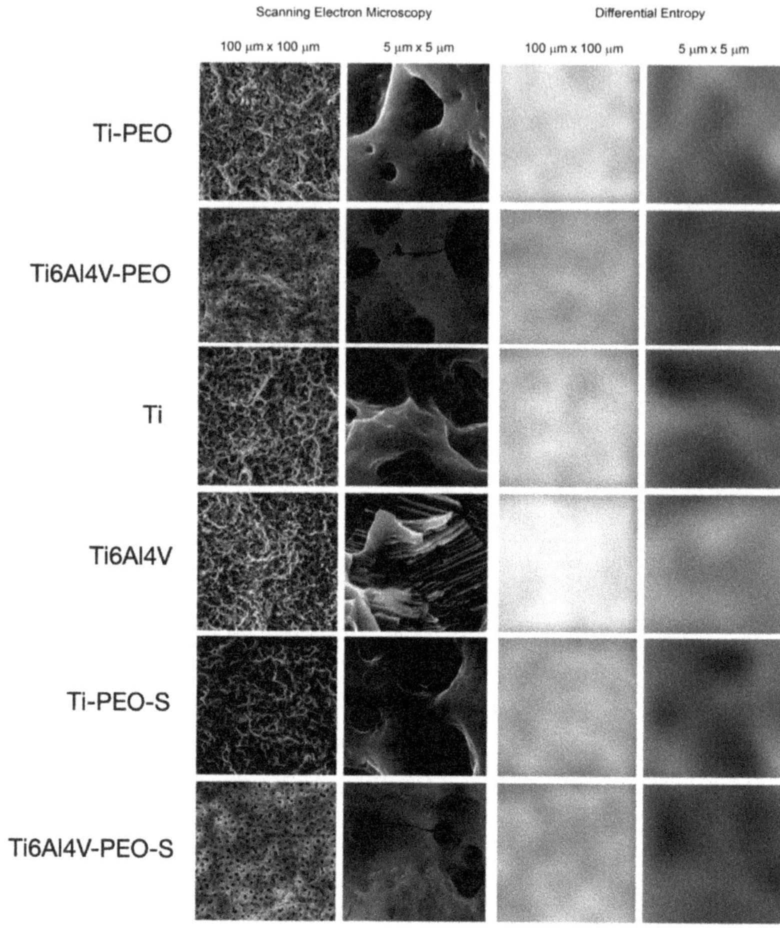

Figure 6. Texture analysis by means of calculation of differential entropy in SEM images on two scales. The two columns on the left show SEM images in the large (100 μm × 100 μm) and small field of view (5 μm × 5 μm). The two columns on the right represent intensity maps of the texture feature studied here in the original SEM image. The whiter areas indicate where the differential entropy is higher (i.e., the surface development is greater), while the darker areas indicate where the differential entropy is low (the implant surface image is more homogeneous). The differences between the tested surfaces are statistically significantly different from one another ($p < 0.05$) in terms of differential entropy (both at low and high magnification).

When examining the differential entropy as a measure of the development of the implant surface as seen in the SEM, significant differences were noted between the analyzed surfaces in the 100 μm × 100 μm field of view. Statistically significant differences are presented in Table 6.

Table 6. Post hoc ANOVA results (least significant difference) for comparison of difference entropy (DifEntrp) for ROI size 100 µm × µm between each examined surface (SD—standard deviation).

	Surface	DifEntrp		$p < 0.05$
		Mean	SD	
1	Ti-PEO	1.2948	0.0070	3,4,5,6
2	Ti6Al4V-PEO	1.2807	0.0134	4,5,6
3	Ti-PEO-S	1.2577	0.0144	1,4,6
4	Ti6Al4V-PEO-S	1.2208	0.0183	1,2,3,5,6
5	Ti	1.2504	0.0062	1,2,4,6
6	Ti6Al4V	1.3252	0.0055	1,2,3,4,5

difference entropy (DifEntrp), standard deviation (SD).

In turn, when examining the differential entropy of the implant surface image at higher magnification (i.e., in a 5 µm × 5 µm field of view), the texture feature increased, as presented in Table 7.

Table 7. Post hoc ANOVA results (least significant difference) for comparison of difference entropy for ROI size 5 µm × 5 µm between each examined surface.

	Surface	DifEntrp		$p < 0.05$
		Mean	SD	
1	Ti-PEO	1.1799	0.0521	5,6
2	Ti6Al4V-PEO	1.2329	0.2890	3,5,6
3	Ti-PEO-S	1.1340	0.0275	2
4	Ti6Al4V-PEO-S	1.1779	0.0326	6
5	Ti	1.1180	0.0738	1,2
6	Ti6Al4V	1.0897	0.0286	1,2,4

difference entropy (DifEntrp), standard deviation (SD).

The increase of Sa was related ($r = 0.41$) with an increase in difference entropy of theimplant surface measured in 100 × 100 µm field of view (Table. 8). DifEntrp (5 µm × 5 µm) was positively correlated with the medium Au and cells [mm^2] ($r = 0.72$ and $r = 0.68$ respectively). On the contrary, a negative correlation was found between DifEntrp (100 µm × 100 µm) and the medium Au and cells [mm^2] ($r = -0.46$ and $r = -0.51$ respectively). A moderate correlation ($r = 0.76$) was found between DiffEntrop (100 µm × 100 µm) and FD. No correlation was found between DiffEntrop (5 µm × 5 µm) and FD ($r = 0.06$). Table 8.

Regarding research hypotheses:

1. The first null hypothesis was accepted. None of the experimental samples expressed cytotoxicity.
2. The second null hypothesis has been rejected. Samples after surface treatment have substantially improved cell growth and cell adherence compared to reference SLA samples.
3. The third null hypothesis has been sustained. We did not reveal a correlation between examined features, except a negative correlation between FD, difference entropy (DifEntrp) (in scale 100 µm × 100 µm), and amount of cells, a positive moderate correlation between DifEntrp and number of cells, and a positive strong correlation between the DifEntrp and FD in scale 100 µm × 100 µm.
4. The fourth null hypothesis has been rejected. Our study revealed statistically significant differences between examined surfaces in the aspect of fractal dimension and texture analysis.

Table 8. The values of the Pearson correlation coefficient (r) between the value of difference entropy (DifEntrp) calculated in different scales (100 μm × 100 μm and 5 μm × 5μm) and the Sa and number of cells per mm^2 and medium Au. No statistically significant relations were found.

Feature	vs.	Feature	r
DifEntrp (100 μm × 100 μm)	vs.	Sa	0.2904
DifEntrp (5 μm × 5 μm)	vs.	Sa	0.4173
DifEntrp (100 μm × 100 μm)	vs.	FD	0.7667
DifEntrp (5 μm × 5 μm)	vs.	FD	0.0606
DifEntrp (100 μm × 100 μm)	vs.	cells [mm^2]	−0.5145
DifEntrp (5 μm × 5 μm)	vs.	cells [mm^2]	0.6813
DifEntrp (100 μm × 100 μm)	vs.	medium Au	−0.4656
DifEntrp (5 μm × 5 μm)	vs.	medium Au	0.7201

difference entropy (DifEntrp).

3. Discussion

The properties of the titanium alloy's surface, including its microtopography and nanoscale modification and chemical composition, as well as strategies and methods for improving biocompatibility to bone tissue, have been contemporarily well investigated [14,25]. Wang et al. [26] have found that titanium disc surfaces treated with low-temperature argon–oxygen plasma are more hydrophilic compared to nontreated surfaces, and the activation with plasma can enhance the attachment, proliferation, and mineralization of osteoblasts on the surfaces. A recent study by Hadzik et al. [27] has shown that the anodization of Ti-6Al-4V alloy and its further low-pressure radiofrequency oxygen plasma treatment is a promising method of implant transgingival parts modification. Cheng et al. [28] have found that, when bioactive oxide film on titanium dental implants is created by the oxygen plasma, bone cells' differentiation and osseointegration are improved. Such modifications provide effective binding to hard tissue and, therefore, promote osteointegration. Our study has confirmed that none of the tested surfaces here have any cytotoxic effect on HGF cell lines, so they can be safe when used as dental implant surfaces. Results of our study prove that PEO as well as PEO-S that were treated with a low-pressure OP represent promising options of Ti surface modification. We have reported a significant increase in cell number and adhesion fusion specifically for the tested experimental PEO and PEO-S surfaces compared to the SLA reference surface.

In our study, the fibroblast model was selected to assess the cytocompatibility of the experimental samples. The adhesion behavior of fibroblasts is known to differ among materials with varying degrees of surface roughness. Generally, maximum adhesion is observed for more rough surfaces. The reason for that is the physiology of fibroblasts and their filopodial structure that extends further into grooves and microstructures of rough surfaces. However, a strong positive relationship between bacterial adhesion and plaque accumulation rate and surface roughness in the supragingival region has been also reported. Rough surfaces in such an application may lead to possible peri-implant mucositis and periimplantitis development [29–31].

Hence, the ideal micro- and nano-scale titanium surface topography for implants should balance the facility for fibroblast adhesion without simultaneously favoring bacterial growth. One of the methods of titanium surface nanotexturing that matches the abovementioned condition of the ideal surface is electrochemical anodization. Due to that process, the hierarchic superstructure of the TiO_2 layer in the shape of nanotubes can be synthesized.

It seems that, in the case of the tested experimental surfaces, their cellular response may be related not only to the roughness, which, reaching Sa values from 0.59 to 2.38 nm for various tested surfaces, is within or slightly exceeds the optimal roughness values described in the literature by Wennerberg and Albrektsson, 1–2 μm [5]. Blinova et al. [32]

were one of the first to report fibroblast ingrowth both on its surface and inside of it when cultured on a porous titanium implant. They also proved that interactions between the cell and titanium implant are more evident in samples with nonuniform porosity. Moreover, Whiteside et al. [33] reported more specific criteria for biologically advantageous porosity, where surfaces with a greater number of finer pores are favored for cell attachment.

The plasma electrochemical oxidation (PEO) process in a solution containing calcium and phosphorus compounds has to cause the natural oxide layer on the titanium to thicken. EDX study has confirmed the incorporation of calcium and phosphorus into the oxide layer, both in titanium and the Ti6Al4V alloy. Additional treatment of the samples with oxygen plasma did not change the surface morphology of the samples. However, of modified titanium and alloy in the oxide layer, the oxygen content increased slightly—around 1%. The significantly better cell adhesion we achieved compared to reference SLA samples is conditioned by the chemical structure that results from the applied PEO and PEO-S treatment of the implant surfaces.

The fractal dimension analysis (FDA) of the samples has provided a mathematical formalism for describing complex spatial and dynamical structures and describing the entropy potential of the surface. The entropy of the surface generally raises with it complicity and roughness. It has been broadly used in many areas of science, including medicine, dentistry, technology, and materials science [34]. However, the contemporary method is rarely employed to evaluate titanium implant surfaces. In this study, we can observe a negative moderate correlation between FD and amounts of cells in scale 100 μm × 100 μm at a similar level of correlation between difference entropy (DifEntrp) and cell number. This observation has a reflection in FD interpretation. The lower FD's value is, the more complex of an analyzed surface is present and the higher number of cells we observed. This observation confirmed that FD and DifEntrp in scale 100 μm × 100 μm and DifEntrp in scale 5 μm × 5 μm can be used as surface features in the aspect of cell ability to colonize. Our results show a lack of correlations between Sa and medium Au and cell growth. We revealed a lack of correlation between FD and Sa in contrast to a weak correlation between DifEntrp and Sa in both scales. It is interesting to observe a strong positive correlation between FD and DifEntrp in scale 100 μm × 100 μm in contrast to a weak negative correlation in scale 5 μm × 5 μm. Skośkiewicz-Malinowska et al. [35] reported a positive moderate correlation between the FDA and the number of the fibroblasts when tested on cement surfaces. In our previous study, a weak correlation (r = 0.38) between the number of fibroblasts and the fractal dimension in the 100 μm × 100 μm scale was observed. Meanwhile, in the 5 μm × 5 μm scale, the correlation coefficient was lower, and r = 0.24 [7].

4. Materials and Methods

For the purpose of this study, titanium plates as a sample of different experimental implant surfaces were delivered and tested. The roughness parameters of the surface were measured. Based on the SEM images, fractal dimension analysis and texture analysis was calculated. All the titanium plates were tested for cytotoxicity and, finally, normal human dermal fibroblasts (NHDF) were used to assess the cell culture adhesion to each Ti sample.

4.1. Titanium Plates Preparation and Surface Modification

Titanium plates with standard SLA (sandblasted and acid-etched) were prepared and delivered by NanoPrime company (NanoPrime, Dębica, Poland) from the Titanium Grade 4 (Ti) and Titanium Grade 23 (Ti6Al4V). Ti and Ti6Al4V sandblasted and acid-etched Titanium dental implant samples used in this study were previously used in our study as a reference to compare against the Laser-Induced Periodic Surface Structures (LIPSS) dental implant surfaces in our other study [7]. In this study, new experimental electrolytic-modified implant surfaces were used for Ti-PEO and Ti6Al4V-PEO (plasma electrolytic oxidation) and Ti-PEO-S and Ti6Al4V-PEO-S treated with additional low-pressure radio-frequency oxygen plasma (low-pressure RF OP). The details of the titanium surface samples are presented in Table 9.

Table 9. Preparation of the titanium plates.

Name	Titanium Grade	Method of Preparation
Ti Ti6Al4V	Grade 4 Grade 23	Sandblasted and acid-etched (SLA) Titanium Dental Implant—Al_2O_3 sandblasting process with a fraction of 30–100 μm. Purified samples were subjected to the etching process (conditions: oxalic acid 100 g L^{-1}, time: 60 min, temperature: boiling). Samples were washed in an ultrasonic cleaner (DEMI water, time: 10 min).
Ti-PEO Ti6Al4V-PEO	Grade 4 Grade 23	SLA surfaces were anodized in a PEO (plasma electrolytic oxidation) regime. Treatment details were presented in previous studies by Simka et al. [17,18]. An electrolyte was composed of Ca and P compounds. Titanium surfaces were oxidized via the PEO process with a high voltage DC power supply, Kikusui PWR400H, (Kikusui Electronics Corporation, Kanagawa, Japan) at 300 V for 5 min. The PEO treatment was realized via DC galvanostatic anodization (anodic current density = 100 mA cm^{-2}) up to limiting voltage. After the process voltage reached the limiting voltage (300 V), the treatment was conducted under a potentiostatic regime. Samples were washed in an ultrasonic cleaner (DEMI water, time: 10 min).
Ti-PEO-S Ti6Al4V-PEO-S	Grade 4 Grade 23	After PEO, samples were treated with low-pressure RF OP and placed in a vacuum chamber for 5 min. Frequency: 40 Mhz, power: 500 Watt. During this time, oxygen was pumped into the chamber (1 L/min).

4.2. *Surface Analysis Surface Topography Ra, Rz, Sa Measurement*

The surface roughness (Sa) parameter was measured with the use of a scanning electron microscope SEM (Thermo Fisher Scientific Inc., Waltham, MA, USA), and 15 keV accelerating voltage was applied. The roughness parameters were measured with Phenom 3D Roughness Reconstruction Software (version 1, Thermo Fisher Scientific Inc., Waltham, MA, USA). Each surface was triplicated, and the measurements were done in four places on each sample. The results were averaged, and a standard deviation (SD) was calculated.

4.3. *Surface Analysis FDA*

We analyzed SEM images under two magnifications: 5000× and 100,000×. Five regions of interest (ROIs) were selected from each image. ROI dimension was 100 μm × 100 μm in case of 5000× magnification and 5 μm × 5 μm for 100,000× magnification. These were applied in the intensity difference fractal dimension counting method to analyze each ROI.

ImageJ version 1.53e (Image Processing and Analysis in Java—Wayne Rasband and contributors, National Institutes of Health, Bethesda, MD, USA, public domain license, https://imagej.nih.gov/ij/, accessed on 1 December 2022) and the FracLac plugin version 2.5 (Charles Sturt University, Bathurs, Australia, public domain license) were applied to do all fractal dimension analyses. The full algorithm of fractal dimension calculation was fully described in our previous study [7].

4.4. *Surface Analysis TA*

MaZda 4.6 (MaZda ver. 4.6, Technical University of Łodź, Institute of Electronics, Łodź, Poland) was used to check how the features were describing analyzed images [36,37]. Primary 8-bit images were reduced to 6 bits. Regions of interest (ROIs) were normalized ($\mu \pm 3\sigma$) to share the same average (μ) and standard deviation (SD) of optical density within the ROI. Difference entropy (DifEntrp) was selected as a texture feature from the co-occurrence matrix to calculate in the ROI:

$$\text{DifEntr} = -\sum_{i=1}^{Ng} p_{x-y}(i) log(p_{x-y}(i)) \quad (1)$$

where Σ is sum, *Ng* is the number of levels of optical density in the microphotograph, *i* and *j* are optical density of pixels 5-pixel distant one from another, *p* is probability, and *log* is common logarithm [36,38,39].

4.5. *Biological Analyses*

4.5.1. Cell Culture

In vitro studies were performed using two models: the Normal Human Dermal Fibroblast (NHDF) cell line (Lonza Group, Basel, Switzerland) and L929 cells, (Sigma-Aldrich, Merck Group, Darmstadt, Germany) (The European Collection of Authenticated Cell Cultures-ECACC). Cells were cultured under standard conditions at 37 °C, 5% CO_2,

95% humidity, in a CO_2 incubator. Cells were always cultured for a minimum of 2 weeks after thawing before starting a series of experiments. Cell cultures were passaged with trypsin/EDTA solution. Cells were counted using a NucleoCounter® NC-200 automatic cell counter (ChemoMetec A/S, Allerod, Denmark). Cells were cultured in Dulbecco's modified Eagle medium (DMEM), 10% fetal bovine serum (FBS), penicillin (10,000 U/mL), streptomycin (10 mg/mL), and L-glutamine (200 mM). All culture reagents were purchased from Biological Industries (Biological Industries, Kibbutz Beit-Haemek, Israel).

4.5.2. Preparation of Samples

The test samples were packaged and autoclaved for sterilization. For the in vitro biological cytotoxicity assessment test, liquid extracts of the test materials were prepared according to the provisions of the standard: EN ISO 10993-5: Biological evaluation of medical devices—Part 5: In vitro cytotoxicity testing. Extraction was carried out in sterile, chemically inert, sealed tubes for 24 h at 37 °C in an incubator. Before the experiments involving direct assessment of the interaction of materials with cells, the prepared sections were wetted with serum culture medium. The test material was incubated in the presence of culture medium at a ratio of 1:10. The resulting extract was sterilized by phytoextraction through a 0.22 µm filter.

4.5.3. Evaluation of the Effects on Growth and Vitality of Cell Cultures

The test was performed according to the guidelines of the standard for testing the cytotoxicity of biomaterials. Cells were obtained from culture NHDF. Tests were performed in 96-well plates at 1×10^4 cells/well. Cells were incubated for 24 h (5% CO_2, 37 °C, 90% humidity) so that the cells formed a monolayer on the plate surface. Before the experiment, each plate was checked under a phase-contrast microscope to ensure that cell growth was relatively uniform across the test plate. After 24 h of incubation, the medium was removed from above the cells. Then, 100 µL of medium containing the appropriate sample extracts, control, or blank only was added to each well. The test plates were incubated for a further 24 h (5% CO_2, 37 °C, 90% humidity). After 24 h of incubation, each plate was viewed under a phase-contrast microscope to assess the growth of control and extract-treated cells. The observed changes in cell morphology may have been due to the cytotoxic effect of the test sample extract. The culture medium was then carefully removed from the plates and 50 µL of 1 mg/mL MTT solution was added to each well. The plates were incubated for a further 2 h in an incubator at 37 °C. After this time, the MTT solution was removed and 100 µL of isopropanol was added to each well. Absorbance was read on a MultiscanGo reader (Thermo Scientific, Waltham, MA, USA) at 570 nm.

4.5.4. Co-Culture of Cells with Materials

In this assay, L929 cells were seeded into 24-well plates alongside previously placed sterile test biomaterials. In this method, the direct interaction between the cells and the test material was checked by assessing the morphology of the cells and the percentage of live and dead cells (Cell Viability Imaging Kit, Green/Red), which was evaluated on a fluorescence microscope EVOS FLoid (Thermo Scientific, Waltham, MA, USA).

4.5.5. Cell Attachment

Cells for the study were obtained according to the methodology described above. Cultures of normal human dermal fibroblasts (NHDF) were used for the study. The cell suspension prepared for the test was counted and suspended in a culture medium. The density of the cell suspension for the adhesion assay was 1×10^6 cells/mL, and 24-well plates were used for the assay. Materials for the test were placed in wells. Cells in the assay were applied to the wetted material using an automatic pipette. After application, the cells were incubated for 2 h (5% CO_2, 37 °C, 90% humidity) to allow the cells to adhere to the test materials. After this time, the wells were replenished with serum culture medium in a volume of 1000 µL. The test plates were incubated for a further 72 h (5% CO_2, 37 °C,

90% humidity). After the incubation time, cells growing on the test surfaces were stained using the Cell Viability Imaging Kit (Blue/Green). Staining involves adding dye to the culture and incubating for 5 to 30 min. After this time, images were taken using a BioTek Lionheart microscope (Agilent Technologies, Santa Clara, CA, USA) using fluorescence excitation with an led illuminator: ex 377 em 447 and ex 469 em 525. Further analysis was performed using GEN5 dedicated image analysis software (Agilent Technologies, Santa Clara, CA, USA). The fluorescence intensity and the number of cells stained with each dye were analyzed. Cells showing blue fluorescence were counted as alive, and green cells as dead.

4.6. Statistical Analysis

Statistica version 13.3 (StatSoft, Cracow, Poland) was applied to calculate all fractal dimension analysis statistical tests. The statistically significant level was set to 0.05. The normality of distribution was confirmed by the Shapiro–Wilk test. Due to normal distribution, we performed parametric tests. Analysis of variance (ANOVA) and post hoc least significant difference was used to show differences in fractal dimensions between each surface in two scales. The correlation matrix was applied to calculate the Pearson correlation coefficient (r) between the fractal dimension of lesions in two scales and Sa, the number of cells in mm^2, medium Au, and Sa vs. amounts of cells in mm^2 and medium Au. The following are the ranges of the r value: r greater than or equal to 0.7—strong correlation; r between 0.7 and 0.5—moderate; and r lower than 0.5—weak correlation. Sample size was calculated on the basis of a power of test. In this study, we used a one way ANOVA for five groups. In this case, the 80% of power of test is achieved for $N = 50$ in each group.

For texture analysis, between-group comparisons were performed with the one way ANOVA or the Kruskal–Wallis test, depending on the presence of normal distribution. Statgraphics Centurion version 18.1.12 (StatPoint Technologies Inc., Warrenton, VA, USA) was used for statistical analyses.

4.7. Study Limitations

A limitation of the study was that a machined titanium sample was not used as a control for the cell culture tests. Furthermore, the wettability and microhardness properties of the modified surfaces should be investigated in the future.

5. Conclusions

The presented study shows that the surface modification by PEO and PEO-S did not affect the sample cytotoxicity. Greater cell growth of HGF cells was observed on PEO and PEO-S samples compared to reference SLA titanium. The number of cells is correlated with the value of fractal dimension and DifEntrp in scale 100 μm × 100 μm. These two parameters can be used to describe the potential of the surface in the aspect of the ability of a cell to grow. A strong positive correlation between fractal dimension value and DifEntrp was found. In the case of our samples, we have not found a correlation between Sa value and cell growth.

Author Contributions: Conceptualization, J.H. and K.J.; methodology, J.H., K.J. and W.S.; software, K.J., W.S., J.H. and M.K.; formal analysis, M.D.; investigation, J.H., T.G. (Tomasz Gębarowski) and W.S.; resources, J.H., A.T., E.S. and A.T.-S.; data curation, J.H. and K.J.; writing—original draft preparation, J.H.; writing—review and editing, J.H., K.J., P.K.-R. and W.S.; supervision, J.H. and W.S.; project administration, J.H.; funding acquisition, M.D. and T.G. (Tomasz Gedrange). All authors have read and agreed to the published version of the manuscript.

Funding: This research was supported by the Wrocław Medical University Statutory Grant no. SUBZ.B040.23.048 and University of Łodź grant no. 503/1-138-01/503-51-001-19-00. Samples of the experimental dental implant surfaces were donated by Nano Prime sp z o.o. (Dębica, Poland).

Institutional Review Board Statement: Not applicable.

Informed Consent Statement: Not applicable.

Data Availability Statement: Data available on request.

Acknowledgments: Authors would like to thank Anna Trzcionka Szajna and Ernest Szajna for preparing and delivering titanium samples for testing.

Conflicts of Interest: Ernest Szajna represented the WEA Techlab sp. z o.o., ul. Perla 10, 41-301 Dąbrowa Górnicza the manufacturer of the experimental dental implant surfaces. Anna Trzcionka Szajna NanoPrime ltd represented the Nano Prime Sp. z o.o. 39-200 Dębica ul. Metalowców 25. The principal investigator in the study Jakub Hadzik and the other subinvestigators responsible for research and testing samples are not related to WEA Techlab ltd. and NanoPrime ltd. and did not receive from either WEA Techlab ltd. or NanoPrime ltd. any remuneration for carrying out research. The data are in the possession of the authors. Ernest Szajna and Anna Trzcionka-Szajna declare that have not interfered with the sample testing process or the publication this manuscript.

References

1. Hoque, M.E.; Showva, N.N.; Ahmed, M.; Bin Rashid, A.; Sadique, S.E.; El-Bialy, T.; Xu, H. Titanium and titanium alloys in dentistry: Current trends, recent developments, and future prospects. *Heliyon* **2022**, *8*, e11300. [CrossRef]
2. Liu, X.; Chen, S.; Tsoi, J.K.H.; Matinlinna, J.P. Binary titanium alloys as dental implant materials-a review. *Regen. Biomater.* **2017**, *4*, 315–323. [CrossRef] [PubMed]
3. Cruz, M.B.; Silva, N.; Marques, J.F.; Mata, A.; Silva, F.S.; Caramês, J. Biomimetic Implant Surfaces and Their Role in Biological Integration—A Concise Review. *Biomimetics* **2022**, *7*, 74. [CrossRef] [PubMed]
4. Dohan Ehrenfest, D.M.; Coelho, P.G.; Kang, B.S.; Sul, Y.T.; Albrektsson, T. Classification of osseointegrated implant surfaces: Materials, chemistry and topography. *Trends Biotechnol.* **2010**, *28*, 198–206. [CrossRef]
5. Wennerberg, A.; Albrektsson, T. Effects of titanium surface topography on bone integration: A systematic review. *Clin. Oral Implants Res.* **2009**, *20*, 172–184. [CrossRef]
6. Jurczyszyn, K.; Kubasiewicz-Ross, P.; Nawrot-Hadzik, I.; Gedrange, T.; Dominiak, M.; Hadzik, J. Fractal dimension analysis a supplementary mathematical method for bone defect regeneration measurement. *Ann. Anat.* **2018**, *219*, 83–88. [CrossRef]
7. Hadzik, J.; Kubasiewicz-Ross, P.; Simka, W.; Gębarowski, T.; Barg, E.; Cieśla-Niechwiadowicz, A.; Szajna, A.T.; Szajna, E.; Gedrange, T.; Kozakiewicz, M.; et al. Fractal Dimension and Texture Analysis in the Assessment of Experimental Laser-Induced Periodic Surface Structures (LIPSS) Dental Implant Surface-In Vitro Study Preliminary Report. *Materials* **2022**, *15*, 2713. [CrossRef]
8. Grzebieluch, W.; Kowalewski, P.; Grygier, D.; Rutkowska-Gorczyca, M.; Kozakiewicz, M.; Jurczyszyn, K. Printable and machinable dental restorative composites for cad/cam application—Comparison of mechanical properties, fractographic, texture and fractal dimension analysis. *Materials* **2021**, *14*, 4919. [CrossRef]
9. Schestatsky, R.; Zucuni, C.P.; Dapieve, K.S.; Burgo, T.A.L.; Spazzin, A.O.; Bacchi, A.; Valandro, L.F.; Pereira, G.K.R. Microstructure, topography, surface roughness, fractal dimension, internal and marginal adaptation of pressed and milled lithium-disilicate monolithic restorations. *J. Prosthodont. Res.* **2020**, *64*, 12–19. [CrossRef] [PubMed]
10. Costa, A.L.F.; de Souza Carreira, B.; Fardim, K.A.C.; Nussi, A.D.; da Silva Lima, V.C.; Miguel, M.M.V.; Jardini, M.A.N.; Santamaria, M.P.; de Castro Lopes, S.L.P. Texture analysis of cone beam computed tomography images reveals dental implant stability. *Int. J. Oral Maxillofac. Surg.* **2021**, *50*, 1609–1616. [CrossRef]
11. Abdollah, V.; Parent, E.C.; Dolatabadi, S.; Marr, E.; Croutze, R.; Wachowicz, K.; Kawchuk, G. Texture analysis in the classification of T2-weighted magnetic resonance images in persons with and without low back pain. *J. Orthop. Res.* **2021**, *39*, 2187–2196. [CrossRef]
12. Kozakiewicz, M.; Wach, T. Exploring the Importance of Corticalization Occurring in Alveolar Bone Surrounding a Dental Implant. *J. Clin. Med.* **2022**, *11*, 7189. [CrossRef]
13. Kozakiewicz, M. Measures of Corticalization. *J. Clin. Med.* **2022**, *11*, 5463. [CrossRef]
14. Makowiecki, A.; Hadzik, J.; Błaszczyszyn, A.; Gedrange, T.; Dominiak, M. An evaluation of superhydrophilic surfaces of dental implants—A systematic review and meta-analysis. *BMC Oral Health* **2019**, *19*, 79. [CrossRef]
15. Hadzik, J.; Kubasiewicz-Ross, P.; Nawrot-Hadzik, I.; Gedrange, T.; Pitułaj, A.; Dominiak, M. Short (6 mm) and regular dental implants in the posterior maxilla–7-years follow-up study. *J. Clin. Med.* **2021**, *10*, 940. [CrossRef] [PubMed]
16. De Fenzo, A.; Scherillo, F.; Astarita, A.; Testani, C.; Squillace, A.; Bellucci, F. Influence of hot deformation on the electrochemical behavior of natural oxides of hipped Ti6Al4V. *Corros. Sci.* **2015**, *94*, 79–87. [CrossRef]
17. Simka, W.; Pogorielov, M.; Mishchenko, O. Method for Modifying the Surface of Implants Made of Titanium or Titanium. Alloys. Patent PL-240205, 23 June 2019.
18. Kyrylenko, S.; Oleshko, O.; Warchol, F.; Husak, Y.; Basiaga, M.; Kazek-Kesik, A.; Dercz, G.; Pogorielov, M.; Simka, W. A New Solution for an Electrochemical Modification of Titanium Surface. *J. Nano-Electron. Phys.* **2020**, *12*, 06038. [CrossRef]
19. Fattah-Alhosseini, A.; Molaei, M.; Babaei, K. The effects of nano- and micro-particles on properties of plasma electrolytic oxidation (PEO) coatings applied on titanium substrates: A review. *Surf. Interfaces* **2020**, *21*, 100659. [CrossRef]
20. Jamali, R.; Bordbar-Khiabani, A.; Yarmand, B.; Mozafari, M.; Kolahi, A. Effects of co-incorporated ternary elements on biocorrosion stability, antibacterial efficacy, and cytotoxicity of plasma electrolytic oxidized titanium for implant dentistry. *Mater. Chem. Phys.* **2022**, *276*, 125436. [CrossRef]

21. Francisca, F.G.S.; Vitoriano, J.D.O.; Alves-Junior, C. Controlling plasma electrolytic oxidation of titanium using current pulses compatible with the duration of microdischarges. *Results Mater.* **2022**, *15*, 100310. [CrossRef]
22. Kyrylenko, S.; Warchoł, F.; Oleshko, O.; Husak, Y.; Kazek-Kęsik, A.; Korniienko, V.; Deineka, V.; Sowa, M.; Maciej, A.; Michalska, J.; et al. Effects of the sources of calcium and phosphorus on the structural and functional properties of ceramic coatings on titanium dental implants produced by plasma electrolytic oxidation. *Mater. Sci. Eng. C* **2021**, *119*, 111607. [CrossRef]
23. Michalska, J.; Sowa, M.; Stolarczyk, A.; Warchoł, F.; Nikiforow, K.; Pisarek, M.; Dercz, G.; Pogorielov, M.; Mishchenko, O.; Simka, W. Plasma electrolytic oxidation of Zr-Ti-Nb alloy in phosphate-formate-EDTA electrolyte. *Electrochim. Acta* **2022**, *419*, 140375. [CrossRef]
24. Chou, W.C.; Wang, R.C.C.; Huang, C.L.; Lee, T.M. The effect of plasma treatment on the osseointegration of rough titanium implant: A histo-morphometric study in rabbits. *J. Dent. Sci.* **2018**, *13*, 267–273. [CrossRef]
25. Liu, Y.; Rath, B.; Tingart, M.; Eschweiler, J. Role of implants surface modification in osseointegration: A systematic review. *J. Biomed. Mater. Res. Part A* **2020**, *108*, 470–484. [CrossRef]
26. Wang, L.; Wang, W.; Zhao, H.; Liu, Y.; Liu, J.; Bai, N. Bioactive Effects of Low-Temperature Argon-Oxygen Plasma on a Titanium Implant Surface. *ACS Omega* **2020**, *5*, 3996–4003. [CrossRef] [PubMed]
27. Hadzik, J.; Kubasiewicz-Ross, P.; Gębarowski, T.; Waloszczyk, N.; Maciej, A.; Stolarczyk, A.; Gedrange, T.; Dominiak, M.; Szajna, E.; Simka, W. An Experimental Anodized Titanium Surface for Transgingival Dental Implant Elements—Preliminary Report. *J. Funct. Biomater.* **2023**, *14*, 34. [CrossRef] [PubMed]
28. Chen, C.S.; Chang, J.H.; Srimaneepong, V.; Wen, J.Y.; Tung, O.H.; Yang, C.H.; Lin, H.C.; Lee, T.H.; Han, Y.; Huang, H.H. Improving the in vitro cell differentiation and in vivo osseointegration of titanium dental implant through oxygen plasma immersion ion implantation treatment. *Surf. Coat. Technol.* **2020**, *399*, 126125. [CrossRef]
29. Kubasiewicz-Ross, P.; Fleischer, M.; Pitułaj, A.; Hadzik, J.; Nawrot-Hadzik, I.; Bortkiewicz, O.; Dominiak, M.; Jurczyszyn, K. Evaluation of the three methods of bacterial decontamination on implants with three different surfaces. *Adv. Clin. Exp. Med.* **2020**, *29*, 177–182. [CrossRef]
30. Romandini, M.; Lima, C.; Pedrinaci, I.; Araoz, A.; Soldini, M.C.; Sanz, M. Prevalence and risk/protective indicators of peri-implant diseases: A university-representative cross-sectional study. *Clin. Oral Implants Res.* **2021**, *32*, 112–122. [CrossRef] [PubMed]
31. Hussain, R.A.; Miloro, M.; Cohen, J.B. An Update on the Treatment of Periimplantitis. *Dent. Clin. North Am.* **2021**, *65*, 43–56. [CrossRef]
32. Blinova, M.I.; Yudintzeva, N.M.; Nikolaenko, N.S.; Potokin, I.L.; Raykhtsaum, G.; Pitkin, M.R.; Pinaev, G.P. Cell cultivation on porous titanium implants with various structures. *Cell Tissue Biol.* **2010**, *4*, 572–579. [CrossRef]
33. Whiteside, P.; Matykina, E.; Gough, J.E.; Skeldon, P.; Thompson, G.E. In vitro evaluation of cell proliferation and collagen synthesis on titanium following plasma electrolytic oxidation. *J. Biomed. Mater. Res. Part A* **2010**, *94*, 38–46. [CrossRef]
34. Kato, C.N.A.O.; Barra, S.G.; Tavares, N.P.K.; Amaral, T.M.P.; Brasileiro, C.B.; Mesquita, R.A.; Abreu, L.G. Use of fractal analysis in dental images: A systematic review. *Dentomaxillofacial Radiol.* **2020**, *49*, 20180457. [CrossRef] [PubMed]
35. Skośkiewicz-Malinowska, K.; Mysior, M.; Rusak, A.; Kuropka, P.; Kozakiewicz, M.; Jurczyszyn, K. Application of texture and fractal dimension analysis to evaluate subgingival cement surfaces in terms of biocompatibility. *Materials* **2021**, *14*, 5857. [CrossRef]
36. Haralick, R.M.; Dinstein, I.; Shanmugam, K. Textural Features for Image Classification. *IEEE Trans. Syst. Man Cybern.* **1973**, *SMC-3*, 610–621. [CrossRef]
37. Szczypiński, P.M.; Strzelecki, M.; Materka, A.; Klepaczko, A. MaZda-A software package for image texture analysis. *Comput. Methods Programs Biomed.* **2009**, *94*, 66–76. [CrossRef] [PubMed]
38. Trafalski, M.; Kozakiewicz, M.; Jurczyszyn, K. Application of fractal dimension and texture analysis to evaluate the effectiveness of treatment of a venous lake in the oral mucosa using a 980 nm diode laser—A preliminary study. *Materials* **2021**, *14*, 4140. [CrossRef] [PubMed]
39. Kołaciński, M.; Kozakiewicz, M.; Materka, A. Textural entropy as a potential feature for quantitative assessment of jaw bone healing process. *Arch. Med. Sci.* **2015**, *11*, 78–84. [CrossRef] [PubMed]

Disclaimer/Publisher's Note: The statements, opinions and data contained in all publications are solely those of the individual author(s) and contributor(s) and not of MDPI and/or the editor(s). MDPI and/or the editor(s) disclaim responsibility for any injury to people or property resulting from any ideas, methods, instructions or products referred to in the content.

MDPI
St. Alban-Anlage 66
4052 Basel
Switzerland
www.mdpi.com

International Journal of Molecular Sciences Editorial Office
E-mail: ijms@mdpi.com
www.mdpi.com/journal/ijms

Disclaimer/Publisher's Note: The statements, opinions and data contained in all publications are solely those of the individual author(s) and contributor(s) and not of MDPI and/or the editor(s). MDPI and/or the editor(s) disclaim responsibility for any injury to people or property resulting from any ideas, methods, instructions or products referred to in the content.

www.ingramcontent.com/pod-product-compliance
Lightning Source LLC
Chambersburg PA
CBHW040812250926
51850CB00007BA/1143
* 9 7 8 3 7 2 5 8 1 0 7 6 5 *